COMPREHENSIVE BIOCHEMISTRY

ELSEVIER PUBLISHING COMPANY

335 Jan van Galenstraat, P.O. Box 211, Amsterdam, The Netherlands

ELSEVIER PUBLISHING COMPANY LIMITED

Barking, Essex, England

AMERICAN ELSEVIER PUBLISHING COMPANY, INC.

52 Vanderbilt Avenue, New York, N.Y. 10017

Library of Congress Card Number 62–10359
ISBN 0–444–40871–1

With 31 illustrations and 24 tables

COMPREHENSIVE BIOCHEMISTRY

COMPREHENSIVE BIOCHEMISTRY

SECTION I (VOLUMES 1–4)

PHYSICO-CHEMICAL AND ORGANIC ASPECTS
OF BIOCHEMISTRY

SECTION II (VOLUMES 5–11)

CHEMISTRY OF BIOLOGICAL COMPOUNDS

SECTION III (VOLUMES 12–16)

BIOCHEMICAL REACTION MECHANISMS

SECTION IV (VOLUMES 17–21)

METABOLISM

SECTION V (VOLUMES 22–29)

CHEMICAL BIOLOGY

HISTORY OF BIOCHEMISTRY (VOLUME 30)

GENERAL INDEX (VOLUME 31)

COMPREHENSIVE BIOCHEMISTRY

EDITED BY

MARCEL FLORKIN

Professor of Biochemistry, University of Liège (Belgium)

AND

ELMER H. STOTZ

*Professor of Biochemistry, University of Rochester, School of Medicine
and Dentistry, Rochester, N.Y. (U.S.A.)*

VOLUME 21

METABOLISM OF VITAMINS AND TRACE ELEMENTS

ELSEVIER PUBLISHING COMPANY

AMSTERDAM · LONDON · NEW YORK

1971

CONTRIBUTORS TO THIS VOLUME

GENE M. BROWN, Ph.D.
Professor of Biochemistry, Department of Biology, Massachusetts Institute
of Technology, Cambridge, Mass. 02139 (U.S.A.)

BENEDICT J. CAMPBELL, Ph.D.
Professor of Biochemistry, Department of Biochemistry, University of Missouri,
Columbia, Mo. 65201 (U.S.A.)

BETTY E. HASKELL, B.Sc., M.A., Ph.D.
Associate Professor of Nutrition, Department of Food Science,
University of Illinois, Urbana, Ill. 61801 (U.S.A.)

DONALD B. McCORMICK, B.A., Ph.D.
Professor of Biochemistry, Section of Biochemistry and Molecular Biology,
Division of Biological Sciences, Graduate School of Nutrition, Cornell University,
Savage Hall, Ithaca, N.Y. 14850 (U.S.A.)

L. MERVYN, B.Sc. (Liverpool), Ph.D. (Liverpool), A.R.I.C.
Section Head, Biochemistry Department, Glaxo Research Ltd., Greenford,
Middlesex (Great Britain)

BOYD L. O'DELL, Ph.D.
Professor of Agricultural Chemistry, Department of Agricultural Chemistry,
University of Missouri, 105 Schweitzer Hall, Columbia, Mo. 65201 (U.S.A.)

GERHARD W. E. PLAUT, Ph.D.
Professor and Chairman, Department of Biochemistry, Building 4127, Kilmer Area,
Rutgers Medical School, Rutgers University, New Brunswick, N.J. 08903 (U.S.A.)

TETSUO SHIOTA, Ph.D.
Associate Professor of Microbiology, Biochemistry and Biology,
Department of Microbiology, Medical Center, University of Alabama in Birmingham,
1919 Seventh Avenue South, Birmingham, Ala. 35233 (U.S.A.)

ESMOND E. SNELL, B.A., M.A., Ph.D.
Professor of Biochemistry, Department of Biochemistry, University of California,
Berkeley, Calif. 94720 (U.S.A.)

LEMUEL D. WRIGHT, B.Sc., M.Sc., Ph.D.
Professor of Biochemistry, Section of Biochemistry and Molecular Biology,
Division of Biological Sciences, Graduate School of Nutrition, Cornell University,
Savage Hall, Ithaca, N.Y. 14850 (U.S.A.)

GENERAL PREFACE

The Editors are keenly aware that the literature of Biochemistry is already very large, in fact so widespread that it is increasingly difficult to assemble the most pertinent material in a given area. Beyond the ordinary textbook the subject matter of the rapidly expanding knowledge of biochemistry is spread among innumerable journals, monographs, and series of reviews. The Editors believe that there is a real place for an advanced treatise in biochemistry which assembles the principal areas of the subject in a single set of books.

It would be ideal if an individual or small group of biochemists could produce such an advanced treatise, and within the time to keep reasonably abreast of rapid advances, but this is at least difficult if not impossible. Instead, the Editors with the advice of the Advisory Board, have assembled what they consider the best possible sequence of chapters written by competent authors; they must take the responsibility for inevitable gaps of subject matter and duplication which may result from this procedure.

Most evident to the modern biochemists, apart from the body of knowledge of the chemistry and metabolism of biological substances, is the extent to which he must draw from recent concepts of physical and organic chemistry, and in turn project into the vast field of biology. Thus in the organization of Comprehensive Biochemistry, the middle three sections, Chemistry of Biological Compounds, Biochemical Reaction Mechanisms, and Metabolism may be considered classical biochemistry, while the first and last sections provide selected material on the origins and projections of the subject.

It is hoped that sub-division of the sections into bound volumes will not only be convenient, but will find favour among students concerned with specialized areas, and will permit easier future revisions of the individual volumes. Toward the latter end particularly, the Editors will welcome all comments in their effort to produce a useful and efficient source of biochemical knowledge.

M. Florkin

E. H. Stotz

Liège/Rochester

PREFACE TO SECTION IV

(VOLUMES 17–21)

Metabolism in its broadest context may be regarded as the most dynamic aspect of biochemistry, yet depends entirely for its advances on progress in the knowledge of the structure of natural compounds, structure–function relationships in enzymes, bioenergetics, and cytochemistry. Approaches to the study of metabolism range from whole organism studies, with a limited possibility of revealing mechanisms, to cytochemical or even purified enzyme systems, sometimes with little attention to physiological conditions. Yet all approaches broaden our understanding of metabolism, and all of them may be recognized in the volumes assembled in Section IV on *Metabolism*. It is not unexpected, then, that previous sections of *Comprehensive Biochemistry* actually deal with some aspects under the broad heading of *Metabolism* and that the succeeding Section V on *Chemical Biology* will certainly draw heavily on a basic understanding of metabolism. Nevertheless Section IV attempts to bring together the broad outlines of the metabolism of amino acids, proteins, carbohydrates, lipids, and their derived products. The currently rapid advances in feed-back, hormonal, and genetic control of metabolism make it particularly difficult that these volumes be current, but the authors, editors, and publishers have made all possible efforts to include the most recent advances.

M. FLORKIN

Liège/Rochester

E. H. STOTZ

CONTENTS

VOLUME 21

METABOLISM OF VITAMINS AND TRACE ELEMENTS

Chapter I. Metabolism of Water-Soluble Vitamins

Section a. The Biosynthesis of Thiamine

by GENE M. BROWN

Chapter I. Metabolism of Water-Soluble Vitamins

Section b. The Biosynthesis of Riboflavin

by G. W. E. PLAUT

Chapter I. Metabolism of Water-Soluble Vitamins

Section c. The Metabolism of Vitamin B_6

by ESMOND E. SNELL AND BETTY E. HASKELL

Chapter I. Metabolism of Water-Soluble Vitamins

Section d. Biosynthesis of Pantothenic Acid and Coenzyme A

by GENE M. BROWN

Chapter I. Metabolism of Water-Soluble Vitamins

Section e. The Metabolism of Biotin and Analogues

by Donald B. McCormick and Lemuel D. Wright

Chapter I. Metabolism of Water-Soluble Vitamins

Section f. The Biosynthesis of Folic Acid and
6-Substituted Pteridine Derivatives

by T. Shiota

Chapter I. Metabolism of Water-Soluble Vitamins

Section g. The Metabolism of the Cobalamins

by L. Mervyn

Chapter II. Trace Elements:

Metabolism and Metabolic Function

by B. L. O'Dell and B. J. Campbell

COMPREHENSIVE BIOCHEMISTRY

Chapter I

Metabolism of Water-Soluble Vitamins

Section a

The Biosynthesis of Thiamine

GENE M. BROWN

Department of Biology, Massachusetts Institute of Technology, Cambridge, Mass. (U.S.A.)

1. Introduction

Thiamine, also known as vitamin B_1 and aneurin, contains pyrimidine and thiazole moieties joined together by a methylene group as shown in Fig. 1. The pyrimidine and thiazole portions of thiamine are formed by independent biosynthetic pathways and then used for the biosynthesis of the vitamin. The

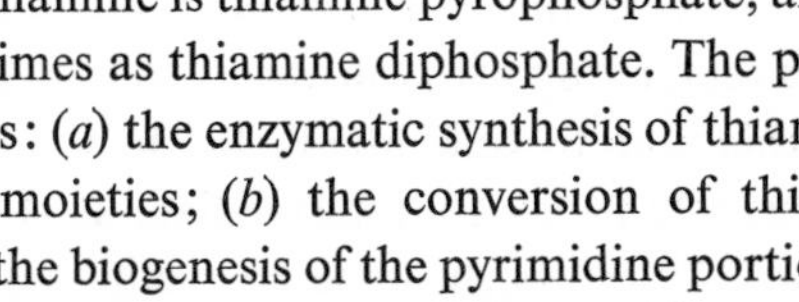

Fig. 1. Formula for thiamine.

coenzyme form of thiamine is thiamine pyrophosphate, also known as cocarboxylase and sometimes as thiamine diphosphate. The purpose of this presentation is to discuss: (*a*) the enzymatic synthesis of thiamine from its pyrimidine and thiazole moieties; (*b*) the conversion of thiamine to thiamine pyrophosphate; (*c*) the biogenesis of the pyrimidine portion of thiamine; and (*d*) the biogenesis of the thiazole moiety. These subjects will be considered in separate sections of the discussion which follows.

2. Enzymatic synthesis of thiamine

The initial enzyme work on the final stages of thiamine biosynthesis was done by Harris and Yavit[1], who found that 2-methyl-4-amino-5-hydroxymethyl-pyrimidine (hereafter referred to as "pyrimidine") and 4-methyl-5-(β-hy-droxyethyl)thiazole (or "thiazole") could be converted to thiamine in the presence of Mg^{2+}, ATP and cell-free extracts of bakers' yeast. This observation was followed by investigations by other groups[2-9] that have led to the conclusion that this synthesis proceeds by the set of reactions shown in Fig. 2.

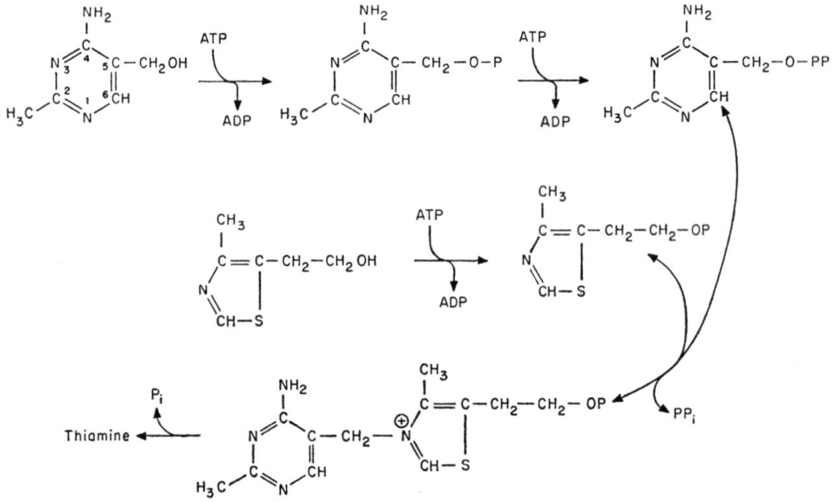

Fig. 2. Enzymatic reactions involved in the synthesis of thiamine from its pyrimidine and thiazole moieties. P_i = inorganic phosphate and PP_i = inorganic pyrophosphate. Phosphate and pyrophosphate esters are indicated as –O–P and –O–PP, respectively.

Leder[2] and Camiener and Brown[6-8] found that thiamine monophosphate (thiamine-P), instead of thiamine, is the product formed by the enzyme system from yeast. This fact, along with the observation that the enzyme system could not convert thiamine to thiamine-P, led to the notion that thiazole monophosphate (thiazole-P) is an intermediate in the over-all process. Camiener and Brown[6-8] established the correctness of this suggestion by demonstrating the presence of an enzyme in yeast that catalyzes the transfer of a phosphate residue from ATP to thiazole to yield thiazole-P.

The requirement for ATP also suggested that the pyrimidine compound had to be "activated" perhaps through the formation of a phosphate ester. Harris and Yavit[1] synthesized pyrimidine monophosphate (pyrimidine-P) and showed that, with crude extracts of yeast, this compound could replace pyrimidine as substrate for thiamine sysnthesis; however Leder[10] found that this monophospho ester could replace pyrimidine only if ATP were also supplied. Nose et al.[4] and Camiener and Brown[6,7] then independently presented evidence for the enzymatic formation of pyrimidine pyrophosphate (pyrimidine-PP) and a role for this compound as an intermediate in the over-all process. The pyrophosphate ester has been synthesized and shown to be converted to thiamine-P in the presence of thiazole-P and the purified enzyme needed for thiamine-P synthesis[3,4]. No ATP was needed for this reaction to take place. The enzyme that catalyzes the synthesis of thiamine-P has been partially purified from yeast by Camiener and Brown[8] and by Leder[3]. Since Leder found that the reaction is reversible, he suggested the enzyme be called "thiamine-P pyrophosphorylase"; Camiener and Brown have suggested "thiamine-P synthetase"[8].

The possibility that pyrimidine-P might be an intermediate in the biosynthesis of pyrimidine-PP was suggested by the observations that (a) extracts of yeast can catalyze the synthesis of pyrimidine-P from pyrimidine and ATP[6,7] and (b) the rate of thiamine-P synthesis is greater from pyrimidine-P than from pyrimidine[1,10]. That this suggestion is correct was established by investigations by Lewin and Brown[9] and by Fujita and coworkers[11,12]. The former investigators found that two enzymes are present in yeast extracts; one of these catalyzes the formation of pyrimidine-P from pyrimidine and ATP and the other uses pyrimidine-P and ATP as substrates for the synthesis of pyrimidine-PP. These two enzymes can be distinguished on the basis of their differences in stability to heating, in susceptibility to inhibition with p-hydroxymercuribenzoate, and abilities to use nucleoside triphosphates other than ATP as a phosphate donor[9]. Fujita and collaborators[11,12] used a different approach to demonstrate that pyrimidine-PP synthesis occurs in two steps. They showed that utilization of ATP labelled with ^{32}P only in the terminal phosphate residue resulted in the formation of pyrimidine-PP labeled equally with radioactivity in both phosphate residues. This could have occurred only by a two-step process with the utilization of two moles of ATP.

The dephosphorylation of thiamine-P to thiamine presumably is catalyzed by intracellular phosphatases. This reaction is a necessary step in the de novo biosynthesis of thiamine-PP, since evidence to be discussed in the succeeding section indicates that thiamine-P cannot be used directly for the formation of thiamine-PP. No evidence has been reported for the existence of specific phosphatases that use thiamine-P as substrate.

3. Enzymatic synthesis of thiamine-PP

The formation of thiamine-PP is known to occur by the transfer of a pyrophosphate group from ATP to thiamine. The initial information to suggest the operation of such a reaction was provided with the findings that thiamine is used more effectively than thiamine-P for the formation of thiamine-PP in the presence of crude-yeast extracts[13] and rat-liver preparations[14]. Support for this view came from the work of Forsander[15] who showed that the use of ATP labeled with ^{32}P only in the terminal phosphate yielded thiamine-PP labeled only in the terminal phosphate. Camiener and Brown[6,8] provided further pertinent evidence with their observation that a partially purified enzyme from yeast could use thiamine but not thiamine-P as substrate for the formation of thiamine-PP. The conclusive evidence came from the investigations of Shimazono and coworkers[16,17] who purified the yeast enzyme and established, with the use of ^{32}P-labeled ATP, that the reaction results in the transfer of a pyrophosphate group. Thus, the enzyme should appropriately be referred to as "thiamine pyrophosphokinase". Investigations by Mano[18,19] have indicated that synthesis of thiamine-PP from thiamine in mammalian tissue also occurs by a one-step transfer of a pyrophosphate group.

4. Biogenesis of the pyrimidine moiety

Although the biosynthetic pathway for the formation of the pyrimidine portion of thiamine has not yet been completely elucidated, enough is known to eliminate the possibility that this pyrimidine is made by the same pathway by which pyrimidines found in nucleic acids are made. Goldstein and Brown[20] found that neither pyrimidines (uracil and orotic acid) nor pyrimidine precursors (aspartic acid and CO_2) were incorporated into the pyrimidine of thiamine, and other work with radioactive compounds (to be discussed below) has indicated that the two pathways are dissimilar.

The first positive evidence concerning the nature of the precursors of this pyrimidine was supplied by the independent reports from several laboratories that exogenous formate is incorporated into this compound, with very little dilution in specific radioactivity, by growing cells of either bacteria[20,21] or yeast[22]. There is some disagreement about which carbon of the pyrimidine is labeled by formate in yeast. David et al.[23] reported that only carbon 4 (see Fig. 2 for the numbering system) is derived from formate, whereas Johnson et al.[24] have concluded that only the methyl group is labeled by formate. However, it should be pointed out that the method used by Johnson et al.[24] could not distinguish between the methyl group and carbon 2 of the ring. There is agreement, however, about the site of incorporation of formate by enteric bacteria. Kumaoka and Brown[25] found that the formate was incorporated exclusively into carbon 2 of the ring by E. coli, and Estramareix and Lesieur[26] have confirmed this with the use of Salmonella typhimurium. These observations leave the possibility open that yeast and enteric bacteria use different pathways for the production of the pyrimidine moiety of thiamine.

Other compounds that have been found to be incorporated into pyrimidine in significant quantities are glycine and acetate. Goldstein and Brown[20] demonstrated that both of these compounds were incorporated by E. coli, but at a dilution in specific radioactivity far greater than that observed for radioactive formate. Johnson et al.[24] have reported that in bakers' yeast only C-2 of acetate is effectively incorporated, and Tomlinson[27] has concluded that C-2 of acetate is the precursor of the methyl group on the 2 position of the pyrimidine. However, some doubt has arisen about the significance of the incorporation of acetate as a result of the findings of Kumaoka and Brown[25] that administration to E. coli of nonradioactive formate dilutes out the incorporation of radioactive acetate to an insignificant value.

Some of the most significant work on the biogenesis of pyrimidine has been done by Newell and Tucker[28–30], who have presented direct evidence for a close biosynthetic relationship between purines and the pyrimidine of thiamine. The stimulation for their work was provided by the isolation of mutants of Salmonella typhimurium that require both a purine and the pyrimidine of thiamine as a result of a single mutation[31]. This suggested that a portion of the biosynthetic pathway for purines also functions in the biosynthesis of pyrimidine. Newell and Tucker provided evidence for this supposition with the demonstration that 4-aminoimidazole ribonucleotide is a common intermediate in the biosynthesis of pyrimidine and purines. The formulation of this conclusion was made possible by the ability of the investigators to develop a secondary mutant that is permeable to aminoimidazole ribonucleoside. They were then able to show that this compound can satisfy simultaneously the requirement for purine and the pyrimidine of thiamine. Furthermore, they could show that radioactive aminoimidazole ribonucleoside was converted by this mutant to pyrimidine without significant dilution of specific radioactivity. Evidence was presented that, although methionine is required for the conversion of aminoimidazole ribonucleoside to pyrimidine, no carbons from methionine are incorporated into pyrimidine. Finally, they established that both carbons of glycine are incorporated into the pyrimidine. This would have to be true if aminoimidazole ribonucleoside (or ribonucleotide) is an intermediate in the biosynthesis of pyrimidine since glycine is known to be a precursor of this compound. The observation by Newell and Tucker that no carbon from methionine (including the methyl group) is incorporated into pyrimidine[29] is consistent with the observations of Goldstein and Brown[20] who earlier had also reported that methionine is not incorporated.

The reactions whereby aminoimidazole ribonucleotide might be converted to pyrimidine remain unknown. Hypothetical schemes that would account for the labeling data with formate are presented in Fig. 3. One possibility (Scheme 1, Fig. 3) is that the imidazole ring is ruptured between carbons 4 and 5. This would be followed by: (a) the insertion of a two-carbon compound to form carbon 5 of the pyrimidine ring and the hydroxymethyl group attached

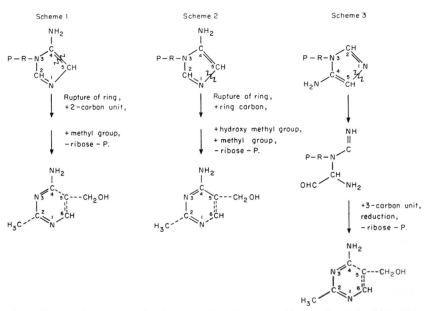

Fig. 3. Hypothetical schemes for the conversion of aminoimidazole ribonucleotide to the pyrimidine of thiamine. R–P represents a ribose-5-phosphate unit.

to the ring at that position; (b) the removal of the ribose phosphate unit; and (c) the addition of a methyl group. The source of the methyl group is particularly puzzling since methionine does not supply this group[20,29] and neither can it be derived from the one-carbon (formate) pool[25]. Another possibility (shown as Scheme 2, Fig. 3) is that the imidazole ring is split between atoms 1 and 5. This would have to be followed by the addition of carbons as follows: (a) to complete the formation of the pyrimidine ring (carbon 6 of the ring); (b) for the formation of the hydroxymethyl group; and (c) for the formation of the methyl group. Finally, in this scheme the ribose phosphate unit would have to be removed. These two schemes are consistent with the fact that in enteric bacteria carbon 2 of the imidazole ring is derived from formate, and in these two schemes carbon 2 of imidazole would become carbon 2 of pyrimidine. It would appear that Scheme 1 might be more feasible since this would require the addition of a 2-carbon unit and a methyl group, whereas Scheme 2 would require the addition of what appears to be three one-carbon units. This seems unlikely since it is known that none of these carbons is derived from formate.

Another possibility that was suggested by Newell and Tucker[30] is that the imidazole ring is ruptured between positions 1 and 5 and this would be followed by reduction of the resultant formyl group to a methyl group and the addition of a 3-carbon compound to complete the formation of the pyrimidine ring and provide the hydroxymethyl group. This scheme is shown in Fig. 3 as Scheme 3. This hypothesis predicts that carbon 2 of imidazole would become carbon 4 of the pyrimidine and carbon 5 would become the methyl group. This is not consistent with the labeling data with formate in E. coli and S. typhimurium, since with these bacteria formate appears in carbon 2 and not carbon 4 of the pyrimidine ring. However, this scheme might be considered as a possibility in yeast if the report is true that in yeast carbon 4 is derived from formate.

Other possibly pertinent information concerning the biogenesis of pyrimidine has come from the work of Diorio and Lewin[32,33]. These investigators have found that certain thiamine-requiring Neurospora mutants produce the 5-aminomethyl and 5-formyl derivatives of pyrimidine and have suggested that these compounds may be intermediates in the biosynthesis of the 5-hydroxymethylpyrimidine compound which is the substrate for the enzymatic synthesis of thiamine. A previous report by Camiener and Brown[7] that the aminomethylpyrimidine can be converted to the hydroxymethylpyrimidine by cell-free extracts of yeast provides some support for roles of these compounds as biosynthetic intermediates.

5. Biogenesis of thiazole

Very little is known about the biosynthesis of the thiazole moiety of thiamine. Investigations in this area have been directed toward the identification of the precursors with radioactive incorporation experiments. Methionine or cysteine has most often been considered as the probable donor of sulfur for thiazole synthesis; however, there is no conclusive evidence for this supposition. Harrington and Moggridge[34] have suggested that methionine might condense with acetaldehyde and ammonia to give α-amino-β-(4-methylthiazole)-5-propionic acid which might then be converted to the thiazole portion of thiamine, but there is no evidence to support this speculative scheme. Johnson et al.[24] have reported that equal quantities of ^{35}S and ^{14}C are incorporated into thiazole by yeast provided with methionine labeled with ^{35}S and with ^{14}C in the methyl group. However, the efficiency of incorporation was low and other workers have found that the methyl of methionine is not incorporated into thiazole[35,36]. Nakayama[37] has suggested that cysteine might provide both sulfur and carbons for thiazole synthesis. The evidence is based on his contention that certain thiazole-requiring mutants of E. coli and Neurospora crassa can utilize cysteine, 4-methylthiazole or 4-thiazolidine carboxylic acid in place of thiazole. He suggested that cysteine and formate might be involved in the synthesis of the other two compounds. However, serious doubts about a role for 4-methylthiazole as an intermediate have arisen as a result of the observation by Korte et al.[38] that 4-methyl-[2-^{14}C]thiazole is not converted to thiazole by a variety of microorganisms. Other workers have found that neither [^{14}C]cysteine nor [^{14}C]formate is incorporated into thiazole by E. coli[35]. Compounds that have been demonstrated to be efficiently incorporated into thiazole by microorganisms are alanine[24,39], acetate (both carbons)[24,39,35], glycine[35,36] and serine[36]. How these compounds are used to make thiazole remains a subject for further investigations.

REFERENCES

1 D. L. HARRIS AND J. YAVIT, *Federation Proc.*, 16 (1957) 192.
2 I. G. LEDER, *Biochem. Biophys. Res. Commun.*, 1 (1959) 63.
3 I. G. LEDER, *J. Biol. Chem.*, 236 (1961) 3066.
4 Y. NOSE, K. UEDA AND T. KAWASAKI, *Biochim. Biophys. Acta*, 34 (1959) 277.
5 Y. NOSE, K. UEDA, T. KAWASAKI, A. IWASHIMA AND T. FUJITA, *J. Vitaminol. (Kyoto)*, 7 (1962) 98.
6 G. W. CAMIENER AND G. M. BROWN, *J. Am. Chem. Soc.*, 81 (1959) 3800.
7 G. W. CAMIENER AND G. M. BROWN, *J. Biol. Chem.*, 235 (1960) 2404.
8 G. W. CAMIENER AND G. M. BROWN, *J. Biol. Chem.*, 235 (1960) 2411.
9 L. M. LEWIN AND G. M. BROWN, *J. Biol. Chem.*, 236 (1961) 2768.
10 I. G. LEDER, *Federation Proc.*, 18 (1959) 270.
11 T. KAWASAKI AND T. FUJITA, *Seikagaku*, 33 (1961) 737; *Chem. Abstr.*, 56 (1962) 7676.
12 T. KAWASAKI AND T. FUJITA, *Seikagaku*, 33 (1961) 742; *Chem. Abstr.*, 56 (1962) 7676.
13 H. WEIL-MALHERBE, *Biochem. J.*, 33 (1939) 1997.
14 F. LEUTHARDT AND H. NIELSEN, *Helv. Chim. Acta*, 35 (1952) 1196.
15 O. FORSANDER, *Soc. Sci. Fennica, Commentationes Phys. -Math.*, 19 (1956) 2.
16 N. SHIMAZONO, Y. MANO, R. TANAKA AND Y. KAZIRO, *J. Biochem. (Tokyo)*, 46 (1959) 959.
17 Y. KAZIRO, R. TANAKA, Y. MANO AND N. SHIMAZONO, *J. Biochem. (Tokyo)*, 49 (1961) 472.
18 Y. MANO, *J. Biochem. (Tokyo)*, 47 (1960) 283.
19 Y. MANO, *J. Biochem. (Tokyo)*, 47 (1960) 24.
20 G. A. GOLDSTEIN AND G. M. BROWN, *Arch. Biochem. Biophys.*, 103 (1963) 449.
21 M. J. PINE AND R. GUTHRIE, *J. Bacteriol.*, 78 (1959) 545.
22 S. DAVID AND B. ESTRAMAREIX, *Biochim. Biophys. Acta*, 42 (1960) 562.
23 S. DAVID, B. ESTRAMAREIX AND H. HIRSHFELD, *Biochim. Biophys. Acta*, 127 (1966) 264.
24 D. B. JOHNSON, D. J. HOWELLS AND T. W. GOODWIN, *Biochem. J.*, 98 (1966) 30.
25 H. KUMAOKA AND G. M. BROWN, *Arch. Biochem. Biophys.*, 122 (1967) 378.
26 B. ESTRAMAREIX AND M. LESIEUR, *Biochim. Biophys. Acta*, 192 (1969) 375.
27 R. V. TOMLINSON, *Biochim. Biophys. Acta*, 115 (1966) 526.
28 P. C. NEWELL AND R. G. TUCKER, *Nature*, 215 (1967) 1384.
29 P. C. NEWELL AND R. G. TUCKER, *Biochem. J.* 106 (1968) 271.
30 P. C. NEWELL AND R. G. TUCKER, *Biochem. J.*, 106 (1968) 279.
31 M. DEMEREC, H. MOSER, R. C. CLOWES, E. L. LAHR, H. OZEKI AND W. VIELMETTER, *Carnegie Inst. Wash., Year Book*, 55 (1955–1956) 309.
32 A. F. DIORIO AND L. M. LEWIN, *J. Biol. Chem.*, 243 (1968) 3999.
33 A. F. DIORIO AND L. M. LEWIN, *J. Biol. Chem.*, 243 (1968) 4006.
34 C. R. HARRINGTON AND R. C. G. MOGGRIDGE, *Biochem. J.*, 34 (1940) 685.
35 M. JULIUS AND G. M. BROWN, unpublished observations (1966).
36 P. E. LINNETT AND J. WALKER, *Biochem. J.*, 109 (1968) 161.
37 H. NAKAYAMA, *Vitamins (Kyoto)*, 11 (1956) 169.
38 F. KORTE, H. WEITKAMP AND J. VOGEL, *Ann. Chem.*, 628 (1959) 159.
39 R. V. TOMLINSON, D. P. KUHLMAN, P. F. TORRENCE AND H. TIECKELMANN, *Biochim. Biophys. Acta*, 148 (1967) 1.
40 P. E. LINNETT AND J. WALKER, *J. Chem. Soc.*, (1967) 796.

Metabolism of Water-Soluble Vitamins

Section b

The Biosynthesis of Riboflavin

G. W. E. PLAUT

Department of Biochemistry, Rutgers Medical School, Rutgers University, New Brunswick, N. J. (U.S.A.)

1. Introduction

Riboflavin is formed by plants and a large number of microorganisms. Some of the latter produce the vitamin in unusually large amounts. A large body of information is available on commercial production of riboflavin by fermentation with the organisms *Clostridium acetobutylicum*, Candida yeasts, *Eremothecium ashbyii*, and *Ashbya gossypii*. Literature describing media and other conditions of growth necessary for optimal riboflavin production has been reviewed previously[1-5].

Studies on the mechanism of biosynthesis of flavins were initiated by the observations that the pyrimidine and pyrazine portions of riboflavin are derived from purine precursors. Thus, the incorporation of radioactive formate, glycine, and CO_2 occur in analogous positions into purines, riboflavin, and pteridines (for review see refs. 5–7). In addition, the observation by McLaren[8] that riboflavin formation by *Eremothecium ashbyii* is enhanced by the addition of certain purines to the growth medium and the

* The experimental work from this laboratory presented here and the preparation of of this article, have been assisted by grants from the National Institute of Arthritis and Metabolic Diseases (AM 10501).

subsequent demonstration by McNutt[9-11] that label in carbon and nitrogen of added purines was recovered in high yield in riboflavin formed by *Eremothecium ashbyii* suggested that the purine skeleton, excepting carbon 8, is incorporated intact into the pyrimidine and pyrazine portions of ribo-flavin. Additional confirmation for the utilization for riboflavin formation of the intact purine molecule came from studies by Goodwin and Brown[12], in which it was shown that the addition of unlabeled adenine (or guanine) led to dilution of the label incorporated from [14C]serine by intact cells of *Eremothecium ashbyii*. The obligatory role of a purine compound in ribo-flavin biosynthesis was demonstrated by Howells and Plaut[13], who found with purineless strains of *E. coli* that radioactivity from labeled glycine is not incorporated into riboflavin while label from [14C]adenine (or guanine) can be recovered in the flavin. However, in the wild-parent strains of these mutants of *E. coli*, transfer of label from glycine to flavin does occur. It is possible that the biosynthesis of the pyrimidine and pyrazine portions of riboflavin occurs by way of a nucleotide by mechanisms analogous to those indicated for the structurally similar pteridines, toxiflavin, etc. Evidence in support of this assumption will be presented in a later portion of this review. The precursor and structural relationship between purines, pteridines, and riboflavin has been reviewed previously[5-7].

The origin of the *o*-xylene portion of the riboflavin molecule was outlined by studies on the incorporation of label from a number of simple precursors with intact cells of *A. gossypii*. A metabolite containing two or four carbon atoms is the probable precursor since it was shown that incorporation of label occurred into the methyl groups from methyl-labeled acetate and glucose labeled in positions 1 or 6, whereas carboxyl-labeled acetate only led to transfer of label into carbons $6+7$ and $4a+8a$ of the aromatic ring[14]. It is quite possible that a 2- or 4-carbon fragment closely related to carbo-hydrate metabolism is the immediate precursor. Thus, it has been observed that more extensive incorporation of radioactivity with intact cells of *E. ashbyii* occurs with labeled glucose or ribose than with [14C]acetate[15]. It may also be pertinent in this connection that in the biosynthesis of the dimethylbenzimidazole portion of vitamin B_{12}, label from [U-14C]erythritol is incorporated efficiently into the ring[16]. In any case, in view of evidence to be presented later, incorporation of the 2- or 4-carbon fragment must occur first into carbons 6 and 7 and the adjacent methyl groups of 6,7-dimethyl-8-ribityllumazine from which riboflavin is formed by action of a specific enzyme.

The origin of the ribityl group of the flavin is still rather uncertain. Studies on the incorporation of ^{14}C into this group from glucose labeled in various positions by intact cells of *A. gossypii* are consistent with the possibility that the ribityl group is derived from a ribosyl or other pentosyl unit[17]. Earlier studies by McNutt and Forrest[18] with intact cells of *E. ashbyii* indicated that the purine and ribosyl portions of uniformly ^{14}C-labeled guanosine were not transferred as a unit into the isoalloxazine and ribityl portions of the riboflavin molecule. However, if formation of riboflavin occurs from a nucleoside triphosphate precursor, this question may be worthy of reexamination.

2. Formation of riboflavin from 6,7-dimethyl-8-ribityllumazine

(a) Isolation and chemical synthesis of 6,7-dimethyl-8-ribityllumazine and derivatives

A green fluorescent and a blue fluorescent compound were isolated from cultures of *E. ashbyii*[19,22] and *A. gossypii*[23,24]. The structure of the green fluorescent compound was identified by chemical synthesis to be 6,7-dimethyl-8-ribityllumazine[23-27] and that of the blue fluorescent substance, 6-methyl-7-hydroxy-8-ribityllumazine[28-30]. A number of procedures have been devised for the chemical synthesis of these compounds varying in certain details, but in all cases 4-chloro-2,6-dihydroxypyrimidine and ribitylamine have been the usual starting materials and a final condensation step between 4-ribitylamino-5-amino-2,6-dihydroxypyrimidine and an α,β-diketone (2,3-butanedione) to form 6,7-dimethyl-8-ribityllumazine (or an ester of pyruvic acid to form 6-methyl-7-hydroxy-8-ribityllumazine) have been used. The methods of chemical synthesis are outlined in Fig. 1. In scheme A, 4-chloro-2,6-dihydroxypyrimidine is condensed with ribitylamine and the resulting 4-ribitylaminopyrimidine is nitrosolated to form 4-ribitylamino-5-nitroso-2,6-dihydroxypyrimidine (scheme A[24,26,31]). In scheme B, chlorouracil is first condensed with a diazo compound and then with ribitylamine to form the corresponding 5-substituted diazo derivative[27]. Either compound (schemes A and B) is then reduced with sodium hydrosulfite or by catalytic hydrogenation to yield 4-ribitylamino-5-amino-2,6-dihydroxypyrimidine. In scheme C[32,33], 4-chloro-2,6-dihydroxypyrimidine is first nitrated. The reactive 4-chloro-5-nitro-2,6-dihydroxypyrimidine condenses under relatively mild conditions with ribitylamine to form 4-ribityl-

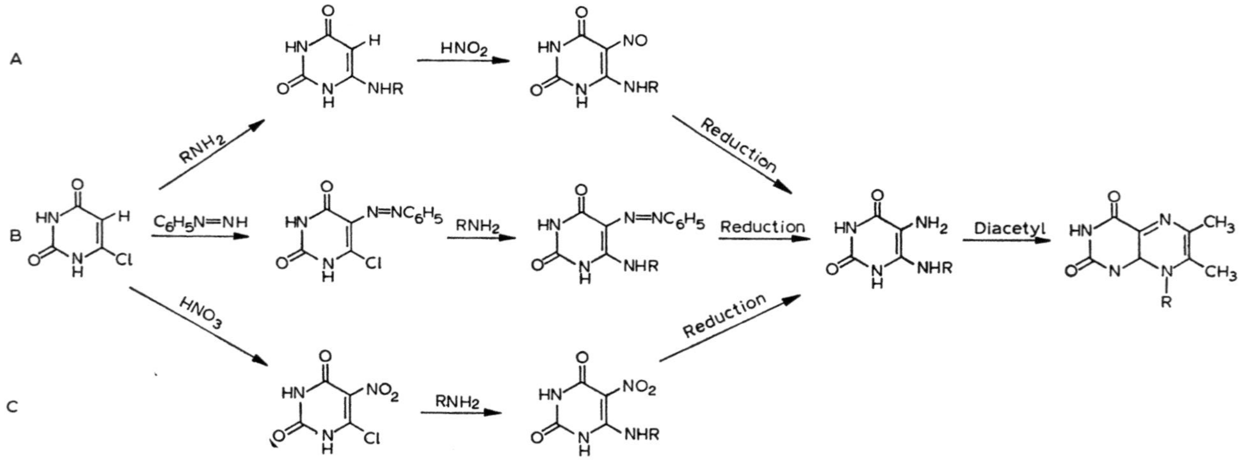

Fig. 1. Chemical synthesis of 6,7-dimethyl-8-ribityllumazine. R, ribityl group.

amino-5-nitro-2,6-dihydroxypyrimidine which is reduced catalytically to 4-ribitylamino-5-amino-2,6-dihydroxypyrimidine.

These methods have been used for the synthesis of 6,7-dimethyl-8-ribityl-lumazine labeled in various positions of the molecule[34]. For example, label has been introduced into the heterocyclic ring of the lumazine by starting with barbituric acid labeled in various positions which was then converted by way of 2,4,6-trichloropyrimidine to 4-chloro-2,6-dihydroxypyrimidine. Lumazine labeled in the ribityl group was prepared from [^{14}C]ribose by reduction of the intermediate ribose oxime to ribitylamine. Labeling of the methyl groups and the adjacent 6- and 7-carbon atoms has been accomplished by the use of labeled 2,3-butanedione[35,36].

A large number of analogues of the lumazine varying in substituents at positions 2, 4, 6, 7 and 8 have been prepared by the methods outlined in Fig. 1. Thus, lumazines bearing different substituents at position 8 have been made by substituting other amines for ribitylamine in the synthesis. Variations in substituents at positions 6 and 7 were accomplished by condensation of 4-ribitylamino-5-amino-2,6-dihydroxypyrimidine with a number of α,β-diketones[27,31-33]. Pteridines bearing an amino instead of an oxygen function at position 2 (including the 2-amino analogue of 6,7-dimethyl-8-ribityllumazine) were prepared by Davoll and Evans[32] using 2-amino-4-chloro-6-hydroxy-5-nitropyrimidine instead of the corresponding 2,6-dihydroxypyrimidine derivative as the starting material in the synthesis.

(b) 6,7-Dimethyl-8-ribityllumazine as precursor of riboflavin

The similarities of the chemical structure of 6,7-dimethyl-8-ribityllumazine and riboflavin suggested to several investigators that the lumazine may be a precursor of the flavin (for review see ref. 7). Evidence in support of this possibility came from the demonstration that incorporation of radioactivity from [^{14}C]formate, [^{14}C]glycine, [2-^{14}C]adenine and [U-^{14}C]adenine (but not [8-^{14}C]adenine) occurred into analogous positions of 6,7-dimethyl-8-ribityllumazine and riboflavin with intact cells of *A. gossypii*. Furthermore, the specific radioactivity of 6,7-dimethyl-8-ribityllumazine formed from the precursors was either the same or greater than riboflavin isolated at the same time periods[23,24].

Direct evidence for the precursor–product relationship came from the demonstration that radioactive 6,7-dimethyl-8-ribityllumazine could be converted to radioactive riboflavin by cell-free extracts of *A. gossypii*[37].

TABLE I

CONVERSION OF 6,7-DIMETHYL-8-RIBITYLLUMAZINE TO RIBOFLAVIN
BY EXTRACTS FROM VARIOUS MICROORGANISMS

Microorganism	B_2 formation[a] (mμmoles/mg protein)
Aerobacter aerogenes[b,c]	4.8
Pseudomonas sp.[b,c]	2.7
Escherichia coli 97[b,c]	3.9
Escherichia coli ATCC 9637[b]	1.8
Ashbya gossypii[b]	4.5
Saccharomyces cerivisiae[b,d]	1.8
Lactobacillus plantarum ATCC 8014[b]	2.4
Lactobacillus casei ATCC 7469[b]	0.0
Neurospora crassa (Rib—) FGSG 83[e]	1.8
Neurospora crassa (wild type) 1A[e]	1.5
Bacillus subtilis[b,c]	2.4
Bacillus subtilis (Rib—) MB 1568[e,f]	2.4
Bacillus subtilis (Rib—) MB 1757[e,f]	8.7
Bacillus subtilis (Rib+) MB 290[e,f]	4.8

[a] The assay medium contained 0.15 mM 6,7-dimethyl-8-ribityllumazine, 0.017 M phosphate at pH 6.9, 0.01 M Na_2SO_3 at pH 6.9, enzyme and water to a final volume of 3.0 ml. The samples were incubated for 60 min at 37°. Riboflavin was determined spectrophotometrically or fluorometrically.
[b] From Plaut[74].
[c] These cultures were obtained from the collection of the Department of Bacteriology, University of Utah College of Medicine, through the courtesy of Dr. Paul S. Nicholes.
[d] "Active dry yeast" from the Red Star Yeast Co., Milwaukee, Wisconsin.
[e] From T. Aogaichi, D. Lees and G. W. E. Plaut, unpublished observations.
[f] These cultures were obtained through the courtesy of Dr. A. L. Demain, Merck and Co., Inc.

This conversion has been demonstrated in extracts from a number of organisms[37-41]. As can be seen in Table I, the rate of conversion of the lumazine to riboflavin by extracts from a number of microorganisms is quite similar, even though the extent of formation of riboflavin by intact cells from different species varies considerably. Thus, production of riboflavin by A. gossypii and E. ashbyii in the gram per ml range has been observed, whereas formation of the flavin by E. coli amounts to less than 1 mg per ml of medium. Lactobacillus casei[5] and Lactobacillus plantarum[41], which require riboflavin for growth, do not exhibit conversion activity. However, requirement of riboflavin for growth does not necessarily imply absence of riboflavin synthetase activity. Thus, as can be seen in Table I, riboflavinless mutants of *Neurospora crassa* (FGSC #83 and 130)[42] and *B. subtilis* (MB #1568 and 1757) do contain riboflavin synthetase activity in extracts[43]. The block in riboflavin synthesis in the latter organisms and riboflavinless mutants of *S. cereviseae*[68] must occur at a step before the enzymatic conversion of 6,7-dimethyl-8-ribityllumazine to riboflavin.

(c) Stoichiometry of the riboflavin synthetase reaction

The conversion of 6,7-dimethyl-8-ribityllumazine to riboflavin requires the addition of 4 carbon atoms to form the *o*-xylene portion of the vitamin. It became apparent, however, in studies with the purified enzyme from a number of sources (*A. gossypii, E. coli*, baker's yeast and spinach) that the 4 carbon atoms were derived from the lumazine itself rather than from an external source. Thus, it could be shown spectrophotometrically that the stoichiometry of the disappearance of 6,7-dimethyl-8-ribityllumazine ($\lambda_{max} = 407$ mμ) and formation of riboflavin ($\lambda_{max} = 450$ mμ) involved the disappearance of two molecules of the lumazine per molecule of flavin formed[35,36,44,45]. Furthermore, the specific *molar* radioactivity of riboflavin formed from 6,7-dimethyl-8-ribityllumazine labeled in the methyl groups[35] or the lumazine labeled in the methyl groups and the adjacent 6- and 7-carbon atoms[36] was twice that of the precursor lumazine. Chemical degradation of riboflavin formed from 6,7-di-[^{14}C]methyl-8-ribityllumazine with purified enzymes from *A. gossypii* or baker's yeast revealed equal amounts of labeling in the methyl groups and carbon atoms 5 and 8 and an absence of the label from positions 6+7 and 4a+8a[34,35]. These findings demonstrated that enzymatic formation of riboflavin occurs by transfer of 4 carbon atoms containing the methyl groups and the adjacent carbons 6 and 7 from one molecule of 6,7-dimethyl-8-ribityllumazine to a second molecule of the lumazine to form the aromatic ring. The stoichiometry of flavin formation implied that a product in addition to riboflavin should be formed from the lumazine donating the 4-carbon atom portion. The formation of such a second product was demonstrated by use of 6,7-dimethyl-8-ribityllumazine labeled in positions 2 and 4+8a[34]. The second product was identified as 4-ribitylamino-5-amino-2,6-dihydroxypyrimidine and its formation from 6,7-dimethyl-8-ribityllumazine in amounts equivalent to riboflavin was demonstrated[46]. The balanced reaction catalyzed by riboflavin synthetase and the transfer of label from specific positions of the lumazine molecule to products is shown in Fig. 2.

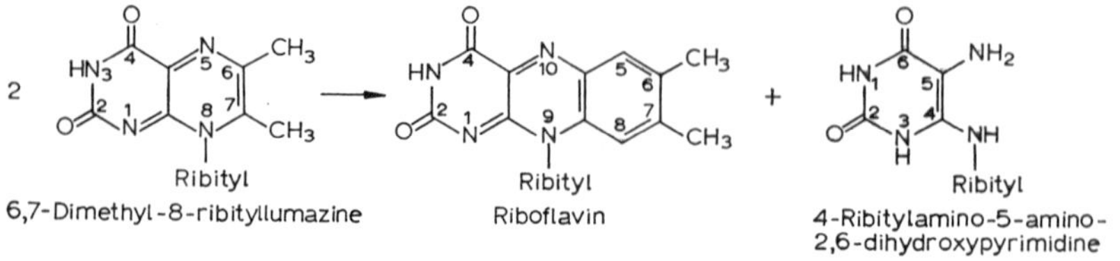

Fig. 2. Stoichiometry of enzymatic conversion of 6,7-dimethyl-8-ribityllumazine to riboflavin.

(d) Substrate specificity of the riboflavin synthetase reaction

A number of derivatives of 6,7-dimethyl-8-ribityllumazines in which the ribityl group at position 8 has been replaced by other substituents have been prepared[27,31-33]. As can be seen in Table II, the enzyme has a very specific requirement for the substituent at this position[47,48], even though the transformation it catalyzes involves the heterocyclic ring system. So far only one other lumazine derivative (6,7-dimethyl-8-[1'-(5'-D-deoxyribityl)]-lumazine) has been found to be converted to a flavin (5'-deoxyriboflavin) with purified enzymes from *A. gossypii* or yeast. 6,7-Dimethyl-8-(1'-D-xylityl)lumazine is not converted to a flavin, but acts as a competitive inhibitor in the conversion of the D-ribityl compound to riboflavin; the corresponding L-xylityl derivative is, however, inactive. A number of pentahydroxyhexityl derivatives were tested (glycityl, galactyl, and mannityl) and found to be inactive, as were the methyl and β-hydroxyethyl derivatives[47]. The D- and L-threityl and erythrityl derivatives show some inhibition of the enzyme from yeast, however, the competitive inhibition constants (K_i) are about 100 times larger than the Michaelis constant (K_m) of the substrate[43]. It would appear from these results that a minimal requirement for the configuration of the side-chain at position 8 requires the presence of a pentyl group bearing hydroxyl groups in the D-configuration at positions 2' and 4'. The latter is emphasized by the recent observation that 6,7-dimethyl-8-(2'-deoxy-D-ribityl)lumazine is inactive and that 6,7-dimethyl-8-(3'-deoxy-D-ribityl)lumazine is a potent inhibitor of the enzyme[49].

A number of lumazine derivatives bearing different substituents at positions 6 and 7 have been prepared and tested with purified enzymes

TABLE II

INHIBITION OF RIBOFLAVIN SYNTHETASE FROM *A. gossypii* AND YEAST BY
6,7-DIMETHYLLUMAZINE DERIVATIVES AND FLAVINS

Compound	Enzyme from	
	A. gossypii Competitive inhibition (K_i) (M)	*Yeast* Competitive inhibition (K_i) (M)
6,7-Dimethyl-8-substituted lumazines (substituent)		
D-Ribityl	$3.1 \cdot 10^{-5}$ [a]	$1 \cdot 10^{-5}$ [a]
D-Arabityl		
L-Arabityl		
D-Lyxityl		
L-Lyxityl		
D-Xylityl	$8.7 \cdot 10^{-5}$	$1 \cdot 10^{-5}$
L-Xylityl		
5′-Deoxy-D-ribityl	$1 \cdot 10^{-4}$ [b]	$1 \cdot 10^{-3}$ [b]
	$4.2 \cdot 10^{-4}$ [a]	$4 \cdot 10^{-5}$ [a]
3′-Deoxy-D-ribityl [c]	—	$8 \cdot 10^{-7}$
2′-Deoxy-D-ribityl [c]	—	
D- and L-Threityl [c]	—	$1 \cdot 10^{-3}$
D- and L-Erythrityl [c]	—	$1 \cdot 10^{-3}$
D-Galactityl		
D-Glucityl		
D-Mannityl		
β-Hydroxyethyl		
Methyl		
Hydrogen		
Flavins		
Riboflavin		$5 \cdot 10^{-6}$
Lumiflavin		
5′-Deoxyriboflavin	$1.9 \cdot 10^{-5}$	—
Riboflavin 2-imine		
FAD		$> 1 \cdot 10^{-3}$
FMN		$> 1 \cdot 10^{-3}$

Data from Winestock *et al.*[47] and Harvey and Plaut[48]. Values for constants are not reported whenever equimolar concentrations of substrate and analogue tested gave less than 10% inhibition. Dashes indicate that the compound was not tested.

[a] This number represents K_m.

[b] This was not strictly competitive inhibition. The value of K_i has been estimated by use of the Dixon plot.

[c] R. L. Beach, T. Aogaichi and G. W. E. Plaut, unpublished observations.

TABLE III

INHIBITION OF RIBOFLAVIN SYNTHETASE FROM *A. gossypii* BY SUBSTITUTED
LUMAZINE DERIVATIVES

Compound	Inhibition[a] (%)	Concentration levels tested (mM)	Competitive inhibition (K_i) (M)
6,7-Substituted 8-(1'-D-ribityl)lumazines (substituent)			
Methyl, ethyl	15	0.15, 0.30	
Methyl, *n*-propyl	15	0.14, 0.28	
Methyl, *n*-butyl	15	0.14, 0.28	
Methyl, *n*-pentyl	20	0.15, 0.32	$2.1 \cdot 10^{-4}$
Diethyl	9	0.14, 0.28	
Di-*N*-propyl	8	0.13, 0.26	
Methyl, phenyl	8	0.13, 0.26	
Diphenyl	6	0.15	
7-(2-Hydroxy-2-methyl-3-oxobutyl)-6-methyl	42	0.15, 0.30	$2.5 \cdot 10^{-4}$ [b]
6-Methyl-7-hydroxy	92	0.003–0.15	$2.0 \cdot 10^{-6}$ $(5 \cdot 10^{-7})$ [c]
6,7-Dihydroxy	100	$2 \cdot 10^{-6}$–0.15	$9 \cdot 10^{-9}$ [d] $(< 3 \cdot 10^{-8})$ [c, d]
Other lumazines			
5,6,7,8-Tetrahydro-9-(1'-D-ribityl)-isoalloxazine	18	0.15, 0.26	$1.6 \cdot 10^{-4}$
6,7-Diethyl-8-(1'-β-hydroxyethyl)lumazine	−3	0.15	
6,7-Diethyl-8-methyllumazine	3	0.22, 0.44	
6,7-Diisopropyl-8-methyllumazine	2	0.20, 0.40	
6-Methyl-7-hydroxy-8-(1'-β-hydroxyethyl)lumazine	5	0.21, 0.42	
6,8-Dimethyl-7-hydroxylumazine	4	0.24, 0.48	

Data from Winestock *et al.*[47].

[a] The average percentage inhibition is reported for approximately equal concentrations of substrate (0.15 mM) and inhibitor with the exception of the last four lumazines, for which the value is reported for the lowest level tested.

[b] This value is an approximation, since the crystalline compound is not available. The concentration was estimated from the value of the molar extinction of the analogous 8-methyllumazine.

[c] Enzyme from yeast[48].

[d] This was not strictly competitive inhibition. The value of K_i has been esitmated by use of the Dixon plot.

from *A. gossypii* and yeast (Table III)[47]. None of these was converted to a flavin derivative; however, a number of these compounds proved to be competitive inhibitors of the riboflavin synthetase reaction. For example, 6-methyl-7-pentyl-8-ribityllumazine* and 5,6,7,8-tetrahydro-9-(1′-D-ribityl)-

TABLE IV

INHIBITION OF RIBOFLAVIN SYNTHETASE FROM *A. gossypii*
BY 2-AMINOPTERIDINES

Compound	Inhibition[a] (%)	Concentration levels tested (mM)	Inhibition (K_i) (M)
2-Amino-4,6-dihydroxy-8-D-ribityl-7(8H)-pteridinone	97	0.1–0.15	$2.4 \cdot 10^{-6}$ [b]
2-Amino-4-hydroxy-8-D-ribityl-7(8H)-pteridinone	3	0.15	
2-Amino-4-hydroxy-6-methyl-8-D-ribityl-7(8H)-pteridinone	0	0.15	
2-Amino-4-hydroxy-6-methyl-8-D-glucityl-7(8H)pteridinone	4	0.15	
2-Amino-4-hydroxy-6,7-dimethyl-8-D-ribityl-(7,8H)-pteridine	4	0.15	
2-Amino-6,7-dimethyl-4-hydroxypteridine	−2	0.15	
2-Amino-4,7-dihydroxy-6-methylpteridine	0	0.15	
2-Amino-6,7-diphenyl-4-hydroxypteridine	−7	0.15	
Xanthopterin	0	0.15	
Leucopterin	0	0.15	
Deaminoleucopterin	−7	0.15	

Data from Winestock *et al.*[47].
[a] The percentage inhibition is reported for 0.15 mM substrate and 0.15 mM inhibitor.
[b] This was not strictly competitive inhibition. The value of the competitive inhibition constant, K_i, has been estimated by use of the Dixon plot.

* The configuration of substituents at the 6- and 7-carbon atoms is probably as indicated, since it was shown recently by NMR spectroscopy that the corresponding 6,7-substituted methyl-, ethyl-derivative has the configuration 6-ethyl-7-methyl-8-ribityllumazine[50].

isoalloxazine had values of K_i about 10 times higher than that of K_m. The natural substance 6-methyl-8-hydroxy-8-ribityllumazine is a potent competitive inhibitor of the reaction and 6,7-dihydroxy-8-ribityllumazine exhibits a value of K_i which is approximately 1/1000th that of K_m of the substrate. It is noteworthy that the presence of a D-ribityl group is necessary for activity of these inhibitory analogues. Thus, replacement of the ribityl group by hydrogen, methyl, or β-hydroxyethyl in 6-methyl-7-hydroxy substituted or in 6,7-dihydroxy substituted lumazines led to loss of activity.

Replacement of the oxygen function in position 2 of a number of lumazines by an amino group leads to compounds with diminished or no activity (Table IV). The 2-amino analogue of the substrate 6,7-dimethyl-8-ribityllumazine, namely, 2-amino-4-hydroxy-6,7-dimethyl-8-ribityl-7(8H)-pteridine, was neither substrate nor inhibitor of the synthetase. 2-Amino-4,6-dihydroxy-8-ribityl-7(8H)-pteridinone, the analogue of the most potent inhibitor tested, 6,7-dihydroxy-8-ribityllumazine, was considerably less active, and 2-amino-4-hydroxy-6-methyl-8-ribityl-7-(8H)-pteridinone was inactive although the corresponding lumazine was an effective competitive inhibitor. A number of other 2-aminopteridines were also found to be inactive, including the potent folic acid inhibitors, 2-amino-4-hydroxy-6,7-dimethylpteridine and 2-amino-4-hydroxy-6,7-diphenylpteridine[47].

(e) Purification and certain properties of riboflavin synthetase

The enzyme has been partially purified from extracts of E. coli, A. gossypii[34] and spinach leaves[45]. Riboflavin synthetase has been purified several thousand fold from extracts of baker's yeast[48] and, in recent improvements of the method of isolation, activity can be recovered from the extract in 30–40% yield[51,52].

The best preparations from yeast had a specific activity of 14 000 mμmoles of riboflavin formed per mg of protein per hour at 37° and showed a single zone of protein possessing activity when examined by cellulose acetate and polyacrylamide disc gel electrophoresis[51]. Preliminary studies in the ultracentrifuge showed a sedimentation constant of approximately 4.3, indicating a molecular weight of around 80 000[48]. The effect of pH on K_m and V_m is summarized in Table V. Although the enzyme is inactive between pH 4.4 and pH 4.8, it is somewhat more stable under acidic conditions than at pH 7.0 where it is maximally active. Oxidizing compounds readily and irreversibly inactivate riboflavin synthetase. Dissolved molecular oxygen is particularly destructive. It is stabilized by a number of reducing agents (Na_2SO_3, mercaptoethanol and dithiothreitol).

p-Chloromercuribenzenesulfonate or p-chloromercuribenzoate at concentrations greater than 10^{-4} M completely inhibit riboflavin synthetase activity (50% inhibition at $3 \cdot 10^{-5}$ M). The inhibition can be almost com-

TABLE V

EFFECT OF pH ON K_m AND V_m OF RIBOFLAVIN SYNTHETASE FROM YEAST

pH	K_m $(M \cdot 10^5)$	Relative V_m[a] (%)
5.8	3.8	41
6.4	4.4	62
7.0	1.0	100
7.6	1.1	61
8.2	4.1	32
8.9	47.0	6

Data from Harvey and Plaut[48].
[a] Maximal velocities are expressed as the percentage of the activity at pH 7.0.

pletely reversed by subsequent addition of an excess of cysteine or thioethanol. Partial protection against the inhibition is afforded by the substrate 6,7-dimethyl-8-ribityllumazine or the competitive inhibitor 6,7-dimethyl-8-D-xylityllumazine when added before the mercurial; however, when the order of addition of components is reversed, the inhibition by p-chloromercuribenzenesulfonate cannot be relieved by these lumazine derivatives[48]. Relatively high concentrations of iodoacetamide inhibit activity[52]. These properties suggest that a sulfhydryl group may play a part in the mechanism of catalysis of the enzyme.

Riboflavin is an effective competitive inhibitor of the enzyme from yeast (approximate K_i, $5 \cdot 10^{-6}$ M)[48], but not of that from A. gossypii[34]. This inhibition accounts for the decrease in reaction velocity after prolonged incubation of the yeast enzyme and substrate and may have a bearing on the fact that large quantities of riboflavin are accumulated by cultures of A. gossypii, but not by baker's yeast. 4-Ribitylamino-5-amino-2,6-dihydroxypyrimidine, the second reaction product, does not contribute appreciably to product inhibition. Riboflavinimine, FAD, or FMN are very poor inhibitors of the conversion of 6,7-dimethyl-8-ribityllumazine to riboflavin[47].

3. Mechanism of riboflavin synthetase from yeast

(a) Complexes of the enzyme with lumazine derivatives and riboflavin[48]

The purified enzyme has a yellow color and upon denaturation of the protein by heat, urea, or trichloroacetic acid, etc., riboflavin is liberated. The spectral properties of the complex are similar to those of other flavoproteins[53]. That is, the visible part of the spectrum is displaced to longer wavelengths in the bound compared to the unbound form of the flavin, Fig. 3. Riboflavin is bound very tightly to the protein and cannot be removed by prolonged dialysis or treatment by chromatography on molecular sieves. *p*-Chloromercuribenzene sulfonate leads to dissociation of the enzyme-riboflavin complex and treatment with charcoal removes the flavin, indicating that binding is not due to covalent linkages. The charcoal-treated protein retains catalytic activity and no induction in rate of riboflavin formation from 6,7-dimethyl-8-ribityllumazine is observed with time of incubation, suggesting that bound riboflavin does not act as a cofactor in the transformation.

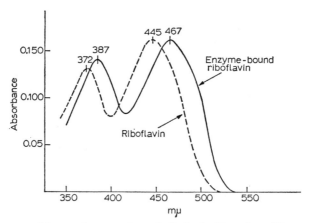

Fig. 3. Spectra of free and enzyme-bound riboflavin. Data from Harvey and Plaut[48].

A number of lumazine derivatives will bind to the protein to form isolatable complexes with the enzyme. These include the substrates, 6,7-dimethyl-8-(1'-D-ribityl)lumazine and 6,7-dimethyl-8-[1'-(5'-deoxy-D-ribityl)]lumazine and the competitive inhibitors 6-methyl-7-hydroxy-8-ribityllumazine and 6,7-dimethyl-8-D-xylityllumazine. Only those substances are bound which

TABLE VI

BINDING OF RIBOFLAVIN AND LUMAZINE DERIVATIVES BY YEAST RIBOFLAVIN
SYNTHETASE

These results were obtained in experiments with aliquots from the same enzyme preparations, specific activity 1500

Enzyme-complex isolated	Buffer		Sephadex column dimensions (cm)	Net binding (mμmoles/mg protein)
	Composition[a]	pH		
Riboflavin	Na$_2$SO$_3$	7.0	1 × 15	6.0
Riboflavin	Na$_2$SO$_3$	7.0	1 × 10	7.5
Riboflavin	Citrate	4.5	1 × 15	6.5
6-Methyl-7-hydroxy-8-ribityllumazine	Na$_2$SO$_3$	7.0	1 × 18	5.4
6-Methyl-7-hydroxy-8-ribityllumazine	Na$_2$SO$_3$	7.0	2 × 20	5.7
6-Methyl-7-hydroxy-8-ribityllumazine	Na$_2$SO$_3$	7.0	2 × 20	5.2
6-Methyl-7-hydroxy-8-ribityllumazine	Citrate	4.5	1 × 15	4.8[b]
6,7-Dimethyl-8-D-xylityllumazine	Na$_2$SO$_3$	7.0	1 × 10	7.0
6,7-Dimethyl-8-L-xylityllumazine	Na$_2$SO$_3$	7.0	1 × 10	0.7

Data from Harvey and Plaut[48].
[a] Na$_2$SO$_3$ and citrate were used at concentrations of 0.01 M and 0.10 M, respectively.
[b] The enzyme–lumazine complex from the preceding experiment was chromatographed again at pH 4.5 as indicated.

have a kinetic effect on the enzyme. For example, 6,7-dimethyl-8-L-xylityllumazine does not inhibit activity and is not bound. The complexes contain riboflavin, substrate or analogues of 6,7-dimethyl-8-ribityllumazine, respectively, in equivalent molecular amounts (Table VI).

The extent of binding of various substances to the protein is relatively constant in buffers of varying composition and pH (Table VI). It was possible, therefore, to form complexes between protein and the substrate 6,7-dimethyl-8-ribityllumazine at pH 4.5 where product formation does not occur; with analogues of the substrate the composition of the complex is the same as that at pH 7.0 where the enzyme catalyzes the conversion of the substrate to riboflavin. As in the case of riboflavin, binding of the lumazine derivatives to the enzyme leads to a shift to longer wavelengths of the absorption maxima (Table VII). The emission maximum of the fluorescence emission spectrum of free and enzyme-bound riboflavin are identical, although there is a marked decrease of fluorescence intensity on

TABLE VII

ABSORPTION SPECTRA OF ENZYME-BOUND RIBOFLAVIN AND LUMAZINES

Compound	pH	Absorption maximum		Absorbance at maximum on binding
		Free ($m\mu$)	Enzyme-bound ($m\mu$)	
Riboflavin	7.0	455	467	Unchanged
6,7-Dimethyl-8-ribityllumazine	4.5	407	412	Unchanged
6,7-Dimethyl-8-D-xylityllumazine	7.0	407	412	
6-Methyl-7-hydroxy-8-ribityllumazine	7.0	340	345	

Data from Harvey and Plaut[48].

binding. With lumazine derivatives there is a shift to shorter wavelength of the emission maxima on binding, however, fluorescence intensity is essentially unchanged (Table VIII).

The complexes of enzyme with riboflavin and lumazine derivatives are very stable thermodynamically, indicated by the inability to demonstrate dissociation by dialysis or by chromatography on the molecular sieve, Sephadex G-25. However, such complexes are kinetically labile since a tightly bound compound can be rapidly displaced by an excess of the same or another substance with which the enzyme can combine. Thus, displacement of enzyme-bound riboflavin by 6-methyl-7-hydroxy-8-ribityllumazine has been demonstrated, resulting in binding of the lumazine to the protein in amounts equivalent to the quantity of riboflavin liberated. The rate of exchange is rapid; it has been shown that about one-half of enzyme-bound riboflavin can be displaced by 6-methyl-7-hydroxy-8-ribityllumazine within 2–5 sec at pH 7.0 and 10°. Rapid equilibrium must also exist between the

TABLE VIII

FLUORESCENCE EMISSION SPECTRA OF ENZYME-BOUND RIBOFLAVIN
AND LUMAZINES

| Compound | pH | Emission maximum[a] | | Fluorescence intensity at maximum on binding |
		Free (mμ)	Enzyme bound (mμ)	
Riboflavin	7.0	522	522	50–70% decrease
6,7-Dimethyl-8-ribityllumazine	4.5	483	473	Unchanged
6,7-Dimethyl-8-D-xylityllumazine	7.0	483	473	Unchanged
6-Methyl-7-hydroxy-8-ribityllumazine	7.0	422	407	Unchanged

Data from Harvey and Plaut[48].
[a] Excitation at 366-mμ line of mercury lamp with Corning filter No. 7-54; temperature, 10°.

free and bound forms of the same substance since exchange of enzyme-bound labeled 6,7-dimethyl-8-ribityllumazine with unlabeled 6,7-dimethyl-8-ribityllumazine has been observed.

(b) Kinetics[48]

The stoichiometry indicates that riboflavin synthetase catalyzes a two-substrate reaction. The reaction can be visualized (Fig. 4) to involve addition of two molecules of 6,7-dimethyl-8-ribityllumazine (L) to sites on the

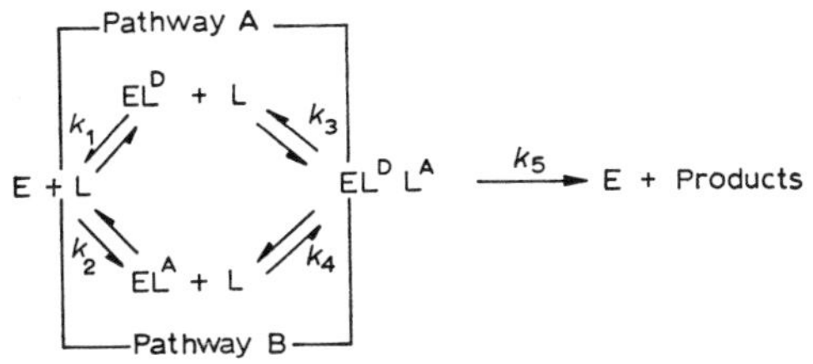

Fig. 4. Random binding scheme of substrate to riboflavin synthetase. L is 6,7-dimethyl-8-ribityllumazine, L^D is 6,7-dimethyllumazine bound to donor site, L^A is 6,7-dimethyl-8-ribityllumazine bound to acceptor site.

enzyme (E) leading to donation (E–L_D) and acceptance (E–L_A) of the 4-carbon moiety from one molecule of lumazine to another in a ternary complex (E–$L_D L_A$) to form products. If one assumes that the reactions described by dissociation constants K_1 to K_4 are in rapid equilibrium, and that the decomposition of the ternary complex (E–$L_D L_A$) is rate-limiting and irreversible (k_5), the system (Fig. 4) can be described by the usual formulation for a two-substrate reaction.

$$\frac{1}{v} = \frac{1}{V_m} \times \left[\frac{K_3 + K_4}{(L)} + \frac{K_1 K_3}{(L)^2} \right]$$

The potential second-order term, $K_1 K_3 / L^2$ (in the case of pathway B the term would be $K_2 K_4 / L^2$) makes a negligible contribution to the reaction rate since it has been shown that a plot of $1/v$ against $1/L$ is linear even at substrate concentrations as low as $5.8 \cdot 10^{-7}$ M, although the apparent K_m is $1 \cdot 10^{-5}$ M. In addition, riboflavin synthetase shows zero to first order kinetics over a wide range of substrate concentrations. The second term of the equation can, therefore, be discarded and the apparent Michaelis constant, K_m, thus becomes equal to the sum of the dissociation constants of the donor and acceptor sites of the ternary enzyme–lumazine complex, $i.e.$, $K_3 + K_4$. The values of dissociation constants of sites binding the first (K_1 or K_2) and second molecules (K_3 or K_4) of the lumazine must, therefore, differ substantially.

(c) Binding of 4-carbon donor and acceptor[48]

The results of the kinetic analysis suggest that binding of the lumazine occurs with different affinities to the donor and acceptor sites on the enzyme. However, it is difficult experimentally to distinguish between these sites because the same substrate is bound to each. An approach to this problem became possible by use of 6,7-dimethyl-8-[1'-(5'-deoxyribityl)]lumazine. The Michaelis constants of the 5'-deoxyribityl derivative and the natural substrate are similar (Table II), however, the maximal velocity of the conversion of the deoxyribityllumazine to 5'-deoxyriboflavin is about 1/100th that of the transformation of the ribityl compound to riboflavin. Nevertheless, the rate of formation of flavin from 6,7-dimethyl-8-ribityllumazine is decreased only slightly by the presence of 6,7-dimethyl-8-[1'-(5'-deoxyribityl)-lumazine (apparent $K_i = 1 \cdot 10^3$ M $vs.$ $K_m = 4 \cdot 10^{-5}$ M). This suggested the possible formation of flavin by transfer of a 4-carbon moiety from one lumazine analogue to the other. This possibility was studied by examining the products arising from a mixture of 6,7-dimethyl-8-ribityllumazine and 6,7-di-[^{14}C]-methyl-8-(5-deoxyribityl)-lumazine incubated with riboflavin synthetase. Of the products which could be expected to be formed in such an experiment, Fig. 5, only labeled riboflavin could be detected (Scheme C). The extent of transfer of radioactivity from 6,7-di-[^{14}C]methyl-8-(5′-deoxyribityl)-lumazine to riboflavin formed from the mixture of substrates indicated that the deoxyribityllumazine derivative is an effective donor of the 4-carbon portion but interacts poorly at the acceptor site.

This observation was confirmed and extended by experiments in which stoichiometric amounts of enzyme–lumazine complexes were permitted to interact with equivalent amounts of various lumazine derivatives in the free

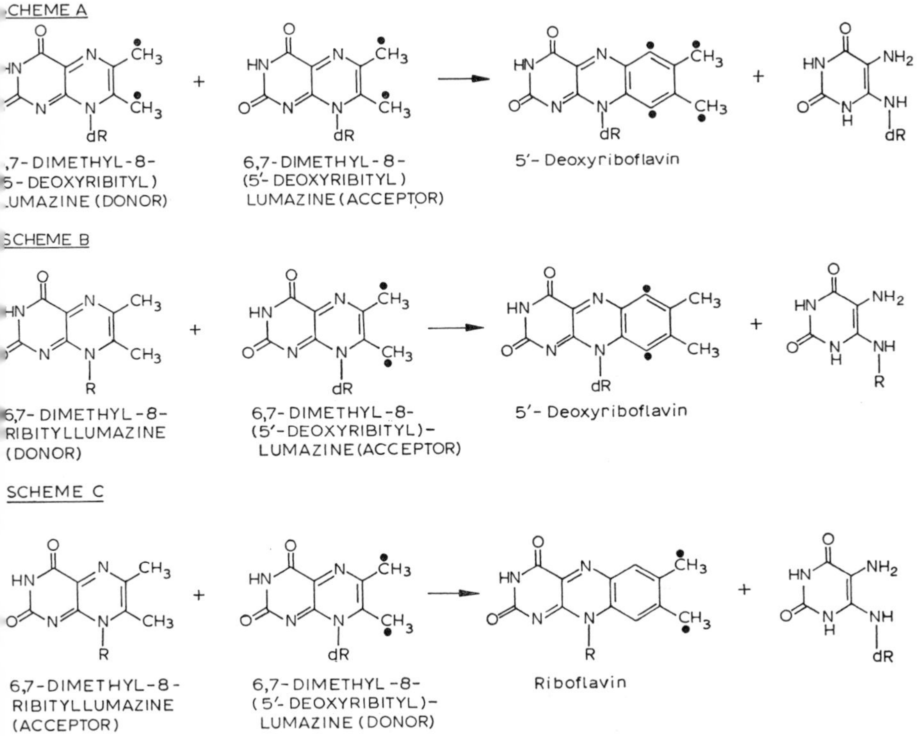

SCHEME A

6,7-DIMETHYL-8-(5-DEOXYRIBITYL)LUMAZINE (DONOR)   6,7-DIMETHYL-8-(5′-DEOXYRIBITYL)LUMAZINE (ACCEPTOR)   5′-Deoxyriboflavin

SCHEME B

6,7-DIMETHYL-8-RIBITYLLUMAZINE (DONOR)   6,7-DIMETHYL-8-(5′-DEOXYRIBITYL)-LUMAZINE (ACCEPTOR)   5′-Deoxyriboflavin

SCHEME C

6,7-DIMETHYL-8-RIBITYLLUMAZINE (ACCEPTOR)   6,7-DIMETHYL-8-(5′-DEOXYRIBITYL)-LUMAZINE (DONOR)   Riboflavin

Fig. 5. Possible reactions between 6,7-dimethyl-8-ribityllumazine and 6,7-di-[^{14}C]methyl-8-(5′-deoxyribityl)lumazine. ●, ^{14}C; dR, 5′-deoxyribityl group; R, ribityl group.

TABLE IX

DONOR–ACCEPTOR SPECIFICITY OF ENZYME-BOUND LUMAZINES

Expt. 1, an aliquot of enzyme–6,7-di-[^{14}C]methyl-8-ribityllumazine containing 2.0 mμ-moles of bound lumazine (188 000 c.p.m. per μmole) was adjusted to pH 7, warmed from 0–37°, and 2.0 mμmoles of free 6,7-dimethyl-8-(5'-deoxyribityl)lumazine were added. Expt. 2, a solution of enzyme–6,7-dimethyl-8-(5'-deoxyribityl)lumazine containing 35 mμmoles of bound lumazine was adjusted to pH 7, warmed from 0–37°, and 34 mμmoles of 6,7-di-[^{14}C]methyl-8-ribityllumazine (81 000 c.p.m. per μmole) were added. Expt. 3, a solution of enzyme–6,7-di-[^{14}C]methyl-8-(5'-deoxyribityl)lumazine containing 5.1 mμ-moles of bound lumazine (25 000 c.p.m. per μmole) was warmed from 0–37° and 5.1 mμmoles of 6,7-dimethyl-8-ribityllumazine were added. After incubation at 37° for 1 h, the reaction mixtures (Expts. 1–3) were deproteinized with 5% trichloroacetic acid, unlabeled riboflavin and deoxyriboflavin were added, and the reaction components were separated by column chromatography on Lloyd's reagent followed by paper chromatography in 1-butanol–ethanol–water (500:175:360, v/v)

Reaction components	Recovery of radioactivity from substrate in reaction product	
	Riboflavin (%)	Deoxyriboflavin (%)
1. Enzyme–6,7-di-[^{14}C]methyl-8-ribityllumazine + 6,7-dimethyl-8-(5'-deoxyribityl)lumazine	1	0
2. Enzyme–6,7-dimethyl-8-(5'-deoxyribityl)lumazine + 6,7-di-[^{14}C]methyl-8-ribityllumazine	33[a]	1
3. Enzyme–6,7-di-[^{14}C]methyl-8-(5'-deoxyribityl)lumazine + 6,7-dimethyl-8-ribityllumazine	27[a]	0

Data from Harvey and Plaut[48].

[a] Essentially quantitative formation of riboflavin occurred in Expts. 2 and 3 as judged by the results of chromatography of the incubation mixture on Lloyd's reagent; no unreacted lumazine could be detected in the appropriate effluents from such columns. The low recovery of radioactivity in these experiments recorded in the table is primarily due to incomplete elution of the flavins from the paper in the subsequent paper-chromatography step.

form (Expts. 1–3, Table IX). Essentially no flavin formation occurred between enzyme–6,7-di-[^{14}C]methyl-8-ribityllumazine and free 6,7-dimethyl-8-(5'-deoxyribityl)lumazine (Expt. 1, Table IX) whereas rapid formation of radioactive riboflavin, but not deoxyriboflavin, occurred in the reaction between enzyme–6,7-dimethyl-8-(5'-deoxyribityl)lumazine and the unbound ribityllumazine (Expts. 2 and 3, Table IX). These experiments suggest that the enzyme–6,7-dimethyl-8-(5'-deoxyribityl)lumazine complex (and possibly the enzyme–6,7-dimethyl-8-ribityllumazine complex) as prepared by chromatography on columns of Sephadex G-25 contains the lumazine bound at the donor site of the enzyme.

(d) Evidence for existence of a ternary complex[48]

The enzyme complexes as isolated contain equivalent amounts of lumazine derivatives. The stoichiometry of the reaction suggests that the ternary complex, however, should contain two molecules of substrate. The examination of the enzyme by fluorescence polarization reveals that titration of an isolated enzyme–6,7-dimethyl-8-ribityllumazine complex with additional free substrate leads to additional polarization of fluorescence equivalent to binding of a second lumazine molecule. A similar interaction has been shown between enzyme–6,7-dimethyl-8-ribityllumazine and 6,7-dimethyl-8-(5'-deoxyribityl)lumazine, and of enzyme–6,7-dimethyl-8-(5'-deoxyribityl)-lumazine with 6,7-dimethyl-8-ribityllumazine. The binding of additional deoxylumazine derivative by enzyme–6,7-dimethyl-8-(5'-deoxyribityl)luma-zine appears to be weak and neither 6,7-dimethyl-8-D-xylityllumazine nor 6,7-dimethyl-8-L-xylityllumazine appears to be bound by the enzyme–6,7-dimethyl-8-ribityllumazine complex. It is pertinent that in the latter reactions flavin formation occurs only slowly or not at all. The low activity of the deoxylumazine as a 4-carbon acceptor cannot be attributed solely to a low affinity to the acceptor site for the substance, since enhancement of polarization of fluorescence occurs upon titration of enzyme–6,7-dimethyl-8-ribityl-lumazine with the deoxy derivative though flavin is not readily formed. It may be that the low rate of flavin formation is due in addition to the slow conversion of this particular ternary complex ($E–L_D–dL_A$) to products.

(e) Studies of nature of reaction intermediates

Suggestions for the possible mode of action of riboflavin synthetase have come from studies on the chemical transformation of lumazine derivatives to flavins. Thus, Birch and Moye[54] and Cresswell and Wood[33] synthesized 7-(2-hydroxy-2-methyl-3-oxobutyl)-6,8-dimethyllumazine and 7-(2-hydroxy-

2-methyl-3-oxobutyl)-6-methyl-8-(1′-D-ribityl)lumazine, respectively, and found that there was a rapid base-catalyzed transformation of these substances to the corresponding flavins (Fig. 6). However, 7-(2-hydroxy-2-methyl-3-oxobutyl)-6-methyl-8-(1′-D-ribityl)lumazine is not a substrate, but rather an inhibitor of the enzyme[47] (Table III).

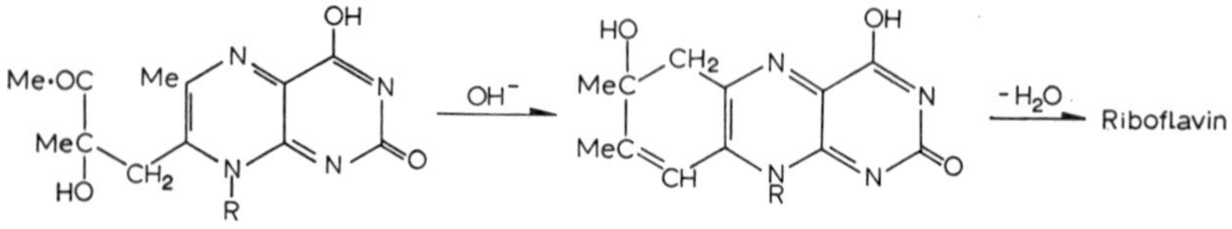

Fig. 6. Chemical transformation of 7-(2-hydroxy-2-methyl-3-oxobutyl)-6-methyl-8-substituted lumazines to flavins. From Birch and Moye[54] and Cresswell and Wood[33].

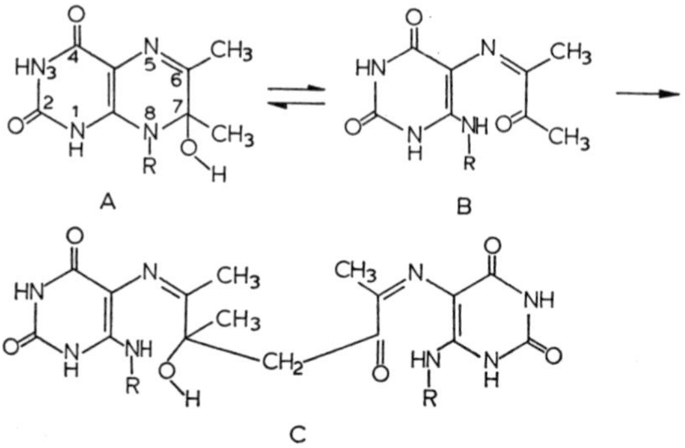

Fig. 7. Mechanism of Rowan and Wood[55,56] for chemical conversion of 6,7-dimethyl-8-substituted lumazine to flavin.

Rowan and Wood[55,56] made the important discovery that non-enzymatic formation of riboflavin occurs when 6,7-dimethyl-8-ribityllumazine is refluxed in phosphate buffer under neutral conditions in an inert atmosphere. They proposed (Fig. 7) that the chemical transformation involves an opening of the pyrazine ring between nitrogen atom 8 and carbon 7 initiated by nucleophilic attack, leading to initial formation of a covalent hydrate of the lumazine (A) followed by reversible ring opening of the pyrazine ring (B). In a subsequent aldol condensation, two molecules of this substance have been suggested to condense to form a dimeric intermediate (C) which cyclizes to riboflavin by a mechanism involving the 6-methyl group and carbon 6 of the initial lumazine. Spectral photometric observations support the occurrence of a ring-opening reaction preceded by hydration of the pyrazine portion of the lumazine under alkaline conditions[57] and measurement of NMR spectra[50] may suggest that a compound (Fig. 7, B) containing a keto group adjacent to the 7-methyl group is formed from certain lumazine derivatives in alkali. Such an α-methyl ketone would participate readily in hydrogen–tritium (or deuterium) exchange with solvent water at the methyl group, a reaction which has been observed to occur more rapidly under alkaline than neutral conditions[50]. However, it is uncertain whether the ingenious mechanism of Rowan and Wood[55,56] explains the enzyme-catalyzed transformation of 6,7-dimethyl-8-ribityllumazine to riboflavin.

Intermediates such as (A) and (B) in Fig. 7 should no longer show light absorption at 407 mμ[57], yet complexes of substrate and enzyme were found to still possess absorption at the long wavelength and the transformation of such complexes to product, as followed by continuous scanning of the spectral change, indicated that essentially all the absorption observed in the visible portion was accounted for by the spectra of 6,7-dimethyl-8-ribityllumazine and riboflavin[48]. Furthermore, attempts to trap the proposed α-methyl ketone intermediate (Fig. 7, B) by inclusion of high concentrations of hydroxylamine in the enzymatic incubation mixture were unsuccessful and the possible formation of a Schiff base from such a keto derivative and an amino group on the enzyme could not be demonstrated[48]. The methods could have been inadequate for the detection of intermediates occurring in small amounts; however, recent studies on the transformation of 6,7-dimethyl-8-ribityllumazine to riboflavin have made it necessary to consider alternative mechanisms for this transformation. Thus, the ring-opening reaction and hydration of the pyrazine portion of the lumazine under alkaline conditions is readily reversible by acidification[50,57]. However, the rate of the chemical exchange of hydrogen at the 7-methyl group of 6,7-dimethyl-8-ribityllumazine and its derivatives is maximal not only in alkali but also in acid[50], and chemical synthesis of riboflavin from the lumazine occurs in good yield upon heating in 0.1 M HCl under a nitrogen atmosphere; a mechanism not involving an initial ring-opening reaction at the level of the lumazine has been proposed to explain these results[58].

Studies by NMR spectroscopy showed that chemical deuterium–hydrogen exchange with solvent D_2O or H_2O occurs at the substituent of carbon 7, but not at carbon 6, of several lumazines. The rate of this exchange is enhanced when a substituent is present at position 8 of the lumazine molecule and occurs most rapidly under either alkaline or acid conditions[50]. In an extension of these investigations to the riboflavin synthetase system[52] it was found that the rates of enzymatic conversion of 6-methyl-7-deutero-methyl-8-ribityllumazine and 6-deuteromethyl-7-methyl-8-ribityllumazine to riboflavin are 80% and 20% of that of the unlabeled substrate, respectively. While the V_{max} of the 6-deuteromethyl-7-methyl-8-ribityllumazine is lower than that of the unlabeled substrate, the values of K_m are nearly identical. The 6-deuteromethyl derivative is a competitive inhibitor of the enzymatic conversion of the unlabeled substrate yielding about equal values of K_i and K_m. The greater discrimination against hydrogen isotopes at the 6-methyl group compared to the 7-methyl group, has also been shown with tritiated lumazines where the predominantly labeled species is likely to contain a single atom of tritium per selectively labeled methyl group. Preferential elimination of protons over tritions into water and retention of the label in riboflavin occurs with 6-tritiomethyl-7-methyl-8-ribityllumazine, whereas there is relatively little isotope discrimination with 6-methyl-7-tritiomethyl-8-ribityllumazine.

The enhancement of the rate of proton elimination from the 7-methyl group of tritiated substrate by riboflavin synthetase has been studied in considerable detail[52]. Under optimal conditions (pH 6.8) rates were comparable for riboflavin formation and the enzymatic elimination of tritions into water from 6-methyl-7-tritiomethyl-8-ribityllumazine. The elimination of tritions must occur at the molecule of the lumazine which acts as acceptor of the 4-carbon moiety, since the specific radioactivity of the methyl group of riboflavin formed is comparable to that of the initial substrate. Catalysis of the hydrogen–tritium exchange reaction between water and the 7-methyl group of lumazines is an inherent part of the riboflavin synthetase reaction. This could be demonstrated with certain analogues of 6,7-dimethyl-8-ribityl-lumazines labeled with tritium in the 7-methyl group which, though bound to the enzyme with about the same affinity, are either converted to flavin at a slower rate than the natural substrate or cannot be transformed at all. With substrates such as 6-methyl-7-tritiomethyl-8-[1'-(5'-deoxy-D-ribityl)]-lumazine and 6-deuteromethyl-7-tritiomethyl-8-ribityllumazine, where K_m is about the same as that of 6,7-dimethyl-8-ribityllumazine but V_{max} is markedly less, the velocities of tritium–hydrogen exchange are faster than flavin formation; the apparent first-order rate constants for tritium–hydrogen exchange are about the same for the substrate 6-methyl-7-tritiomethyl-8-ribityllumazine and the analogues. 6-Methyl-7-tritiomethyl-8-(1'-D-xylityl)-lumazine, which is not converted to flavin but is a competitive inhibitor

TABLE X

COMPARISON OF ENZYMATIC RATES OF FLAVIN SYNTHESIS AND OF EXCHANGE
REACTION BETWEEN 7-TRITIOMETHYL GROUP OF VARIOUS LUMAZINE DERIVA-
TIVES AND WATER

Compound[a]	Relative rates	
	Exchange reaction (%)	Flavin formation (%)
6,7-Dimethyl-8-D-ribityllumazine	100	100
6-Deuteromethyl-7-methyl-8-D-ribityllumazine	80	20
6,7-Dimethyl-8-[1'-(5'-deoxy-D-ribityl)]lumazine	39	1
6,7-Dimethyl-8-D-xylityllumazine	27	0
6,7-Dimethyl-8-L-xylityllumazine	1	0

Data from Plaut et al.[52].
All incubations were carried out in 0.05 M imidazole·HCl at pH 6.8 and 25°.
[a] All compounds listed were labeled with tritium at the 7-methyl group.

and forms complexes with the enzyme, participates in tritium–hydrogen exchange. The relationship between substrate specificity and the exchange reaction is shown by the observation that the enzyme does not increase liberation of tritions into water from 6-methyl-7-tritiomethyl-8-[(1'-L-xylityl)]-lumazine, a compound which does not interact with the enzyme (Table X). Furthermore, proportional inhibition of riboflavin synthesis and trition elimination from 6-methyl-7-tritiomethyl-8-ribityllumazine has been obtained with certain sulfhydryl binding reagents (p-chloromercuribenzene sulfonate, 5,5'-dithio-bis-(2-nitrobenzoate), iodoacetamide) further indicating that the exchange reaction and flavin formation are catalyzed by the same protein.

A mechanism of action of riboflavin synthetase has been proposed[49,52] (Fig. 8) which may account for the experimental findings with the enzyme. Binding of the substrate is visualized to occur at two sites, one leading to donation of a 4-carbon unit (donor site) and the other to its acceptance (acceptor site) (I). Binding of substrate to enzyme is relatively insensitive to pH occurring, for example, at pH 4.5 without formation of riboflavin.

References p. 44

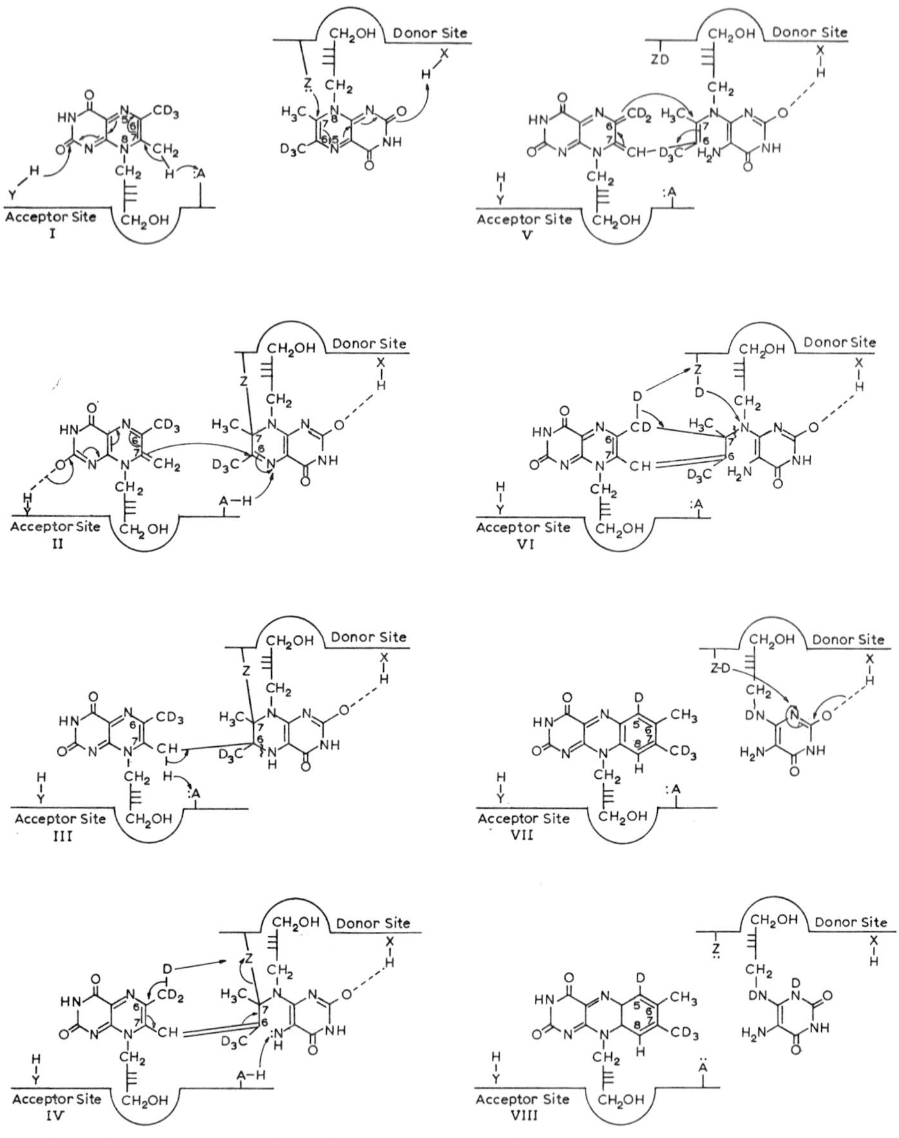

Fig. 8. Proposed mechanism of enzymatic conversion of 6,7-dimethyl-8-ribityllumazine to riboflavin. From Beach and Plaut[49].

The relatively pH-insensitive binding of the lumazine to what appears to be the donor site may occur at a hydrophobic region on the protein and may account for the bathochromic shifts of absorption maxima in the visible region of the spectrum which have been observed when riboflavin and lumazine derivatives are bound to the enzyme (Fig. 3, Table VII). The dependence of activity on pH may be due to interaction at the acceptor site. It is, therefore, suggested that hydrogen bonding may occur between YH on the protein and the 2-oxo group of the lumazine, facilitating extraction of a proton from the 7-methyl group of the substrate by a nucleophilic center (A) on the enzyme (I and II). Interaction of a group Z could occur simultaneously with carbon 7 of the lumazine at the donor site. A dimeric substance (III) could be formed by subsequent carbon–carbon condensation between the 7-methylene group of the lumazine at the acceptor site and the electrophilic center developed at carbon 7 of the lumazine bound to the donor site (II)*. Ring opening of III would lead to the formation of IV which, upon removal of a proton from the 6-methyl group at the acceptor site by group Z leads to formation of a methylene group at this position (V). Attack of the latter group at carbon 7 leads to ring closure (VI), followed by the loss of a second proton from the 6-methyl group, cleavage of the carbon–nitrogen bond of the lumazine residue at the donor site (VII) and formation of riboflavin and 4-ribitylamino-5-amino-2,6-dihydroxypyrimidine (VIII). The enzyme is regenerated (cf. I and VIII).

The rate-limiting effect of the presence of deuterium in the 6-methyl group of 6,7-dimethyl-8-ribityllumazine on riboflavin formation does not influence the velocity of hydrogen–tritium exchange at the 7-methyl group of the substrate. This suggests that the carbon–carbon condensations necessary for the formation of the aromatic ring occur first at the 7-methyl and then at the 6-methyl group. This sequence is reflected in the mechanism shown in Fig. 8. Two groups on the protein (A and Z) are shown as acceptors and donors of protons in the various reactions. Whether or not these groups are the same is not certain, although the proportional inhibition of riboflavin synthesis and protium elimination from the 7-methyl group may suggest that at least one of these is a thiol group. However, in the absence of definitive experimental data, it is also possible that sulfhydryl-binding agents interact

---

* The recent chemical studies of McAndless and Stewart[82] are in agreement with the suggestion here that an electrophilic center is developed at carbon 6 of the donor lumazine when a covalent bond (with group Z) is formed at carbon 7 (II). They found with 1,7-dihydro-6,7,8-trimethyllumazine that chemical deuterium exchange with solvent occurred at the 6-methyl but not the 7-methyl group of this compound, indicating that an electrophilic center develops at carbon 6 of the pyrazine ring when, in this case, carbon 7 is covalently bonded to hydrogen.

References p. 44

with a group on the protein which is not at the catalytic site, but influence enzyme action by steric effects or changes in protein conformation. This is consistent with the experimental observation that the rates of hydrogen–tritium exchange at the 7-methyl group and riboflavin formation are equal at pH 6.8, though neither the exchange or flavin synthesis is catalyzed at pH 4.5, where the group with pK_a 6–7 would be protonated, and the finding that raising the pH above that optimal for riboflavin formation (pH 8.8) leads, nevertheless, to an increased exchange rate.

The initial interactions between substrate and enzyme (Fig. 8, I) involves tautomeric forms of the bound lumazine which should retain absorption at long wavelengths. Opening of the pyrazine ring in the subsequent dimeric intermediates (Fig. 8, III–VI), possibly resulting in loss of light absorption above 400 mμ, is confined to one of the two lumazine components of the structures. This might be consistent with the failure to find significant reduction in color at around 400 mμ in the enzymatic reaction. Furthermore, none of the intermediates contain an α-methyl ketone group and, in agreement with experimental observations, inhibition by high concentrations of hydroxylamine of the riboflavin synthetase reaction would not be expected[48].

The enzyme-catalyzed hydrogen exchange at the 7-methyl group could occur (Fig. 8) between steps I and II and, if the reactions were reversible, between II and IV.

The positioning of ribityl groups on the enzyme in the scheme is intended to imply that sterically exact attachment of the side-chain to the protein may be of importance for the precise positioning of the heterocyclic ring portions for the reaction to occur. The exacting requirements for the configuration of the substituent at nitrogen 8 of the molecule would be in accord with such a possibility.

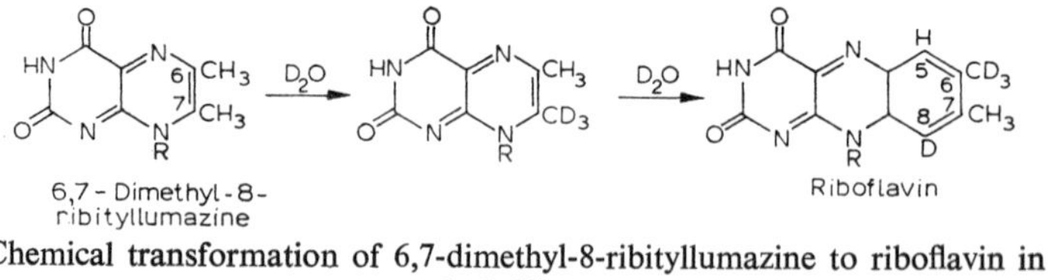

6,7-Dimethyl-8-
ribityllumazine   Riboflavin

Fig. 9. Chemical transformation of 6,7-dimethyl-8-ribityllumazine to riboflavin in D$_2$O
(ref. 59).

Paterson and Wood[59] have proposed recently a mechanism different from that of Rowan and Wood[55,56] and more similar to that described in Fig. 8. In experiments on the chemical transformation of **6,7-dimethyl-8-**

substituted lumazines to flavins at neutral pH in D_2O it was found that deuterium is recovered in carbon 8 and the 6-methyl group of riboflavin (Fig. 9). These investigators[59], therefore, suggested a sequential condensation of the 7-methyl and 6-methyl group of the lumazine acting as the 4-carbon acceptor with carbon 6 and carbon 7, respectively, of the donor molecule.

The same stereospecific mode of transfer of the 4-carbon portion occurs enzymatically since Beach and Plaut[49] have shown that 6-deuteromethyl-7-methyl-8-ribityllumazine (Fig. 8, I) is converted to riboflavin labeled with deuterium in the 7-methyl group and at carbon 5 (Fig. 8, VIII) in the presence of purified yeast riboflavin synthetase.

4. Biosynthesis of 6,7-dimethyl-8-ribityllumazine

The obligatory role of a purine derivative as a precursor in the biosynthesis of the pyrimidine portion of riboflavin and of 6,7-dimethyl-8-ribityllumazine has been established (for review see refs. 5–7, 60, 61). Pyrimidines *per se* do not participate in flavin synthesis since addition of compounds such as uracil and thymine have not been found to stimulate production of the vitamin by *E. ashbyii*[8] and since the radioactivity of added [6-^{14}C]orotic acid could not be recovered in riboflavin biosynthesized by *A. gossypii*[61,62] and other organisms[63].

In a number of compounds related in chemical structure and biosynthesis to riboflavin, it has been shown that the *o*-diaminopyrimidine-containing portion of the molecules are derived from GTP. This has been especially well documented for the synthesis of the pteridine portion of folates (for review see refs. 7, 64). Thus, enzymatic opening of the imidazole ring of GTP followed by elimination of carbon 8 as formate and subsequent ring closure involving carbons 1 and 2 of the ribosyl portion of the nucleotide to form carbons 6 and 7 of 2-amino-4-hydroxy-6-trihydroxypropyldihydropteridine phosphate have been demonstrated. If a similar ring opened form of a guanosine nucleotide were a precursor of riboflavin, one would anticipate a deamination reaction at some intermediary stage to introduce the oxygen function at position 2 of the flavin molecule; alternatively a xanthosine derivative might serve as the most immediate purine precursor of riboflavin. Indications that a guanosine, rather than a xanthosine, compound may be a direct precursor came from studies on the biosynthesis of toxoflavin which contains an oxo group in position 2, since a potent enzymatic activity leading to elimination of carbon 8 from GTP was demonstrated in extracts of *P. cocovenenans*[65].

The possibility that the oxygen function at position 2 of riboflavin is introduced into an intermediate subsequent to the purine derivative is supported by recent work of Bacher and Lingens[66], who found that a mutant of *Aerobacter aerogenes* exhibiting a genetic block between xanthosine monophosphate and guanosine monophosphate was able to convert [2-^{14}C]-xanthine to xanthosine, but not to labeled riboflavin; [2-^{14}C]guanine was transformed to riboflavin without significant dilution. Furthermore, these investigators[67] reported that a riboflavinless mutant of *Saccharomyces cerevisiae* (HK 849) in a minimal medium containing diacetyl forms 2-amino-6,7-dimethyl-4-oxo-8-ribityl-4,8-dihydropteridine. This fluorescent substance is not formed in the absence of diacetyl and it is thought that the compound trapped by the diketone is 2,5-diamino-6-hydroxy-4-ribitylamino-pyrimidine. These investigators[68] observed that other riboflavinless mutants of *S. cerevisiae* accumulate 4-ribitylamino-5-amino-2,6-dihydroxypyrimidine as demonstrated by trapping with glyoxal or diacetyl. With glyoxal, 8-ribityllumazine was identified as the product accumulated, whereas with diacetyl, 6,7-dimethyl-8-ribityllumazine *and* riboflavin were recovered. Furthermore, these mutants (HK645 and HK693) grew in a riboflavinless medium supplemented with low concentrations of diacetyl, indicating that these organisms contain riboflavin synthetase. 4-Ribitylamino-5-amino-2,6-dihydroxypyrimidine has also been reported to occur in a riboflavinless mutant of *Aspergillus nidulans* (W ribo)[69]. These investigators claim that 4-ribitylamino-5-amino-2,6-dihydroxypyrimidine promotes growth in 5 riboflavinless mutants of *A. nidulans*. These studies suggest that a purine precursor, after elimination of formate from position 8 (and certain other likely rearrangements), may yield 2,5-diamino-6-hydroxy-4-ribitylamino-pyrimidine which upon deamination forms 4-ribitylamino-5-amino-2,6-dihydroxypyrimidine and then 6,7-dimethyl-8-ribityllumazine (Fig. 10). Bacher and Lingens[67] have mentioned the possibility that the phosphorylated

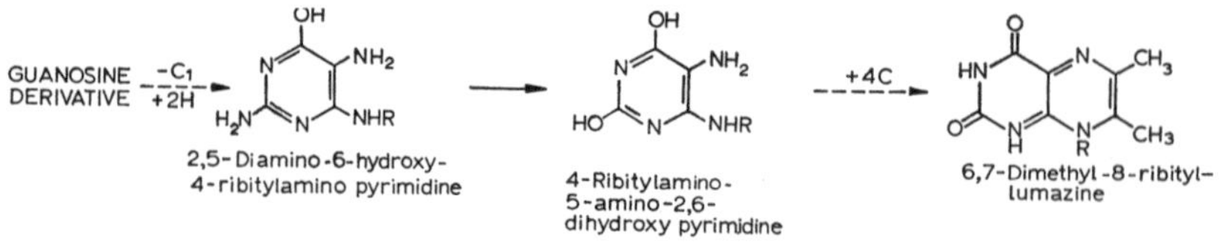

Fig. 10. Scheme of biosynthesis of 6,7-dimethyl-8-ribityllumazine from a purine precursor compound.

forms of these ribityl diaminopyrimidines may be the true intermediates, even though the substances as isolated did not contain phosphate.

The immediate source of carbons 6 and 7 and the adjacent methyl groups of 6,7-dimethyl-8-ribityllumazine is still uncertain. Earlier studies on the incorporation of labeled glucose and acetate into the *o*-xylene portion of riboflavin[14] suggested that a 2-carbon or 4-carbon fragment could be a precursor. It was recognized even then that it seems unlikely that an acetyl unit is the direct precursor of these carbons since the distribution of isotope from [1-14C]glucose and [6-14C]glucose in the aromatic ring was rather specific, while that from [2-14C]acetate was more random. In view of the ease of the chemical condensation between diacetyl and 4-ribitylamino-5-amino-2,6-dihydroxypyrimidine, it is tempting to believe that the diketone or acetoin are the direct precursors. Goodwin and Treble[70] reported incorporation of radioactivity from [1-14C]3-hydroxy-2-butanone by intact cells of *E. ashbyii* into ring A of riboflavin exclusively, and noted that the incorporation of isotope from this precursor was more efficient than that from 2-14C. In retrospect it is difficult to know, however, whether the incorporation of radioactivity from acetoin in this experiment might not have been due to condensation with 4-ribitylamino-5-amino-2,6-dihydroxypyrimidine generated as a second product[46] in the riboflavin synthetase reaction. Katagiri *et al.*[26,30,40], and Kishi *et al.*[71] reported that the formation of 6,7-dimethyl-8-ribityllumazine from 4-ribitylamino-5-amino-2,6-dihydroxypyrimidine and acetoin occurred in the presence of extracts of *E. ashbyii* and *Aerobacter aerogenes*. The reaction appears to be non-specific with regard to the pyrimidine *ortho*diamine, since 4-ribitylamino-5-amino-2,6-dihydroxypyrimidine can be effectively replaced by 4,5-diamino-2,6-dihydroxypyrimidine, 4-methylamino-5-amino-2,6-dihydroxypyrimidine, and 4-(2'-hydroxyethyl)-amino-5-amino-2,6-dihydroxypyrimidine leading to the formation of 6,7-dimethyllumazine, 6,7,8-trimethyllumazine, and 6,7-dimethyl-8-(2'-hydroxyethyl)lumazine, respectively. The extracts of the microorganisms may, therefore, contribute the well known enzyme system catalyzing oxidation of acetoin to 2,3-butanedione[72]; the latter can condense readily in the absence of enzyme with *o*-diaminopyrimidines. Masuda's proposal[73] for an analogous mechanism of formation of 6-methyl-7-hydroxy-8-ribityllumazine from 4-ribityllumazine from 4-ribitylamino-5-amino-2,6-dihydroxypyrimidine and pyruvate is probably untenable since it has been shown that the addition of [1-14C]pyruvate to growing cultures of *A. gossypii* produces no radioactivity in the methyl group and in the carbons

6 and 7 moiety of 6-methyl-7-hydroxy-8-ribityllumazine isolated from the fermentation mixture[74]. It is more likely that 6-methyl-7-hydroxy-8-ribityllumazine is formed by oxidation of 6,7-dimethyl-8-ribityllumazine[38,75,76]. Although there is good evidence that such a conversion occurs chemically, conclusive experiments are needed to prove that such a mechanism prevails under physiological conditions.

The role of acetoin as a rather immediate precursor of the *o*-xylene portion of riboflavin and, therefore, of the methyl groups at carbons 6 and 7 of 6,7 dimethyl-8-ribityllumazine is not favored by the studies of Ali and Al-Khalidi[15] who found with resting cells of *E. ashbyii* that incorporation of radioactivity from [14C]acetoin into the lumichrome portion of riboflavin was even lower than that from [2-14C]pyruvate or [1-14C]acetate. In such experiments, [14C]glucose and [14C]ribose were found to be more effective precursors of the lumichrome portion of riboflavin than acetate, pyruvate, or acetoin. Ali and Al-Khalidi[15] are of the opinion that a derivative of an intermediate of the pentose-phosphate cycle may be a precursor of the *o*-xylene ring of riboflavin. Such a possibility is not inconsistent with previous experiments[17] and finds particular support in the finding of Ali and Al-Khalidi[15] that radioactivity from [1-14C]ribose is incorporated at least as effectively into the xylene portion of riboflavin as that from [14C]glucose. However, a definitive answer to this problem is difficult to obtain with a complex system such as a suspension of resting cells since it was also found in these experiments that, whereas incorporation of radioactivity from [6-14C]glucose into the lumichrome portion of riboflavin was unaffected in the presence of nonradioactive ribose, the incorporation of 14C from [1-14C]ribose was severely depressed when unlabeled glucose was added to the medium. Erythritol has been reported to be an effective precursor for the analogous dimethylbenzimidazole portion of vitamin B_{12}[16]; however, Ali and Al-Khalidi[15] have found that transfer of 14C into riboflavin was much less effective with [U-14C]erythritol than with labeled glucose, ribose, glycerol, or xylose.

The studies of Ali and Al-Khalidi[15] with intact cells of *E. ashbyii* indicate that 20–30% of the label from [1-14C]glucose, [6-14C]glucose or [1-14C]-ribose recovered in riboflavin is located in the ribityl group. These results are consistent with those obtained in similar experiments with *A. gossypii* in which it was concluded on the basis of label distribution within the ribityl group that it could be derived from a pentosyl precursor[17]. The accumulation of 2,5-diamino-6-hydroxy-4-ribitylaminopyrimidine by riboflavinless mutants of *S. cerevisiae*[67] may suggest that the reduction of the ribosyl to a ribityl group may occur during the transformation of a guanosine phosphate derivative to the ribityl-pyrimidine compound, if the ring-opening reaction of the purine is similar to that found in the synthesis of folate derivative from GTP[77,78]. However, purine nucleosides or purine nucleotides may not be direct precursors in the case of riboflavin bio-synthesis since experiments with whole organisms have shown that purine ribonucleosides are no more effective than free purines in stimulating flavin production[9,12] and Forrest and McNutt[79] found that [U-^{14}C]guanosine is incorporated more efficiently into the ring system of riboflavin than into the ribityl substituent by intact cells of *E. ashbyii*. It is, therefore, still possible that reduction of the ribosyl moiety to a ribityl group[5] *e.g.*:

$$\text{XDP-ribose} \xrightarrow[\text{PNH}]{\text{PN}^+} \text{XDP-ribitol}$$

occurs before attachment to the precursor of 6,7-dimethyl-8-ribityllumazine. The 4-amino group of an *o*-diaminopyrimidine derivative could act as an acceptor of the ribityl group from an activated precursor and it may be pertinent that the occurrence of 4,5-diaminouracil has been reported in cultures of *E. ashbyii*[80] and in certain riboflavinless strains of *A. nidulans*[81]. However, if, as suggested by Lingens *et al.*[68], 2,5-diamino-6-hydroxy-4-ri-bitylaminopyrimidine precedes 4-ribitylamino-5-amino-2,6-dihydroxypyrim-idine as a precursor of 6,7-dimethyl-8-ribityllumazine, one would expect a derivative of 2,4,5-triamino-6-hydroxypyrimidine, rather than 4,5-diamino-2,6-dihydroxypyrimidine, to be an acceptor of the ribityl group in such a mechanism.

REFERENCES

1 W. H. Peterson and M. S. Peterson, *Bacteriol. Rev.*, 9 (1945) 49.
2 T. G. Pridham, *Econ. Botany*, 6 (1952) 185.
3 J. M. van Lanen and F. W. Tanner Jr., *Vitamins Hormones*, 6 (1948) 1963.
4 R. J. Hickey, in L. A. Underkofler and R. J. Hickey (Eds.), *Industrial Fermentations*, Vol. 2, Chemical Publ. Co., New York, 1954, p. 157.
5 G. W. E. Plaut, in D. M. Greenberg (Ed.), *Metabolic Pathways*, Vol. II, Academic Press, New York, 1961, p. 673.
6 G. M. Brown, *Physiol. Rev.*, 40 (1960) 331.
7 C. H. Winestock and G. W. E. Plaut, in J. Bonner and J. E. Varner (Eds.), *Plant Biochemistry*, Academic Press, New York, 1965, p. 391.
8 J. A. McLaren, *J. Bacteriol.*, 63 (1952) 233.
9 W. S. McNutt, *J. Biol. Chem.*, 210 (1954) 511.
10 W. S. McNutt, *J. Biol. Chem.*, 219 (1956) 365.
11 W. S. McNutt, *Federation Proc.*, 19 (1960) 157.
12 T. W. Goodwin and S. Pendlington, *Biochem. J.*, 57 (1954) 631.
13 D. J. Howells and G. W. E. Plaut, *Biochem. J.*, 94 (1965) 755.
14 G. W. E. Plaut, *J. Biol. Chem.*, 211 (1954) 111.
15 S. N. Ali and U. A. S. Al-Khalidi, *Biochem. J.*, 98 (1966) 182.
16 W. L. Alworth, H. N. Baker, D. A. Lee and B. A. Martin, *J. Am. Chem. Soc.*, 91 (1969) 5662.
17 G. W. E. Plaut and P. L. Broberg, *J. Biol. Chem.*, 219 (1956) 131.
18 W. S. McNutt and H. S. Forrest, *J. Am. Chem. Soc.*, 80 (1958) 951.
19 T. Masuda, *Pharm. Bull. (Tokyo)*, 4 (1956) 71.
20 T. Masuda, *Pharm. Bull. (Tokyo)*, 4 (1956) 375.
21 T. Masuda, *Pharm. Bull. (Tokyo)*, 5 (1957) 136.
22 T. Masuda, T. Kishi and M. Asai, *Pharm. Bull. (Tokyo)*, 6 (1957) 598.
23 G. F. Maley and G. W. E. Plaut, *Federation Proc.*, 17 (1958) 268.
24 G. F. Maley and G. W. E. Plaut, *J. Biol. Chem.*, 234 (1959) 641.
25 R. M. Cresswell and H. C. S. Wood, *Proc. Chem. Soc.*, (1959) 387.
26 H. Katagiri, I. Takeda and K. Imai, *J. Vitaminol. (Kyoto)*, 5 (1959) 287.
27 W. Pfleiderer and G. Nübel, *Chem. Ber.*, 93 (1960) 1406.
28 G. W. E. Plaut and G. F. Maley, *Arch. Biochem. Biophys.*, 80 (1959) 219.
29 G. W. E. Plaut and G. F. Maley, *J. Biol. Chem.*, 234 (1959) 3010.
30 H. Katagiri, I. Takeda and K. Imai, *J. Vitaminol. (Kyoto)*, 5 (1959) 81.
31 C. H. Winestock and G. W. E. Plaut, *J. Org. Chem.*, 26 (1961) 4456.
32 J. Davoll and D. D. Evans, *J. Chem. Soc.*, (1960) 5041.
33 R. M. Cresswell and H. C. S. Wood, *J. Chem. Soc.*, (1960) 4768.
34 G. W. E. Plaut, *J. Biol. Chem.*, 238 (1963) 2225.
35 G. W. E. Plaut, *J. Biol. Chem.*, 235 (1960) PC 41.
36 T. W. Goodwin and A. A. Horton, *Nature*, 191 (1961) 772.
37 G. F. Maley and G. W. E. Plaut, *J. Am. Chem. Soc.*, 81 (1959) 2025.
38 S. Kuwada, T. Masuda, T. Kishi and M. Asai, *Chem. Pharm. Bull. (Tokyo)*, 6 (1958) 618.
39 H. Katagiri, I. Takeda and K. Imai, *J. Vitaminol. (Kyoto)*, 4 (1958) 211.
40 H. Katagiri, I. Takeda and K. Imai, *J. Vitaminol. (Kyoto)*, 4 (1958) 278.
41 F. Korte and H. V. Aldag, *Ann. Chem.*, 628 (1959) 144.
42 L. Garnjobst and E. L. Tatum, *Am. J. Botany*, 43 (1956) 149.
43 R. L. Beach, *Ph.D. Thesis*, Rutgers University, New Brunswick, N.J., 1970.
44 M. Asai, T. Masuda and S. Kuwada, *Chem. Pharm. Bull. (Tokyo)*, 9 (1961) 496, 503.

45 H. Mitsuda, *Vitamins*, 28 (1963) 465.
46 H. Wacker, R. A. Harvey, C. H. Winestock and G. W. E. Plaut, *J. Biol. Chem.*, 239 (1964) 3493.
47 C. H. Winestock, T. Aogaichi and G. W. E. Plaut, *J. Biol. Chem.*, 238 (1963) 2866.
48 R. A. Harvey and G. W. E. Plaut, *J. Biol. Chem.*, 241 (1966) 2120.
49 R. L. Beach and G. W. E. Plaut, *J. Am. Chem. Soc.*, 92 (1970) 2913.
50 R. L. Beach and G. W. E. Plaut, *Biochemistry*, 9 (1970) 760.
51 G. W. E. Plaut and R. A. Harvey, *Methods in Enzymology*, in the press.
52 G. W. E. Plaut, R. L. Beach and T. Aogaichi, *Biochemistry*, 9 (1970) 771.
53 H. Beinert, in P. D. Boyer, H. Lardy and K. Myrbäck (Eds.), *The Enzymes*, Academic Press, New York, 1960, p. 340.
54 A. J. Birch and C. J. Moye, *J. Chem. Soc.*, (1958) 2622.
55 T. Rowan and H. C. S. Wood, *Proc. Chem. Soc.*, (1963) 21.
56 T. Rowan and H. C. S. Wood, *J. Chem. Soc.*, (*C*) (1968) 452.
57 W. Pfleiderer, J. W. Bunting, D. D. Perrin and G. Nübel, *Chem. Ber.*, 99 (1966) 3503.
58 R. L. Beach and G. W. E. Plaut, *Tetrahedron Letters*, 40 (1969) 3489.
59 T. Paterson and H. C. S. Wood, *Chem. Commun.*, (1969) 290.
60 T. W. Goodwin, *The Biosynthesis of Vitamins and Related Compounds*, Academic Press, New York, 1963.
61 G. W. E. Plaut and J. J. Betheil, *Ann. Rev. Biochem.*, 25 (1956) 463.
62 P. L. Broberg, *Studies on the Biosynthesis of Riboflavin in Ashbya gossypii*, Master's thesis, Univ. Winconsin, Madison, Wisc., 1954.
63 F. Korte, H. V. Aldag, G. Ludwig, W. Paulus and K. Storiko, *Ann. Chem.*, 619 (1958) 70.
64 G. M. Brown and J. J. Reynolds, *Ann. Rev. Biochem.*, 32 (1963) 419.
65 B. Levenberg and S. N. Linton, *J. Biol. Chem.*, 241 (1966) 846.
66 A. Bacher and F. Lingens, *Angew. Chem., Intern. Ed.*, 8 (1969) 371.
67 A. Bacher and F. Lingens, *Angew. Chem., Intern. Ed.*, 7 (1968) 219.
68 F. Lingens, O. Oltmanns and A. Bacher, *Z. Naturforsch.*, 22B (1967) 755.
69 J. Sadique, R. Shanmugasumdaran and E. R. B. Shanmugasumdaran, *Biochem. J.*, 101 (1966) 2C.
70 T. Goodwin and D. H. Treble, *Biochem. J.*, 70 (1958) 14P.
71 T. Kishi, M. Asai, T. Masuda and S. Kuwada, *Chem. Pharm. Bull. (Tokyo)*, 7 (1959) 515.
72 A. Kling, *Compt. Rend.*, 133 (1901) 231.
73 T. Masuda, T. Kishi, M. Asai and S. Kuwada, *Chem. Pharm. Bull. (Tokyo)*, 6 (1958) 523.
74 G. W. E. Plaut, *Ann. Rev. Biochem.*, 30 (1961) 409.
75 G. W. E. Plaut, *Federation Proc.*, 19 (1960) 312.
76 M. Asai and S. Kuwada, *Chem. Pharm. Bull. (Tokyo)*, 10 (1962) 241.
77 J. J. Reynolds and G. M. Brown, *J. Biol. Chem.*, 257 (1962) 2713.
78 J. J. Reynolds and G. M. Brown, *J. Biol. Chem.*, 239 (1964) 317.
79 H. S. Forrest and W. S. McNutt, *J. Am. Chem. Soc.*, 80 (1958) 739.
80 T. W. Goodwin and D. H. Treble, *Biochem. J.*, 67 (1957) 10P.
81 J. Sadique, R. Shanmugasundaran and E. R. B. Shanmugasundaran, *Naturwissenschaften*, 53 (1966) 282.
82 J. M. McAndless and R. Stewart, *Can. J. Chem.*, 48 (1970) 263.

Metabolism of Water-Soluble Vitamins

Section c

The Metabolism of Vitamin B₆

ESMOND E. SNELL and BETTY E. HASKELL

Department of Biochemistry, University of California, Berkeley, Calif.,
and
Department of Food Science, University of Illinois, Urbana, Ill. (U.S.A.)

1. Introduction

An account of the discovery, synthesis, general properties, and metabolic functions of vitamin B_6 appears elsewhere in this treatise[1,2]. The three naturally-occurring forms of this vitamin, pyridoxine (I, also called pyridoxol), pyridoxal (II), and pyridoxamine (III), occur free to only a limited extent in tissues, most of the vitamin being present as the corresponding 5'-phosphate esters (IV–VI, Fig. 1). Of these, only pyridoxamine 5'-phosphate and pyridoxal 5'-phosphate have recognized coenzymatic functions. This chapter deals with the metabolism of these various forms of vitamin B_6, *i.e.*, with their biosynthesis, absorption and transport, interconversion, and catabolic degradation in various organisms. Certain aspects of these processes have been reviewed elsewhere[3-5]. We shall consider these processes primarily as they occur in microorganisms and in higher animals; very little information concerning this topic is available for higher plants.

References p. 68

2. Biosynthesis of vitamin B_6

Vitamin B_6 is synthesized by all higher plants so far studied, although isolated portions of the plant, e.g., root cultures, may require the preformed vitamin[6]. The vitamin is also synthesized by a great many microorganisms, but is required preformed by many fastidious fungi, protozoa, and bacteria[7]. All higher animals so far studied require the vitamin in the diet; these include man[8,9], rat[10], mouse[11], chick[12], pig[13], turkey[14], duck[15], hamster[16], rabbit[17], rainbow trout[18], dog[19], cat[20], and monkey[21].

There is still no definitive information concerning the biosynthetic steps through which vitamin B_6 arises from acyclic precursors in those organisms that synthesize it; as a consequence, the metabolic defect(s) which makes animals and some microorganisms dependent on an external supply of this vitamin is unknown. Progress in unravelling the biosynthetic pathway is hampered by the extremely small quantities of the vitamin produced by organisms studied to date. Even mutant cultures of yeast which "over-produce" this vitamin excrete only 1–3 mg (mostly as pyridoxine) per liter of culture medium[22]; most organisms, including *Escherichia coli*, excrete little or no vitamin[23,24]. This probably reflects feedback inhibition of the biosynthetic process. For example, vitamin B_6 formation by exponentially growing cultures of *E. coli* reportedly ceases within 14 sec on addition of pyridoxine, and is rapidly reestablished when the cells are transferred from vitamin B_6-rich media even under conditions that should prevent protein synthesis[24]. Several vitamin B_6-dependent mutants of this organism have been isolated[25], and appropriate cross-feeding and genetic studies indicate that at least 6–7 distinct steps—all unidentified—are involved in the biosynthesis[26]. Pyridoxine is the form of the vitamin that appears first[27], and inconclusive data suggest that 3-phosphoserine may be one of its early precursors[28].

3. Absorption and transport of vitamin B_6

For organisms that cannot synthesize vitamin B_6, its absorption from external sources is required. Few studies of this process have appeared. Nutritional experiments[12,19,29] in which animals were shown to grow equally well on limiting amounts of pyridoxine, pyridoxamine, or pyridoxal indicate that all 3 forms probably are absorbed with equal efficiency from the gastrointestinal tract. In auxotrophic microorganisms, on the other hand, different forms of the vitamin sometimes show markedly different growth-promoting activities[29], but it is not always clear whether such differences arise from differences in absorption or in intracellular metabolism. It seems clear, however, that vitamin B_6 phosphates are not readily transported across cell membranes of many organisms. For example, pyridoxal-P and pyridoxamine-P are only slightly active in promoting growth of a yeast, *Saccharomyces carlsbergensis*, for which the unphosphorylated compounds are fully active[30]. Similarly, pyridoxal-P has relatively slight activity in promoting growth of *Streptococcus faecalis* under conditions where either free pyridoxal or pyridoxamine-P is highly active[30]. That this results from reduced permeability of the cells to pyridoxal-P is evident from the fact that free pyridoxal, but not pyridoxal-P, activates the tyrosine apodecarboxylase of resting intact cells of this organism, whereas only pyridoxal-P is fully effective in acetone-dried cells, where phosphorylative mechanisms are partially inactivated and the selective permeability of the cell membrane has been destroyed[31].

Studies on the transport of vitamin B_6 phosphate esters into Ehrlich ascites tumor cells[32] and into erythrocytes[33] also show that phosphorylated derivatives are concentrated to a much smaller extent than are free forms of the vitamin. When tumor cells[32] are incubated with 0.05–2 mM pyridoxal at 37° for 30 min, the ratio of intracellular to extracellular pyridoxal remains almost constant at about 1.4 to 1.0; in comparable experiments with pyridoxal-P, the ratio was less than 0.2 to 1. Erythrocytes[33] concentrate free pyridoxal, pyridoxine, and pyridoxamine against a concentration gradient from media containing 0.2 μg of vitamin B_6 per ml. The rate of transport was pyridoxal > pyridoxine > pyridoxamine, and the ratio of intracellular to extracellular vitamin after 30 min was 15 to 1 for pyridoxal and 7 to 1 for pyridoxine and pyridoxamine. Saturation of the transport process occurred at relatively low concentrations of pyridoxine, but was less evident with pyridoxal and pyridoxamine. Transport was favored by glycolysis. Erythrocytes showed no comparable tendency to concentrate phosphorylated derivatives of the vitamin. However, the amount of pyridoxal phosphate incorporated into erythrocytes could be increased by increasing the pyridoxal-P concentration in the medium. In view of the apparent preference of erythrocytes for unphosphorylated forms of the vitamin, it is interesting that erythrocyte ghosts can hydrolyze pyridoxal-P to pyridoxal[33].

The foregoing studies seem to indicate either an active transport process or transport by diffusion coupled with rapid intracellular fixation of the vitamin. In this respect, they contrast with studies of intestinal absorption, carried out with higher concentrations of pyridoxine, which have been interpreted[34,35] as indicating that absorption takes place by simple diffusion rather than active transport. Booth and Brain[34], for example, found that absorption of labeled pyridoxine in rats increased linearly with increasing size of an oral dose. They observed no tendency for pyridoxine uptake to plateau, even after an oral dose of 5 mg, which exceeds by 500-fold the daily requirement of a 200-g rat. Nor did they find evidence that large doses of pyridoxine altered the site of absorption of pyridoxine from the gastrointestinal tract. The preferred site of pyridoxine absorption is from the jejunum, although some absorption also occurs when pyridoxine is introduced into the ileum or caecum. Had a 5-mg dose of pyridoxine saturated absorption sites in the jejunum, absorption from adjoining lower segments of the intestine might have been expected to increase. Such a phenomenon, common in active transport processes, was not observed. Further indication that absorption of these levels of pyridoxine may occur by passive diffusion comes from studies with everted intestinal sacs in which no tendency of pyridoxine to be transported against a concentration gradient was observed[35].

Absorption of pyridoxine is rapid, about half of an oral dose of ^{3}H-labelled vitamin disappearing from the gastrointestinal tract within 10 min after administration[34]. When labeled pyridoxine is introduced into a cannulated loop of rat jejunum with an intact blood supply, after 1 h 81 % of the isotope is recovered from liver, 10% from the gut wall, and 4% from fluid within the gut sac[35]. Unfortunately, similar studies of the absorption of pyridoxal and pyridoxamine or their 5′-phosphates (which comprise most of the vitamin B$_6$ of mixed diets) have not been made. It is not known whether the phosphate esters of vitamin B$_6$ must be hydrolyzed prior to absorption. However, these compounds are readily hydrolyzed by intestinal alkaline phosphatase, as discussed in a later section.

A striking characteristic of vitamin B$_6$ absorption is the ability of the gastrointestinal tract to take up quantities of pyridoxine which exceed greatly the physiological need of the animal. For example, Scudi et al.[36] showed that dogs weighing 7–15 kg excrete within 6 h about 20% of an oral dose of pyridoxine varying in size from 100 to 500 mg. The vitamin B$_6$ requirement of growing dogs of this size is 0.3–0.66 mg per day[37]. Large doses of pyridoxal and pyridoxamine appear to be absorbed by human male adults as readily as pyridoxine as judged by recoveries of metabolic products from the urine[38].

4. Interconversion of various forms of vitamin B$_6$

As noted earlier, pyridoxal, pyridoxamine, and pyridoxine are approximately equally active in supporting animal growth[12,19,29]. Although minor differences in growth response to the 3 forms of vitamin B$_6$ can be observed in rats and chicks fed diets high in soluble carbohydrate, these differences appear to result from effects on the intestinal flora and are not detectable when animals are fed more normal diets containing purified starch as a source of carbohydrate[40]. Since pyridoxal-P is the only generally utilizable coenzymatic form of the vitamin, these observations emphasize the ease of interconversion of the various dietary forms in growing animals. Some vitamin B$_6$-dependent microorganisms, in contrast, exhibit marked preferences for one or another form of the vitamin. For example, *Lactobacillus casei* requires pyridoxal specifically for growth, whereas *Streptococcus faecalis* requires either pyridoxal or pyridoxamine; neither organism utilizes

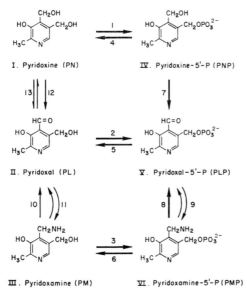

Fig. 1. Metabolic interconversion of various forms of vitamin B$_6$. Reactions 1, 2 and 3 are catalyzed by *pyridoxal kinase* (EC 2.7.1.35), reactions 4, 5 and 6 by various *phosphatases*, reactions 7, 8 and 10 by *pyridoxine-P oxidase*; reactions 9 and 11 by certain *transaminases*, and reactions 12 and 13 by various *pyridoxine dehydrogenases* (EC 1.1.1.65). Predominant reaction pathways at low vitamin concentrations are indicated by heavy arrows. See text for discussion.

References p. 68

pyridoxine[29]. A few strains of lactic acid bacteria are known which require pyridoxamine-P for growth, other forms of the vitamin being utilized only at much higher concentrations[40a]. Vitamin B_6-dependent yeasts and fungi, on the other hand, resemble animals in utilizing all three forms of the free vitamin almost equally well[29].

Our present knowledge of the interconversion of various forms of vitamin B_6 is summarized in Fig. 1. Heavy arrows indicate the preferred reaction pathways (see Section 5, p. 58); transformations indicated by the light arrows have been demonstrated to occur in one organism or another, but their metabolic significance at low substrate concentrations is doubtful. The individual reactions shown in this figure are considered in the following sections.

(a) Phosphorylation of pyridoxine, pyridoxal and pyridoxamine (reactions 1, 2 and 3, Fig. 1)

Pyridoxal kinase (ATP:pyridoxal 5-phosphotransferase, EC 2.7.1.35) catalyzes the formation of pyridoxine-P, pyridoxal-P, and pyridoxamine-P from ATP and the corresponding free form of vitamin B_6 (eqn. 1). The enzyme is present in all tissues tested, but these vary widely in their phos-

$$RCH_2OH + ATP \xrightarrow{\;Zn^{2+}\; or\; Mg^{2+}\;} RCH_2OPO_3^{2-} + ADP \qquad (1)$$

phorylating activity[41]. Kidney, liver, brain and some bacteria (*e.g.*, *Lactobacillus casei*) are particularly rich sources; skeletal muscle, fungi and bacteria such as *Mycobacterium tuberculosis* are poor sources[41]. The enzyme has been partially purified from several sources (Table I). At saturating substrate concentrations, the enzyme phosphorylates all 3 forms of the vitamin at about the same rate[41]. However, kinases from rat liver and beef brain phosphorylate pyridoxal at substantially lower substrate concentrations than they do pyridoxine or pyridoxamine, a characteristic which is even more marked in the kinases from *L. casei* and *S. faecalis*, where the K_m value for pyridoxal is less than 0.01 that for pyridoxine or pyridoxamine. This preference is reversed in yeast, and especially in *E. coli* both of which, in contrast to the other sources examined, can synthesize vitamin B_6. The difference may reflect the fact that pyridoxine is the first form of vitamin B_6 to appear in organisms that synthesize it[27], whereas pyridoxal (and pyridoxamine) is a more abundant source of the vitamin in the diet of auxotrophic organisms.

TABLE I

PROPERTIES OF PYRIDOXAL KINASES FROM VARIOUS SOURCES

Source	Preferred metal ion	pH optimum	K_m values, mM^f			
			ATP	PL	PN	PM
Brain tissue						
Beef[a]	Zn^{2+}	6.0	0.1	0.05	0.20	0.50
Human[b]	Zn^{2+}	6.5	0.05	0.05	(g)	(g)
Mouse[c]	Zn^{2+}	5.5[e]	—	0.086	0.05	0.20
Liver (rat)[a]	Zn^{2+}	5.7	0.08	0.015	0.025	0.15
Bacteria						
Lactobacillus casei[a]	Mg^{2+}	5.0	0.064	0.03	4.0	5.0
Streptococcus faecalis[a]	Mg^{2+}	5.2	—	0.015	1.5	2.5
Yeast[a,d]	Zn^{2+}	6.8	0.33	0.40	0.025	0.015

[a] From McCormick, Gregory and Snell[41]. In the presence of Zn^{2+}, the beef brain kinase is further activated by K^+. This relationship has not been tested for the other kinases.
[b] From McCormick and Snell[42].
[c] From Tsubosaka and Makino[43].
[d] The yeast used was *Saccharomyces carlsbergensis*. Hurwitz[44] has extensively studied the kinase from another strain of yeast.
[e] For phosphorylation of pyridoxal and pyridoxamine; pH 6.6 for pyridoxine.
[f] K_m values vary with the pH; those reported are at the pH optimum.
[g] Not determined; values were similar to those for pyridoxal.

All pyridoxal kinases purified from mammalian tissue are activated most effectively by Zn^{2+}, and less effectively by Mg^{2+}; this relationship is reversed in the bacterial kinases[41]. The mammalian enzymes differ somewhat in pH and temperature optima; whether these apparent differences are due to differing degrees of purification or are species differences is not known. The possibility that pyridoxal kinases in different tissues may be different proteins is interesting in connection with the occurrence of metabolic disorders such as vitamin B_6-dependent convulsions in which vitamin B_6 utilization appears to be altered primarily in brain.

Pyridoxal kinase is inhibited by high concentrations of substrates[41] and by carbonyl reagents such as hydrazine, semicarbazide, isonicotinyl hydrazide, etc.[45]. The actual inhibitors *in vitro* (and presumably also *in vivo*) are the condensation products of these reagents with pyridoxal, which have affinities for the kinase 100–1000 times that of pyridoxal[45-47]. Among the most potent inhibitors of the kinase from brain is the hydrazone of pyridoxal with isonicotinylhydrazide, a drug used in treatment of tuberculosis. This hydrazone has been isolated from urine and tissues of animals treated with isonicotinylhydrazide[48]. It is postulated[41,42] that this high inhibitory potency, coupled with the very low kinase levels found in *M. tuberculosis*, may be a factor in the therapeutic value of isonicotinylhydrazide in human tuberculosis, as well as in the neurological disorders which occur during prolonged isonicotinylhydrazide therapy in man[49].

An enzyme that forms pyridoxine-P from pyridoxine and *p*-nitrophenyl-phosphate is widely distributed in yeasts and fungi[49a]. Rather high concentrations of substrates are required, suggesting the possibility that this reaction may represent an instance of a transphosphorylation reaction catalyzed by a phosphatase. The physiological significance of the reaction, if any, is unknown. Pyridoxal is not a substrate[49a].

(b) Interconversion of pyridoxine 5'-phosphate, pyridoxamine 5'-phosphate, and pyridoxal 5'-phosphate (reactions 7, 8, 9, Fig. 1)

(i) Pyridoxine phosphate oxidase

Pyridoxamine-P is not oxidized by amine oxidase (EC 1.4.3.4) or by D-amino acid oxidase (EC 1.4.3.3)[50]. An enzyme that catalyzes the oxygen-dependent transformation of this compound to pyridoxal-P (pyridoxamine-5'-phosphate: oxygen oxidoreductase (deaminating), E.C. 1.4.3.5) was partially purified from rabbit liver by Pogell[50]; a similar enzyme that oxidizes pyridoxine-5'-P to pyridoxal-P was later discovered by Ichihara and his colleagues[51-53]. Wada and Snell[54] purified these activities 65-fold from rabbit liver, and concluded that a single oxidase (termed pyridoxine-P oxidase) catalyzed both reactions (eqns. 2a and 2b). Thus, activity of pyridoxine-P oxidase for the two substrates increases at exactly the same rate during purification and decreases at the same rate during heat inactivation at two different pH values. Curves for the elution of pyridoxine-P oxidase activity and pyridoxamine-P oxidase activity from DEAE cellulose

$$\text{Pyridoxine-P} + O_2 \rightarrow \text{pyridoxal-P} + H_2O_2 \qquad (2a)$$

$$\text{Pyridoxamine-P} + H_2O + O_2 \rightarrow \text{pyridoxal-P} + NH_3 + H_2O_2 \quad (2b)$$

are superimposable. FMN is required as prosthetic group in each case, and both activities are inhibited by the product, pyridoxal-P. However, because of its lower affinity for pyridoxamine-P, the enzyme is more sensitive to product inhibition when pyridoxamine-P is substrate. The maximum velocity of oxidation of the two substrates under optimal conditions is identical. In addition, pyridoxal-P formation is not additive when pyridoxine-P oxidase is incubated with a mixture of pyridoxine-P (K_m=0.031 mM)

and pyridoxamine-P $(K_m=0.14$ m$M)$; instead, the preferred substrate, pyridoxine-P, strongly inhibits product formation from pyridoxamine-P. Although the pH optima for the two substrates is the same, activity with pyridoxamine-P falls off much more rapidly with pH than does that with pyridoxine-P, so that at pH 7–8, pyridoxine-P is a very much better substrate than pyridoxamine-P. In view of these relative affinities and activities on the two substrates, the enzyme is more appropriately termed pyridoxine-P oxidase (pyridoxine-5'-phosphate:oxygen oxidoreductase) than pyridox-amine-P oxidase. Enzymes with almost identical properties have been purified from *Alkaligenes faecalis* and *Azotobacter agilis* by Ogata *et al.*[55,56]. A similar enzyme is found in mammalian liver[52], and in aerobic micro-organisms[57,58], but is not present in the strictly anaerobic *Clostridia*[58].

Pyridoxine-P oxidase exhibits a strong preference for phosphorylated substrates. Free pyridoxine is not oxidized by the enzyme. At the stage of purity achieved, free pyridoxamine is oxidized (Fig. 1, reaction 10) by the mammalian enzyme[54] (but not by the *Alkaligenes* enzyme[55]) at a rate comparable to that of pyridoxamine-P but only at substrate concentrations 20 times higher and at pH 10. Phosphorylated derivatives of vitamin B_6 (*e.g.* pyridoxal-P-oxime, pyridoxal-P-semicarbazone, etc.) and its analogues are potent inhibitors of the oxidase; the unphosphorylated compounds are relatively ineffective[54–56].

(ii) Reversible transamination between pyridoxamine-P and pyridoxal-P (reaction 9, Fig. 1)

Interconversion of enzyme-bound pyridoxamine-P and enzyme-bound pyridoxal-P occurs as an essential feature of the catalytic action of amino acid transaminases (see eqn. 3, below)[7,59,60], over 30 of which are known[7]. To the extent that coenzyme exchange (eqn. 4) occurs *in vivo* these reactions offer a means for the net conversion of pyridoxamine-P to pyridoxal-P (or *vice versa*), as shown in eqn. 5. The extent of such reactions and their physiological importance as a mechanism for interconversion of pyridoxal-P and pyridoxamine-P have not been assessed.

$$\text{Apotransaminase} \cdot \text{PLP} + \alpha\text{-amino acid} \rightleftharpoons \text{apotransaminase} \cdot \text{PMP}$$
$$+ \alpha\text{-keto acid} \quad (3)$$

$$\text{Apotransaminase} \cdot \text{PMP} + \text{PLP} \rightleftharpoons \text{apotransaminase} \cdot \text{PLP} + \text{PMP} \quad (4)$$

$$\textit{Sum}: \text{PLP} + \alpha\text{-amino acid} \rightleftharpoons \text{PMP} + \alpha\text{-keto acid} \quad (5)$$

An abortive reaction closely analogous to eqns. 3 and 4 occurs infrequently during the action of microbial aspartate β-decarboxylase and explains its activation by either α-keto acids or pyridoxal-P[61].

The strictly anaerobic *Clostridia*, which apparently do not contain pyridoxine-P oxidase[58], do contain a true transaminase (pyridoxamine-P: α-ketoglutarate aminotransferase) which catalyzes eqn. 5 (*i.e.*, reaction 9, Fig. 1). This enzyme has been purified about 55-fold from *Clostridium kainantoi*[62]; pyridoxamine is inactive as a substrate. Glutamate aspartate transaminase also catalyzes this reaction (although at a negligibly slow rate) following treatment with *p*-chloromercuribenzoate, which reduces its affinity for pyridoxal-P[63].

(c) Hydrolysis of vitamin B$_6$ 5'-phosphates to free vitamin B$_6$ (reactions 4, 5 and 6, Fig. 1)

Non-specific phosphatases capable of hydrolyzing pyridoxal-P and pyridoxamine-P are widely distributed in bacterial[64], fungal[65], plant[66] and animal tissues. Alkaline phosphatases which hydrolyze vitamin B$_6$ phosphates have been detected in mammalian liver[50,51], in human brain[67], and in rat[68], monkey[68], pig[68], sheep[68], bovine[69], and human[68] intestine. Pyridoxal-P is not hydrolyzed by an acid phosphatase preparation from *E. coli*[64] or from human brain[67]; however, it is a preferred substrate of prostatic acid phosphatase[66].

Two enzymes which hydrolyze vitamin B$_6$ phosphates have been investigated in some detail. One is a bovine intestinal phosphatase preparation of unspecified purity[69], the other an alkaline phosphatase purified 70-fold from human brain[67]. Both are activated by Mg^{2+}, inhibited by inorganic phosphate, and exhibit a pH optimum of about 9. The intestinal enzyme hydrolyzes pyridoxal-P at lower substrate concentrations ($K_m = 5 \cdot 10^{-5}$ M) than does the brain enzyme ($K_m = 4.3 \cdot 10^{-4}$ M); the brain phosphatase hydrolyzes pyridoxal-P more rapidly than any other substrate tested.

(d) Interconversion of free pyridoxine, pyridoxal and pyridoxamine (reactions 10–13, Fig. 1)

Most of the enzymes concerned in these reactions have been studied primarily in connection with the catabolism of vitamin B$_6$ in bacteria or animals and will be considered in greater detail under that heading (section 6b, p. 61). Since these enzymes show rather low affinities for their substrates, their metabolic significance at the low concentrations of vitamin B_6 normally accessible to the organism is probably limited.

Reaction 10 (Fig. 1), as noted earlier, is catalyzed at high pyridoxamine concentrations and pH 10 by partially purified preparations of pyridoxine-P oxidase from rabbit liver[54] (but not by more highly purified preparations from *Alkaligenes faecalis*[55]); the reaction is oxygen-dependent, and the products are pyridoxal and ammonia[54].

Reaction 11 (Fig. 1) proceeds according to eqn. 6. It is catalyzed inefficiently

$$PM + \alpha\text{-keto acid} \rightleftharpoons PL + \alpha\text{-amino acid} \qquad (6)$$

at high pyridoxamine (or pyridoxal) concentrations by the extramitochondrial aspartate–glutamate *apo*transaminase and shows the same specificity for amino and keto acids as that enzyme[70,71]. In this reaction, pyridoxal and pyridoxamine act as loosely bound analogues of the firmly bound coenzymes, pyridoxal-P and pyridoxamine-P. Addition of these coenzymes completely inhibits transamination of the unphosphorylated compounds[70]. A weakly active pyridoxamine–oxaloacetate transaminase partially purified from both rabbit liver and *E. coli* was similarly inhibited by pyridoxal-P[70]; α-keto-glutarate does not serve as an amino group acceptor. Despite this fact these enzymes have been identified as apoenzymes derived from the aspartate–glutamate transaminases of these tissues[72], the rabbit-liver enzyme being derived from the mitochondrial form of this enzyme. Presumably, other apotransaminase preparations can act inefficiently in a similar fashion. Whether such reactions play a significant role *in vivo* is not known. The affinity of the apoenzymes for free pyridoxamine and pyridoxal is relatively low and at the high concentrations necessary to achieve activity pyridoxal-P formation would almost certainly be sufficient to saturate all apoenzymes fully and thus prevent this reaction. Nevertheless transfer of ^{15}N from the amino group of pyridoxamine to α-ketoglutarate does occur in pig-heart homogenates without intermediate formation of ammonia[73]; the route is unknown.

An efficient transaminase that acts according to eqn. 6 with pyruvate as the amino-group acceptor functions in the utilization of pyridoxamine (at high concentrations) as sole carbon and nitrogen source by a soil pseudo-monad[74]. It does not contain and is not inhibited by PLP. The enzyme (pyridoxamine:pyruvate aminotransferase) has been crystallized and studied extensively as a model for the more complex amino acid trans-aminases[75,76].

Reaction 12 (Fig. 1) is catalyzed efficiently and irreversibly by an FAD-dependent oxidase from *Pseudomonas* MA-1 grown with pyridoxine as sole carbon source; oxygen or 2,6-dichloroquinone chlorimide served as the electron acceptor[77]. An NADP-specific pyridoxine dehydrogenase from yeast also catalyzes reaction 12; however, the equilibrium position greatly favors reduction of pyridoxal (by NADPH) to pyridoxine[78]; *i.e.*, the enzyme normally catalyzes *Reaction 13*. An enzyme from rabbit liver also oxidizes pyridoxine to pyridoxal, but only in the presence of aldehyde oxidase, which apparently pulls the reaction in the direction of oxidation by converting the pyridoxal formed to 4-pyridoxic acid[53]. This suggests that the equilibrium point of the reaction must strongly favor pyridoxine; however, for reasons not yet apparent the purified enzyme did not catalyze reduction of added pyridoxal even on addition of NADH or NADPH.

5. Preferred pathways for pyridoxal-P formation and control of vitamin B_6 metabolism

From the enzymatic specificities discussed in Section 4 and related findings we may postulate a tentative preferred route leading to pyridoxal-P from the relatively small amounts of vitamin B_6 normally available from dietary sources. Since the unphosphorylated forms of the vitamin are generally transported across cell membranes more efficiently than the phosphorylated forms, phosphatase action appears important in preparing the latter forms for absorption from the intestinal tract and also for transfer between cells and cell compartments. The preferred intracellular route from pyridoxine to pyridoxal-P almost certainly involves its phosphorylation by pyridoxal kinase to pyridoxine-P followed by the action of pyridoxine-P oxidase (*i.e. via* reactions 1 + 7, Fig. 1, rather than reactions 12 (or 13) + 2). Such a route is consistent with the observations (*a*) that pyridoxine is more effective than pyridoxal in counteracting the inhibitory effects of 4-deoxypyridoxine in test organisms such as yeast[79] and chick embryos[80], (*b*) that in crude rabbit-liver extracts the rate of formation of 4-pyridoxic acid from pyridoxine through the coupled action of the enzyme catalyzing reaction 13 and excess aldehyde oxidase is only 5 % of the rate of pyridoxal-P formation[53], (*c*) that pyridoxal-P formation in *E. coli* occurs at very low concentrations from pyridoxine, but only at substantially higher concentrations from pyridoxal[27], and (*d*) that pyridoxine-P accumulates in mutants of *E. coli* lacking pyridoxine-P oxidase[27].

Recent experiments on the metabolism of ^{3}H-labeled vitamin B_6 in mice also support this view[81]. When [^{3}H]pyridoxine is injected into mice and the animals are killed after intervals of up to 60 min, the initial high concentrations of radioactive pyridoxine drop rapidly with a concomitant rise at about 12 min of radioactive pyridoxine-P. The pyridoxine-P peak begins to subside after about 15 min with an accompanying rise in liver levels of pyridoxal-P. Pyridoxal-P was the major form of the vitamin present in liver after 60 min. Although some isotope was detectable as pyridoxal and as pyridoxamine, the amounts were very small.

Similar experiments on the distribution of isotope in carcass[81] indicated the same precursor–product relationship observed in liver. However, the formation of pyridoxal-P from pyridoxine-P was much slower, consistent with the observation[52] that pyridoxine-P oxidase activity is low in muscle. One to 7 days after administration of [^{3}H]pyridoxine, most of the radioactivity in both liver and carcass was present as pyridoxal-P or as pyridoxamine-P, in accordance with other data indicating that these two forms of the vitamin predominate in animal tissue[48,82]. Carcass but not liver contained relatively large amounts of free pyridoxal and pyridoxamine. No free [^{3}H]pyridoxine was detectable in liver after 1 h or in carcass after 2 days. There are no comparable data to indicate the relative significance of reactions 3+8 (or 9) as opposed to reactions 10 (or 11)+2 as routes from free pyridoxamine to pyridoxal-P (Fig. 1).

Present knowledge of the control of vitamin B$_6$ metabolism is also meager. On the basis of present information, however, it seems likely that pyridoxine-P oxidase may play an important role in regulating tissue levels of pyridoxal-P. In pregnant and newly-born rats, the increases in pyridoxal-P which occur during pregnancy and post-natal development parallel an increase in pyridoxine-P oxidase activity[83]; pyridoxal kinase levels are unchanged. Purified pyridoxine-P oxidase is highly sensitive to inhibition[54,55] by pyridoxal-P so that accumulation of this product could, in principle, prevent its further formation from pyridoxine or pyridoxamine, although not from pyridoxal. The physiological significance of this inhibition needs further investigation, as does the possible control of tissue levels of pyridoxal-P by a combination of phosphatase and aldehyde oxidase activity. However, since neither phosphatases nor aldehyde oxidase are specific for the metabolism of vitamin B$_6$, these enzymes would represent much less effective control points. 4-Pyridoxic acid (10^{-3} M) has no inhibitory effects on purified preparations of the pyridoxine-P oxidase[54] or (at a concentration of 10^{-4} M) on pyridoxal kinase[45].

In *E. coli*, which synthesizes vitamin B$_6$, addition of pyridoxine effectively stops further synthesis by unknown mechanisms[24].

6. Catabolism of vitamin B$_6$

(a) In animals

4-Pyridoxic acid has been recognized as the principal urinary excretion product derived from vitamin B$_6$ in both man[38,84] and rats[84]. In addition, smaller amounts of pyridoxal, pyridoxamine and pyridoxine (and their phosphorylated derivatives) are excreted when either large test doses of the vitamin are administered[38,84], or when "physiological" amounts are ingested[85]. For example, following administration of a 100-mg dose of pyridoxal to human subjects, 59.7 mg of 4-pyridoxic acid and 1.9 mg of pyridoxal were recovered from urine during a 24-h period; the corresponding figures following administration of pyridoxine (25.3 and 1.7 mg) or pyridoxamine (24.6 and 1.9 mg) reflect the direct or indirect conversion of these compounds to pyridoxal *via* reactions of Fig. 1, discussed earlier, and the efficient oxidation of pyridoxal to 4-pyridoxic acid[38]. The latter oxidation is catalyzed by the relatively non-specific, FAD-dependent general aldehyde oxidase (aldehyde:oxygen oxidoreductase, E.C. 1.2.3.1) of liver according[39] to eqn. 7. The K_m value[39] for pyridoxal is $6.3 \cdot 10^{-7}$ M at the optimal pH of about 7.0.

4-Pyridoxolactone is not an intermediate, and pyridoxal phosphate is not oxidized by the enzyme[86]. The closely related enzyme, xanthine oxidase (EC 1.2.3.2), does not oxidize pyridoxal[86]. Neither 4-pyridoxic acid nor its lactone has vitamin B$_6$ activity[19,29], *i.e.*, reaction 7 is irreversible. 4-Pyridoxic acid disappears from the urine as vitamin B$_6$ deficiency develops[87]; its estimation is therefore of value in assessing the nutritional status.

Several studies[38,88] have shown that only about half of an oral dose of pyridoxine is recovered from urine within 3 days of its administration as vitamin B$_6$ plus 4-pyridoxic acid. Similarly, only about half of an oral dose of 4-pyridoxic acid was excreted in human urine within 24 h of its administration[87]. Nonetheless, it appears unlikely that animals can degrade 4-pyridoxic acid, for subcutaneously administered 4-pyridoxic acid is recovered almost quantitatively (80–111%) from the urine of human subjects[88]. Similarly, Bernett and Pearson[89] observed that no radioactive carbon dioxide was released when [4,5-^{14}C]pyridoxine was injected into vitamin B_6-deficient rats in amounts from 1 to 10 times the daily requirement of the animals. They concluded that any vitamin B_6 oxidation products formed probably retained the integrity of the pyridine ring. Recent investigations with radioactive pyridoxine indicate that the "missing" vitamin B_6 in earlier excretion studies may have been undetected because of its extremely slow rate of excretion. Johansson et al.[90] found that when small amounts of [^{3}H]pyridoxine are administered to human subjects, only about 15–20% of the isotope is excreted in urine within the first day. The remainder of the dose is excreted with a half-life of 18–38 days.

(b) In microorganisms

By use of selective culture techniques, several aerobic microorganisms have been isolated from soil that utilize one or another form of vitamin B_6 as a sole source of carbon and nitrogen[91]. For growth to occur under these conditions, degradation of the vitamin must occur via intermediates that can be used both for synthesis of cellular material and to provide the energy that drives such syntheses. The reaction pathways followed by two such organisms, both pseudomonads, have been traced from vitamin B_6 to open chain compounds, and are shown in Fig. 2. Characterization and properties of the several intermediates, all of which were isolated from partially spent culture media, have been described[91–93], and enzymes concerned in most of the transformations have been partially characterized. Both of these pathways are inducible; i.e., enzymes catalyzing the indicated transformation are found only in cells grown with vitamin B_6. These individual enzymatic steps are described briefly below.

(i) Oxidation of pyridoxamine, pyridoxine and pyridoxal by Pseudomonas MA and MA-1 (Pathway I, Fig. 2)

(a) Reaction 1 (Fig. 2): Pyridoxamine + pyruvate $\rightleftharpoons$ pyridoxal + L-alanine. This reaction is catalyzed by pyridoxamine pyruvate transaminase, which is readily isolated in crystalline form from cell extracts. The enzyme has a

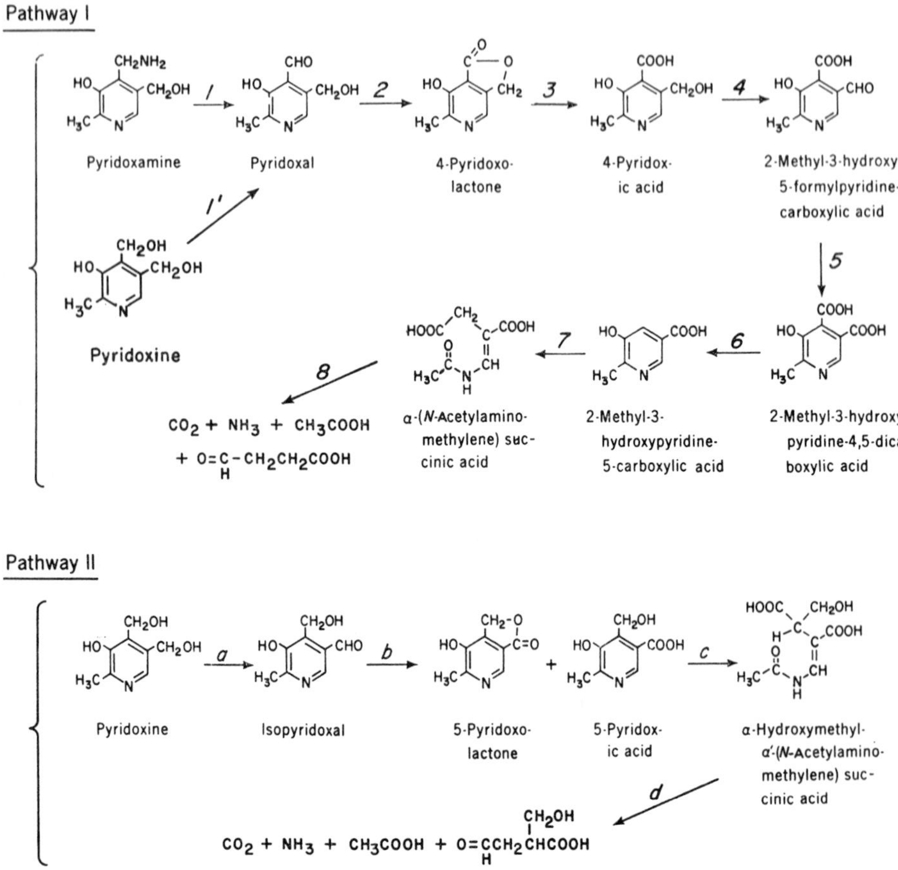

Pathway I

Pyridoxamine → Pyridoxal → 4-Pyridoxolactone → 4-Pyridoxic acid → 2-Methyl-3-hydroxy-5-formylpyridine-4-carboxylic acid

Pyridoxine

$CO_2 + NH_3 + CH_3COOH + O=C-CH_2CH_2COOH$ (H)

α-(N-Acetylaminomethylene) succinic acid

2-Methyl-3-hydroxypyridine-5-carboxylic acid

2-Methyl-3-hydroxypyridine-4,5-dicarboxylic acid

Pathway II

Pyridoxine → Isopyridoxal → 5-Pyridoxolactone → 5-Pyridoxic acid + ... → α-Hydroxymethyl-α'-(N-Acetylaminomethylene) succinic acid

$CO_2 + NH_3 + CH_3COOH + O=CCH_2CHCOOH$ (CH_2OH, H)

Fig. 2. Degradation of various forms of vitamin B_6 to utilizable open-chain metabolic intermediates by *Pseudomonas* MA-1 (pathway 1) and *Pseudomonas* 1A (pathway II). See text for discussion.

molecular weight of 150 000, contains 4 apparently identical subunits, and two binding sites for pyridoxal (or pyridoxamine)[75,94]. It does not contain pyridoxal-P as a prosthetic group[74]. Extensive kinetic studies[75] with various substrates and inhibitors have established kinetic parameters for each of these substances, and show that in the forward reaction pyridoxamine is bound first, then pyruvate; alanine is then released followed by pyridoxal. The reaction is thus fully analogous to a half-reaction catalyzed by pyridoxal-P-dependent transaminases such as aspartate aminotransferase, and has special usefulness in mechanistic investigations[75,76].

(b) *Reaction 1' (Fig. 2)*: Pyridoxine + 2,6-dichloroindophenol → pyridoxal + reduced 2,6-dichloroindophenol. This reaction is catalyzed by an FAD-dependent enzyme, *pyridoxal 4-dehydrogenase*, that utilizes either oxygen or 2,6-dichloroindophenol as the hydrogen acceptor. The partially purified enzyme has high affinity for pyridoxine (K_m 0.043 mM), acts optimally between pH 7 and 8, and shows only slight activity with pyridoxamine or pyridoxine-P as substrates[77]. FMN, which is the required prosthetic group for pyridoxine-P oxidase[54], does not replace FAD as a coenzyme for pyridoxine 4-dehydrogenase[77], thus clearly differentiating the two enzymatic reactions.

(c) *Reaction 2 (Fig. 2)*: Pyridoxal (hemiacetal) + NAD$^+$ → 4-pyridoxolactone + NADH + H$^+$. This enzyme, termed *pyridoxal dehydrogenase*, is specific for NAD$^+$ and does not oxidize benzaldehyde, salicylaldehyde, acetaldehyde, or 5-deoxypyridoxal[95]. Its substrate specificity, coenzyme requirement, and the product of the reaction thus differentiate it from the general aldehyde oxidase, which catalyzes oxidation of pyridoxal to 4-pyridoxic acid in animal tissues. Its failure to oxidize either pyridoxal-P or 5-deoxypyridoxal, together with formation of 4-pyridoxolactone as the product, shows that the internal hemiacetal of pyridoxal is its true substrate[95]. Under the conditions tested, no reduction of 4-pyridoxolactone could be observed. The enzyme is extremely sensitive to inhibition by *m*- and *p*-phenanthroline ($K_I = 1.0$ μM; *cf.* the K_m value for pyridoxal = 76 μM, both at pH 9.3)[95].

(d) *Reaction 3 (Fig. 2)*: 4-Pyridoxolactone + H$_2$O → 4-pyridoxic acid. This enzyme, *4-pyridoxolactonase*, is highly active in crude extracts of *Pseudomonas* MA-1. It acts optimally at pH 7.7, and has a K_m value of 3.1 μM[95].

(e) *Reaction 4 (Fig. 2)*: 4-Pyridoxic acid + X → 2-methyl-3-hydroxy-5-formylpyridine-4-carboxylic acid + XH$_2$. This reaction has been studied only in resting cells of *Pseudomonas* MA-1, which readily oxidize 4-pyridoxic acid on shaking in air. The immediate oxidation product, the corresponding 5-aldehyde, does not accumulate unless a trapping agent such as sodium bisulfite is added[93]. The hydrogen acceptor (X) for the reaction is unknown.

(f) *Reaction 5 (Fig. 2)*: 2-Methyl-3-hydroxy-5-formylpyridine-4-carboxylic acid + X → 2-methyl-3-hydroxypyridine-4,5-dicarboxylic acid + XH$_2$. If resting cells of *Pseudomonas* MA are incubated with 4-pyridoxic acid in the absence of a trapping agent, 4-pyridoxic acid disappears and the corresponding dicarboxylic acid appears[93]. As shown in the preceding section, the 5-aldehyde is intermediate in this process. The second reaction is presumably catalyzed by an aldehyde dehydrogenase; its specificity, the nature of the hydrogen acceptor, and other details of its action are unknown.

(g) *Reaction 6 (Fig. 2)*: 2-Methyl-3-hydroxypyridine-4,5-dicarboxylic acid → CO_2 + 2-methyl-3-hydroxypyridine-5-carboxylic acid. A decarboxylase that catalyzes this reaction has been partially purified from cell-free extracts of *Pseudomonas* MA-1. It shows an obligatory requirement for Mn^{2+} and a reducing agent such as CN^- or mercaptans, a broad pH optimum centered near pH 7.8, and a K_m value of $3.6 \cdot 10^{-5} M$[96].

(h) *Reaction 7 (Fig. 2)*: 2-Methyl-3-hydroxypyridine-5-carboxylic acid + $NADH + O_2$ → α-(N-acetylaminomethylene)succinic acid. It is *via* this interesting reaction that the pyridine ring derived from vitamin B_6 is cleaved to an open-chain product. A single FAD-dependent, crystalline oxygenase catalyzing this reaction has been isolated from *Pseudomonas* MA-1[97]. The enzyme appears homogeneous on electrophoresis and ultracentrifugation, has a molecular weight of 166 000, and contains two moles of FAD per mole of enzyme. Activity is lost on resolution with acidic ammonium sulfate and is restored by adding back FAD; FMN is inactive. Appropriate experiments showed that NADH serves only to reduce the FAD enzyme; the reduced enzyme interacts with oxygen and substrate to yield reoxidized enzyme and product. Two oxygen atoms are incorporated into the product during the cleavage reaction, one of which is partially displaced by water during the course of the reaction. NADPH is almost as effective a hydrogen donor as NADH; 5-pyridoxic acid also is cleaved (*cf.* reaction c, Fig. 2) but at less than 5% of the rate with the true substrate. This reaction is the first representative of a unique class of dioxygenase reactions in which reduction of a ring compound accompanies ring cleavage by oxygenation[97].

(i) *Reaction 8 (Fig. 2)*: α-(N-Acetylaminomethylene)succinic acid + 2 H_2O → $CH_3COOH + CO_2 + OHC\text{-}CH_2CH_2COOH + NH_3$. A partially purified hydrolase from *Pseudomonas* MA-1 catalyzes this reaction, which degrades the open-chain product formed by the preceding enzyme to compounds readily utilizable for growth. The enzyme acts optimally at pH 6–7.5, depending upon the buffer, and has a high apparent affinity (K_m 0.058 mM) for its substrate. No cofactor or metal ion requirements were observed, and no intermediate compounds accumulate during the reaction[98].

(ii) Oxidation of pyridoxine by Pseudomonas IA (Pathway II, Fig. 2)

Although distinct at every step, the reactions of Pathway II are closely analogous to those of Pathway I. *Reaction a,* the oxidation of pyridoxine to isopyridoxal, is catalyzed by pyridoxine-5-dehydrogenase, an FAD enzyme that uses 2,6-dichloroindophenol but not oxygen as hydrogen acceptor[77]. The physiological oxidant is not known. *Reaction b* appears analogous to reaction 2+3 of Pathway I; however, the aldehyde oxidase concerned has not been studied, and it is not known whether the reaction actually proceeds *via* the lactone or directly to 5-pyridoxic acid. *Reaction c* is catalyzed by an oxygenase which is distinct from that catalyzing reaction 7, but appears to act in a fully analogous manner; however, NADPH is specifically required, NADH being less than 5% as effective[97]. Finally, *reaction d* is catalyzed by a specific hydrolase, distinct from but entirely analogous in its action[98] to that catalyzing reaction 8.

(iii) Comments on the pathways for bacterial degradation of vitamin B_6

The oxidative pathways shown in Fig. 2 provide reduced coenzymes which can presumably be reoxidized to provide energy to the organism. However, no reduced carbon compounds or nitrogen become available for growth until the final reactions (reactions 8 and d) have occurred. With these reactions, compounds widely available for biosynthetic purposes and for energy formation *via* recognized metabolic pathways become accessible to the organism.

Pseudomonas MA was originally selected for its ability to grow on pyridoxamine as carbon source[91]; only later was a culture (*Pseudomonas* MA-1) derived from it that utilized pyridoxine by a similar pathway[77]. *Pseudomonas* IA, on the other hand, was originally selected for its ability to grow on pyridoxine[91]; it does not grow with pyridoxamine as carbon source, but does convert this compound in part to an unidentified blue pigment. Since the only two organisms studied in detail utilize different pathways for degradation of pyridoxine, it is quite probable that additional organisms might be found that utilize still different degradative pathways. Both of these organisms are aerobic; anaerobic or facultatively anaerobic bacteria which utilize vitamin B_6 as a source of carbon have not been found as yet.

7. Abnormalities in vitamin B_6 metabolism

Rare disorders of vitamin B_6 metabolism in which symptoms can be partially or completely suppressed by frequent large doses of vitamin B$_6$ are known to occur in humans. In some, but not all, cases, it has been possible to demonstrate that the disease is familial. An interesting feature of these metabolic disorders is that general symptoms of vitamin B$_6$ deficiency are lacking. Instead, each disease is characterized by a single manifestation of abnormal vitamin B$_6$ metabolism — *i.e.* anemia in the macrocytic or microcytic pyridoxine-responsive anemias[99-102]; convulsions in vitamin B$_6$-dependent infants[103-111]; elevated cystathionine levels in cystathioninuria[112]; elevated homocystine and methionine levels in homocystinuria[113-116]; and elevated excretion of xanthurenic acid and certain other tryptophan metabolites in xanthurenic aciduria[117-119]. The amount of vitamin B$_6$ supplied by normal diet is inadequate to reverse symptoms. When large doses of pyridoxine are administered, either orally or by injection, evidence of metabolic abnormality disappears temporarily. Increased dietary supplements of vitamin B$_6$ are known to lead to increased tissue levels[120] of pyridoxal-P. Pyridoxine supplementation, however, does not reverse certain serious features of these metabolic diseases which—in the case of vitamin B$_6$-dependent convulsions, homocystinuria, cystathioninuria and xanthurenic aciduria—commonly include mental retardation.

In cystathioninuria and in xanthurenic aciduria, the elevated vitamin B$_6$ requirement apparently is due to enzymes which function normally only in the presence of levels of pyridoxal phosphate higher than those achieved by normal dietary intake. The defective enzyme in cystathioninuria is cystathioninase (EC 4.2.1.15)[112]; in xanthurenic aciduria, kynureninase (EC 3.7.1.3)[117]. In both disorders, experiments with liver biopsy tissue showed that enzyme activity was low but could be restored to near normal levels by additions of pyridoxal phosphate[112,117]. An apparent explanation of these results is that the apoenzyme in each case had undergone a structural alteration which resulted in reduced affinity for the co-factor. These results are reminiscent of studies with *Neurospora crassa* which indicated that the affinity of tryptophan synthetase (EC 4.2.1.20) for pyridoxal phosphate could be altered by mutation[121].

Metabolic abnormalities associated with homocystinuria suggest that cystathionine synthetase (EC 4.2.1.13) is the missing or defective enzyme. Mudd *et al.*[114] analyzed liver tissue obtained from one homocystinuric patient and found no detectable cystathionine synthetase activity, although assays were carried out in the presence of pyridoxal phosphate. Other investigators have observed, however, that the metabolic abnormalities associated with homocystinuria disappear partially or completely after pyridoxine therapy[113,116]. If cystathionine synthetase is completely lacking, then the effect of pyridoxine therapy is difficult to explain. It has been suggested that the cases observed to date may differ in etiology[115].

The enzymatic causes of vitamin B_6-dependent convulsions and of the pyridoxine-responsive anemias are unknown.

Elevated requirements for vitamin B_6 which result in unusual sensitivity to vitamin B_6 depletion have been reported both in humans[8,122] and in inbred strains of mice[123]. The metabolic basis for the quantitative differences in vitamin B_6 requirements is unknown. It might lie in any of the areas in this review (*i.e.*, faulty absorption of vitamin B_6, faulty conversion to pyridoxal-P or excessive excretion of either the vitamin itself or of 4-pyridoxic acid). Further investigations in this fascinating area undoubtedly will provide additional instances of genetic abnormalities in the metabolism and utilization of both vitamin B_6 and other vitamins.

REFERENCES

1 E. E. SNELL, In M. FLORKIN AND E. H. STOTZ (Eds.), *Comprehensive Biochemistry*, Vol. 11, Elsevier, Amsterdam, 1963, p. 48.
2 B. M. GUIRARD AND E. E. SNELL, in M. FLORKIN AND E. H. STOTZ (Eds.), *Comprehensive Biochemistry*, Vol. 15, Elsevier, Amsterdam, 1964, p. 138.
3 T. W. GOODWIN, *The Biosynthesis of Vitamins and Related Compounds*, Academic Press, New York, 1963, p. 159.
4 E. E. SNELL, *Proc. 4th International Congress of Biochemistry*, Vol. 11, Pergamon, Oxford, 1959, p. 250.
5 E. E. SNELL, *Proc. 5th International Congress of Biochemistry*, Vol. 4, Pergamon, Oxford, 1963, p. 268.
6 W. J. ROBBINS, *Am. J. Botany*, 29 (1942) 241.
7 B. M. GUIRARD AND E. E. SNELL, in I. C. GUNSALUS AND R. Y. STANIER (Eds.), *The Bacteria*, Vol. 4, Academic Press, New York, 1962, p. 55.
8 O. A. BESSEY, D. J. D. ADAMS AND A. E. HANSON, *Pediatrics*, 20 (1957) 33.
9 N. YESS, J. M. PRICE, R. R. BROWN, P. B. SWAN AND H. LINKSWILER, *J. Nutr.*, 84 (1964) 229.
10 P. GYORGY, *Biochem. J.*, 29 (1935) 741.
11 J. B. LYON, H. L. WILLIAMS AND E. A. ARNOLD, *J. Nutr.*, 66 (1958) 261.
12 P. S. SARMA, E. E. SNELL AND C. A. ELVEHJEM, *J. Biol. Chem.*, 165 (1946) 55.
13 M. M. WINTROBE, R. H. FOLLIS, M. H. MILLER, H. J. STEIN, R. ALCAYAGA, S. HUMPHREYS, A. SUKSTA AND G. E. CARTWRIGHT, *Bull. Johns Hopkins Hosp.*, 72 (1943) 1.
14 F. H. BIRD, F. H. KRATZER, V. S. ASMUNDSON AND S. LEPKOVSKY, *Proc. Soc. Exptl. Biol. Med.*, 52 (1943) 44.
15 D. M. HEGSTED AND M. N. RAO, *J. Nutr.*, 30 (1945) 367.
16 G. SCHWARTZMAN AND L. STRAUSS, *J. Nutr.*, 38 (1949) 131.
17 E. L. HOVE AND J. F. HERNDON, *J. Nutr.*, 61 (1957) 127.
18 B. A. MCLAREN, E. KELLER, D. J. O'DONNELL AND C. A. ELVEHJEM, *Arch. Biochem. Biophys.*, 15 (1947) 169.
19 P. S. SARMA, E. E. SNELL AND C. A. ELVEHJEM, *Proc. Soc. Exptl. Biol. Med.*, 63 (1946) 284.
20 A. C. DA SILVA, A. B. FAJER, R. C. DE ANGELIS, M. A. PONTES, A. M. GIESBRECHT AND R. FRIED, *J. Nutr.*, 68 (1959) 213.
21 K. B. MCCALL, H. A. WAISMAN, C. A. ELVEHJEM AND E. S. JONES, *J. Nutr.*, 31 (1946) 685.
22 R. S. PARDINI AND C. J. ARGOUDELIS, *J. Bacteriol.*, 96 (1968) 672.
23 V. NURMIKKO AND R. RAUNIO, *Acta Chem. Scand.*, 15 (1961) 856.
24 W. B. DEMPSEY, *J. Bacteriol.*, 90 (1965) 431.
25 W. B. DEMPSEY AND P. F. PACHLER, *J. Bacteriol.*, 91 (1966) 642.
26 W. B. DEMPSEY, *J. Bacteriol.*, 97 (1969) 1403; 100 (1969) 295.
27 W. B. DEMPSEY, *J. Bacteriol.*, 92 (1966) 333.
28 W. B. DEMPSEY, *J. Bacteriol.*, 100 (1969) 1114.
29 E. E. SNELL AND A. N. RANNEFELD, *J. Biol. Chem.*, 157 (1945) 475.
30 J. C. RABINOWITZ AND E. E. SNELL, *J. Biol. Chem.*, 169 (1947) 643; *Anal. Chem.*, 19 (1947) 277.
31 G. H. SLOANE-STANLEY, *Biochem. J.*, 44 (1949) 567.
32 P. R. PAL AND H. N. CHRISTENSEN, *J. Biol. Chem.*, 236 (1961) 894.
33 K. YAMADA AND M. TSUJI, *J. Vitaminol. (Kyoto)*, 14 (1968) 282.
34 C. C. BOOTH AND M. C. BRAIN, *J. Physiol. (London)*, 164 (1962) 282.

35 H. A. Serebro, H. M. Solomon, J. H. Johnson and T. R. Hendrix, *Bull. Johns Hopkins Hosp.*, 119 (1966) 166.
36 J. V. Scudi, K. Unna and W. Antopol, *J. Biol. Chem.*, 135 (1940) 371.
37 National Academy of Sciences, National Research Council, *Nutrient Requirements of Dogs*, (Rev. 1962) (Publ. No. 989) pp. 22–23.
38 J. C. Rabinowitz and E. E. Snell, *Proc. Soc. Exptl. Biol. Med.*, 70 (1949) 235.
39 R. Schwartz and N. O. Kjelgaard, *Biochem. J.*, 48 (1951) 333.
40 P. E. Waibel, W. W. Cravens and E. E. Snell, *J. Nutr.*, 48 (1952) 531.
40a W. S. McNutt and E. E. Snell, *J. Biol. Chem.*, 173 (1948) 801; 182 (1950) 557.
41 D. B. McCormick, M. E. Gregory and E. E. Snell, *J. Biol. Chem.*, 236 (1961) 2076.
42 D. B. McCormick and E. E. Snell, *Proc. Natl. Acad. Sci. (U.S.)*, 45 (1959) 1371.
43 M. Tsubosaka and K. Makino, *J. Vitaminol. (Kyoto)*, 15 (1969) 131.
44 J. Hurwitz, *J. Biol. Chem.*, 205 (1953) 935; 217 (1955) 513.
45 D. B. McCormick and E. E. Snell, *J. Biol. Chem.*, 236 (1961) 2085.
46 D. B. McCormick and E. E. Snell, *Proc. Natl. Acad. Sci. (U.S.)*, 45 (1959) 1371.
47 D. B. McCormick, B. M. Guirard and E. E. Snell, *Proc. Soc. Exptl. Biol. Med.*, 104 (1960) 554.
48 J. A. Bain and H. L. Williams, *Inhibition in the Nervous System and Gamma Aminobutyric Acid*, Pergamon, New York, 1960, pp. 275–293.
49 L. S. Goodman and A. Gilman (Eds.), *The Pharmacological Basis of Therapeutics*, MacMillan, New York, 1965, pp. 1324–1325.
49a K. Ogata, T. Tochikura, Y. Tani and S. Yamamoto, *Agr. Biol. Chem. (Tokyo)*, 30 (1966) 829.
49b I. B. Wilson, J. Dayan and K. Cyr, *J. Biol. Chem.*, 239 (1964) 4182.
50 B. M. Pogell, *J. Biol. Chem.*, 232 (1958) 761.
51 H. Wada, T. Morisue, Y. Nishimura, Y. Morino, Y. Sakamoto and K. Ichihara, *Proc. Imp. Acad. Japan*, 35 (1959) 299.
52 T. Morisue, Y. Morino, Y. Sakamoto and K. Ichihara, *J. Biochem. (Tokyo)*, 48 (1960) 28.
53 Y. Morino, H. Wada, T. Morisue, Y. Sakamoto and K. Ichihara, *J. Biochem. (Tokyo)*, 48 (1960) 18.
54 H. Wada and E. E. Snell, *J. Biol. Chem.*, 236 (1961) 2089.
55 S. Yamamoto, T. Tochikura and K. Ogata, *Agr. Biol. Chem. (Tokyo)*, 29 (1965) 315.
56 S. Yamamoto, T. Tochikura and K. Ogata, *Agr. Biol. Chem. (Tokyo)*, 29 (1965) 597.
57 H. M. Henderson, *Biochem. J.*, 95 (1965) 775.
58 S. Yamamoto, T. Tochikura and K. Ogata, *Agr. Biol. Chem. (Tokyo)*, 29 (1965) 200.
59 E. E. Snell, *Vitamins and Hormones*, 16 (1958) 77.
60 A. E. Braunstein, in P. D. Boyer, H. Lardy and K. Myrbäck (Eds.), *The Enzymes*, Vol. 4, Academic Press, New York, 1960, p. 113.
61 A. Novogrodsky and A. Meister, *J. Biol. Chem.*, 239 (1964) 879.
62 S. Yamamoto, T. Tochikura and K. Ogata, *Agr. Biol. Chem. (Tokyo)*, 29 (1965) 605.
63 C. Turano, A. Giartosio and P. Fasella, *Arch. Biochem. Biophys.*, 104 (1964) 524.
64 J. M. Turner and F. C. Happold, *Biochem. J.*, 78 (1961) 364; F. C. Happold, *Nature*, 194 (1962) 580.
65 S. D. Wainwright, *Can. J. Biochem. Physiol.*, 37 (1959) 1429; E. Wikberg, *Physiol. Plantarum*, 13 (1960) 616.
66 M. J. Andrews and J. M. Turner, *Nature*, 210 (1966) 1159.

67 S. Saraswathi and B. K. Bachhawat, *J. Neurochem.*, 10 (1963) 127.
68 M. V. Srikantaiah, E. V. Chandrasekaran and A. N. Radhakrishnan, *Indian J. Biochem.*, 4 (1967) 9.
69 J. M. Turner, *Biochem. J.*, 80 (1961) 663.
70 H. Wada and E. E. Snell, *J. Biol. Chem.*, 237 (1962) 127.
71 H. L. C. Wu and M. Mason, *J. Biol. Chem.*, 239 (1964) 1492.
72 H. Wada and Y. Morino, *Vitamins and Hormones*, 22 (1964) 411.
73 S. W. Tanenbaum, *J. Biol. Chem.*, 218 (1956) 733.
74 H. Wada and E. E. Snell, *J. Biol. Chem.*, 237 (1962) 133.
75 J. E. Ayling and E. E. Snell, *Biochemistry*, 7 (1968) 1616, 1626.
76 J. E. Ayling, H. C. Dunathan and E. E. Snell, *Biochemistry*, 7 (1968) 4537.
77 T. K. Sundarum and E. E. Snell, *J. Biol. Chem.*, 244 (1969) 2577.
78 Y. Morino and Y. Sakamoto, *J. Biochem. (Tokyo)*, 48 (1960) 733.
79 J. C. Rabinowitz and E. E. Snell, *Arch. Biochem. Biophys.*, 43 (1952) 399.
80 W. W. Cravens and E. E. Snell, *Proc. Soc. Exptl. Biol. Med.*, 71 (1949) 73.
81 S. Johansson, S. Lindstedt and H. Tiselius, *Biochemistry*, 7 (1968) 2327.
82 J. B. Lyon, J. A. Bain and H. L. Williams, *J. Biol. Chem.*, 237 (1962) 1989.
83 I. Nakahara, Y. Morino, T. Morisue and Y. Sakamoto, *J. Biochem. (Tokyo)*, 49 (1961) 339.
84 J. W. Huff and W. A. Perzweig, *J. Biol. Chem.*, 155 (1944) 345.
85 J. Kelsay, A. Baysal and H. Linkswiler, *J. Nutr.*, 94 (1968) 490.
86 J. Hurwitz, *J. Biol. Chem.*, 212 (1955) 757; *Natl. Vitamin Found. Nutr. Symp. Ser.*, 13 (1956) 49.
87 A. Baysal, B. A. Johnson and H. Linkswiler, *J. Nutr.*, 89 (1966) 19.
88 S. K. Reddy, M. S. Reynolds and J. M. Price, *J. Biol. Chem.*, 233 (1958) 691.
89 G. E. Bernett and W. N. Pearson, *Fed. Proc.*, 27 (1968) 553.
90 S. Johansson, S. Lindstedt, U. Register and L. Wadstrom, *Am. J. Clin. Nutr.*, 18 (1966) 185.
91 V. W. Rodwell and E. E. Snell, *J. Biol. Chem.*, 233 (1958) 1548.
92 M. Ikawa, V. W. Rodwell and E. E. Snell, *J. Biol. Chem.*, 233 (1958) 1555.
93 R. W. Burg, V. W. Rodwell and E. E. Snell, *J. Biol. Chem.*, 235 (1960) 1164.
94 H. Kolb, R. D. Cole and E. E. Snell, *Biochemistry*, 7 (1968) 2946.
95 R. W. Burg and E. E. Snell, *J. Biol. Chem.*, 244 (1969) 2585.
96 E. E. Snell, A. A. Smucker, A. Ringelmann and F. Lynen, *Biochem. Z.*, 341 (1964) 109.
97 L. G. Sparrow, P. P. K. Ho, T. K. Sundarum, D. Zach, E. J. Nyns and E. E. Snell, *J. Biol. Chem.*, 244 (1969) 2590.
98 E. J. Nyns, D. Zach and E. E. Snell, *J. Biol. Chem.*, 244 (1969) 2601.
99 J. W. Harris and D. L. Horrigan, *Vitamins and Hormones*, 22 (1964) 721.
100 J. D. Hines and J. W. Harris, *Am. J. Clin. Nutr.*, 14 (1964) 137.
101 S. S. Bottomley, *J. Am. Med. Ass.*, 180 (1962) 653.
102 M. S. Bourne, M. W. Elves and M. C. G. Israels, *Brit. J. Haematol.*, 11 (1965) 1.
103 A. D. Hunt, J. Stokes, W. W. McCrory and H. H. Stroud, *Pediatrics*, 13 (1954) 140.
104 L. Sokoloff, N. A. Lassen, G. M. McKhann, D. B. Tower and W. Albers, *Nature*, 183 (1959) 751.
105 M. Bejsovec, Z. Kulenda and E. Ponca, *Arch. Dis. Childhood*, 42 (1967) 201.
106 M. M. Robins, *J. Am. Med. Ass.*, 195 (1966) 491.
107 C. Waldinger and R. B. Berg, *Pediatrics*, 32 (1963) 161.
108 R. Garty, Z. Yonis, J. Braham and K. Steinitz, *Arch. Dis. Childhood*, 37 (1962) 21.
109 C. R. Scriver, *Pediatrics*, 26 (1960) 62.

110 C. R. SCRIVER AND A. M. CULLEN, *Pediatrics*, 36 (1965) 14.
111 J. GENTZ, A. HAMFELT, S. JOHANSSON, S. LINDSTEDT, B. PERSSON AND R. ZETTERSTROM, *Acta Paediatr. Scand.*, 56 (1967) 17.
112 G. FRIMPTER, *Science*, 149 (1965) 1095.
113 G. W. BARBER AND G. L. SPAETH, *Lancet, i* (1967) 337.
114 S. H. MUDD, J. D. FINKELSTEIN, F. IRREVERRE AND L. LASTER, *Science*, 143 (1964) 1442.
115 J. D. FINKELSTEIN, S. H. MUDD, F. IRREVERRE AND L. LASTER, *Science*, 146 (1964) 785.
116 S. KELLY AND W. COPELAND, *Metab. Clin. Exptl.*, (1968) 794.
117 K. TADA, Y. YOKOYAMA, H. NAKAGAWA, T. YOSHIDA AND T. ARAKAWA, *Tohoku J. Exptl. Med.*, 93 (1967) 115.
118 D. O'BRIEN AND C. B. JENSEN, *Clin. Sci.*, 24 (1963) 179.
119 A. KNAPP, *Clin. Chim. Acta*, 5 (1960) 6.
120 W. D. BELLAMY, W. W. UMBREIT AND I. C. GUNSALUS, *J. Biol. Chem.*, 160 (1945) 461.
121 D. M. BONNER, Y. SUYAMA AND J. A. DEMOSS, *Fed. Proc.*, 19 (1960) 926.
122 C. R. SCRIVER AND J. H. HUTCHINSON, *Pediatrics*, 31 (1963) 240.
123 J. B. LYON, H. L. WILLIAMS AND E. A. ARNOLD, *J. Nutr.*, 66 (1958) 261.

Chapter I

Metabolism of Water-Soluble Vitamins

Section d

Biosynthesis of Pantothenic Acid and Coenzyme A

GENE M. BROWN

Department of Biology, Massachusetts Institute of Technology, Cambridge, Mass. (U.S.A.)

1. Introduction

Pantothenic acid is one of the B vitamins and thus cannot by synthesized by animals. However, most microorganisms are able to make this vitamin from its precursors, pantoic acid and β-alanine. Both animals and bacteria can convert pantothenic acid to coenzyme A, the metabolically active form of the vitamin. The material included in this chapter will contain what is known about the metabolic origins of pantoic acid and β-alanine as well as information on the enzymatic synthesis of pantothenic acid and coenzyme A.

2. Biogenesis of pantoic acid

The set of enzymatic reactions that is thought to result in the biosynthesis of pantoic acid is shown in Fig. 1. The position of α-keto-β,β-dimethyl-γ-hydroxybutyric acid ("ketopantoate") as a precursor of pantoic acid was first suggested by an observation by Kuhn and Wieland[1] that this compound could be reduced to pantoate by yeast. Confirmatory evidence was supplied by Lansford and Shive[2] who described a class of pantoate-requiring mutants of *Escherichia coli* that could utilize ketopantoate in place of pantoate. The enzymatic reduction of ketopantoate to pantoate has not been studied in enough detail to establish the nature of the necessary reducing agent or the characteristics of the enzyme.

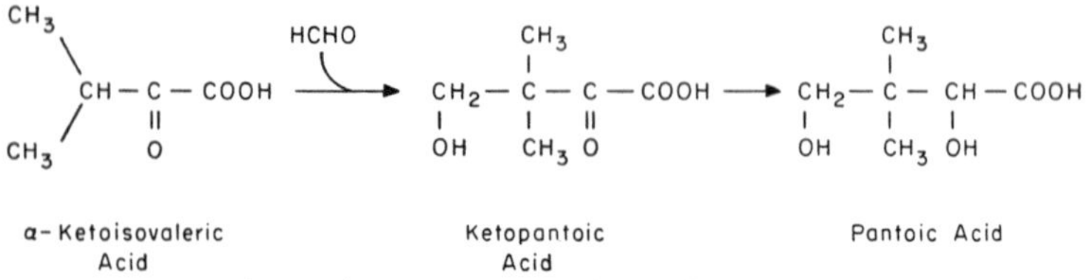

α- Ketoisovaleric
Acid   Ketopantoic
Acid   Pantoic Acid

Fig. 1. The enzymatic synthesis of pantoic acid.

That the precursor of ketopantoic acid is α-ketoisovaleric acid was first suggested by the report by Maas and Vogel[3] that whole cells of a strain of *E. coli* could synthesize ketopantoate and pantoate from α-ketoisovaleric acid, whereas a certain pantoate-requiring mutant could not carry out these transformations. McIntosh et al.[4] reported that α-ketoisovaleric acid could be converted to ketopantoate in the presence of cell-free extracts of *E. coli* only if formaldehyde were also supplied as substrate. These workers were stimulated to do this work by their prior discovery that a strain of *Bacterium linens* needs either pantoic acid or *p*-aminobenzoic acid as a growth factor[5,6], an observation that suggested a role for *p*-aminobenzoate in the synthesis of pantoate. The discovery that formaldehyde is needed to make ketopantoate from α-ketoisovaleric acid provided apparent confirmation for this suggestion since use of a one-carbon unit (formaldehyde) in a metabolic process usually requires the participation of tetrahydrofolic acid, the metabolically active form of *p*-aminobenzoate. Addition of tetrahydrofolate to crude extracts of *E. coli* stimulated the synthesis of ketopantoate[7], but after the enzyme had been partially purified no effect of tetrahydrofolate could be observed[4].

The K_m values for formaldehyde (0.01 M) and for α-ketoisovaleric acid (0.1 M) in this tetrahydrofolate-independent synthesis of ketopantoate are so high that it seems questionable whether this process is of physiological significance. Teller and Snell[8] have reinvestigated ketopantoate synthesis in *E. coli* and have found that two enzymes exist; one similar to that reported by McIntosh et al.[4] which is not stimulated by tetrahydrofolate, and a second enzyme whose activity is dependent on the presence of tetrahydrofolate. This second enzyme requires Mg^{2+} and the K_m values are 10^{-3} M for both formaldehyde and α-ketoisovaleric acid. That the tetrahydrofolate-dependent enzyme is the one that functions in the synthesis of ketopantoate was indicated by the observation that it was missing in certain ketopantoate-requiring mutants of *E. coli*, whereas these mutants retained the tetrahydrofolate-independent enzyme. The physiological role, if any, for the latter enzyme remains unclear.

3. Biogenesis of β-alanine

The view that β-alanine is made in pantothenate-synthesizing microorganisms by the decarboxylation of aspartic acid is based primarily on the observations (*a*) that the growth inhibitory properties of β-hydroxyaspartic acid for *E. coli* can be reversed by either aspartate, β-alanine or pantothenic acid[9] and (*b*) that whole cells of a variety of organisms apparently can make β-alanine when aspartic acid is supplied to the cells[10-12]. However, no tracer or enzyme studies have yet been reported to confirm the suggestion that β-alanine is formed by the decarboxylation of aspartate. A possible alternate method for β-alanine production would terminate with a transamination whereby an amino group could be transferred to malonic semialdehyde to yield β-alanine. Such transaminases are known to exist in *Aspergillus*[13], *Neurospora crassa*[14] and *Pseudomonas fluorescens*[15]. Malonic semialdehyde is known to be made by a number of bacteria[16,17] from acrylyl-coenzyme A; however, whether such reactions can account for the synthesis of β-alanine remains to be established.

In summary, the method of enzymatic synthesis of β-alanine is not well-understood, although there is indirect evidence to indicate that this is accomplished by the decarboxylation of aspartic acid.

4. Enzymatic synthesis of pantothenic acid

The enzyme that catalyzes the synthesis of pantothenic acid from pantoic acid and β-alanine has been purified from *E. coli* by two independent groups[18-24]. ATP, Mg^{2+} or Mn^{2+}, and K^+ or NH_4^+ are known to be required for this process[18] and Maas and Novelli[19] established that during the reaction AMP and inorganic pyrophosphate are produced each in a molar amount equal to the moles of pantothenate formed. Information concerning the mechanism of the reaction was provided by Maas who reported that: (*a*) ^{32}P-inorganic pyrophosphate is incorporated into ATP only in the presence of pantoate[22]; (*b*) [^{14}C]AMP is not incorporated into ATP[22]; (*c*) incubation of hydroxylamine with pantoate, ATP, and enzyme results in the formation of pantoylhydroxam- ic acid[22] and (d) during pantothenate synthesis ^{18}O is transferred from the carboxyl group of pantoate to AMP[23]. These observations led to the formu- lation of the mechanism of the reaction as follows:

$$\text{pantoate} + \text{ATP} + \text{enzyme} \rightleftharpoons \text{enzyme–pantoyl–AMP} + \text{PP}_i$$

$$\text{enzyme–pantoyl–AMP} + \beta\text{-alanine} \rightarrow \text{pantothenate} + \text{AMP} + \text{enzyme}$$

5. Enzymatic synthesis of coenzyme A from pantothenic acid

Early work on the biosynthesis of coenzyme A from pantothenic acid was directed toward the identification of the precursor of β-mercaptoethylamine, the sulfur-containing fragment of coenzyme A. Experiments with whole cells of *Lactobacillus arabinosus*[24] and *Proteus morganii*[25] indicated that this fragment is derived from cysteine since synthesis of coenzyme A from panto- thenate was dependent on the presence of this amino acid. The concept began to take shape that the biosynthesis of coenzyme A occurs *via* the following reactions:

$$\text{pantothenate} + \text{cysteine} \rightarrow \text{pantothenylcysteine}$$

$$\text{pantothenylcysteine} \rightarrow \text{pantetheine} + \text{CO}_2$$

$$\text{pantetheine} \xrightarrow{\text{ATP}} \text{phosphopantetheine}$$

$$\text{phosphopantetheine} \xrightarrow{\text{ATP}} \text{dephospho coenzyme A}$$

$$\text{dephospho coenzyme} \xrightarrow{\text{ATP}} \text{coenzyme A}$$

The conversion of phosphopantetheine to coenzyme A *via* the intermediate, dephospho coenzyme A, has been shown to proceed in the presence of enzymes from hog liver[26] and rat liver[27]. The conversion of pantetheine to phos- phopantetheine can occur by kinases known to be present in bacterial species[28] and in rat tissues[26,28].

The view that pantothenate is converted to pantothenylcysteine which is then decarboxylated to pantetheine was based on the following observations: (a) pantothenylcysteine is several times more effective, on a molar basis, than pantothenic acid as a growth factor for *Acetobacter suboxydans*[25,29,30]; (b) cell-free preparations of *A. suboxydans*[25] and *Lactobacillus helveticus*[31] can decarboxylate pantothenylcysteine to pantetheine; and (c) pantothenyl- cysteine appeared to be an intermediate in coenzyme A synthesis carried out by rat-liver extracts[26]. However, some doubts began to arise as to the general importance of this set of reactions with the findings that bacteria other than *A. suboxydans* and *L. helveticus* cannot decarboxylate pantothenylcysteine[31] and that pantothenylcysteine apparently cannot replace pantothenate as a nutritional factor in rats and chicks[32]. These doubts led to a reexamination of bacterial and animal systems by Brown[33] in 1959. The results of these investigations indicated that the "pantothenylcysteine–pantetheine" pathway does not exist in rat tissues and in bacteria such as *P. morganii* and *E. coli*. Evidence was presented that the following reactions account for the synthesis of phosphopantetheine in these systems:

$$\text{pantothenic acid} \xrightarrow{\text{ATP}} 4'\text{-phosphopantothenic acid}$$

$$4'\text{-phosphopantothenic acid} + \text{cysteine} \xrightarrow[\text{ATP}]{\text{CTP or}} \text{phosphopantothenylcysteine}$$

$$\text{phosphopantothenylcysteine} \rightarrow \text{phosphopantetheine} + CO_2$$

This evidence was based on the findings that each of the above reactions was shown to occur in the presence of cell-free extracts and that, on the other hand, no synthesis of pantothenylcysteine nor decarboxylation of this compound could be detected. Brown found that β-mercaptoethylamine could be used as substrate in place of cysteine to yield phosphopantetheine directly, but there is no evidence that such a reaction is of biosynthetic importance since the enzymatic decarboxylation of cysteine to β-mercaptoethylamine has not been shown to occur. One difference that was observed between the bacterial and rat systems is that only CTP can be used as an energy source for the synthesis of phosphopantothenylcysteine by the bacterial enzyme, whereas the rat enzyme can utilize either CTP or ATP.

As a result of the observations described above, the reactions that result in the synthesis of coenzyme A from pantothenate are thought to be as shown in Fig. 2. Abiko[34,35], in 1967, confirmed and extended the work of Brown with rat-liver preparations. He separated and partially purified the enzymes that catalyze the individual reactions shown in Fig. 2 and confirmed that pantothenylcysteine does not function as an intermediate in coenzyme A biosynthesis in rat liver. Abiko *et al.*[36] have studied the reaction catalyzed by purified phosphopantothenylcysteine synthetase from rat liver and have concluded that ADP and inorganic phosphate are produced from ATP during the course of the reaction. These same workers have also investigated the latter stages of coenzyme A biosynthesis by partially purified rat-liver enzymes and

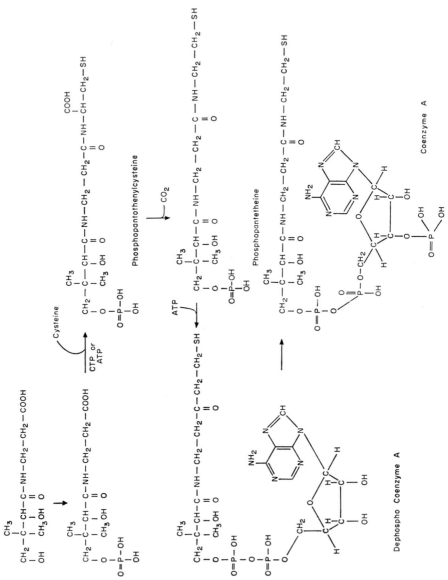

Fig. 2. Enzymatic reactions involved in the formation of coenzyme A from pantothenic acid

have confirmed that the reactions proceed from phosphopantetheine as shown[27] in Fig. 2.

The conclusion in 1954 that rat liver makes coenzyme A by the "pantothenylcysteine–pantetheine" pathway[26] was based on analytical procedures which could not distinguish between this pathway and the "phosphopantothenylcysteine" pathway. Misinterpretation of the data arose primarily because it was not recognized that the kinase can use both pantothenic acid and pantetheine as substrate. All kinases that have been examined from both bacterial and animal sources possess this common property. Some disagreement that has yet to be resolved has arisen with regard to the ability of animal kinases to phosphorylate pantothenylcysteine. Brown could find little evidence for such a reaction in rat liver[33], an observation that is consistent with the findings of Thompson and Bird[32] that pantothenylcysteine is inert as a nutritional factor for rats. On the other hand, Abiko[34] and Cavallini et al.[37] have reported that kinases from rat liver[34] and pigeon liver[37] can catalyze the phosphorylation of pantothenylcysteine.

The "pantothenylcysteine–pantetheine" pathway for production of coenzyme A would seem to be limited possibly to A. suboxydans and L. helveticus of all the organisms that have been studied. The view that this pathway operates even in these organisms must remain tentative, however, until positive evidence is obtained for the enzymatic synthesis of pantothenylcysteine from pantothenate and cysteine.

REFERENCES

1 R. KUHN AND T. WIELAND, *Chem. Ber.*, 75B (1942) 121.
2 E. M. LANSFORD JR. AND W. SHIVE, *Arch. Biochem. Biophys.*, 38 (1952) 353.
3 W. K. MAAS AND H. J. VOGEL, *J. Bacteriol.*, 65 (1953) 338.
4 E. N. MCINTOSH, M. PURKO AND W. A. WOOD, *J. Biol. Chem.*, 228 (1957) 499.
5 M. PURKO, W. O. NELSON AND W. A. WOOD, *J. Bacteriol.*, 66 (1953) 561.
6 M. PURKO, W. O. NELSON AND W. A. WOOD, *J. Biol. Chem.*, 207 (1954) 51.
7 M. PURKO, W. O. NELSON AND W. A. WOOD, *Abstr. Am. Chem. Soc.*, *126th Meeting, New York, September, 1954*, 57c.
8 J. H. TELLER AND E. E. SNELL, *Abstr. Am. Chem. Soc.*, *158th Meeting, New York, September, 1969*, 40.
9 W. SHIVE AND J. MACOW, *J. Biol. Chem.*, 162 (1946) 451.
10 A. I. VIRTANEN AND T. LAINE, *Enzymologia*, 3 (1937) 266.
11 D. BILLEN AND H. C. LICHSTEIN, *J. Bacteriol.*, 58 (1949) 215.
12 E. A. GRULA AND M. M. GRULA, *Biochim. Biophys. Acta*, 74 (1963) 776.
13 E. ROBERTS, P. AYENGAR AND I. POSNER, *J. Biol. Chem.*, 203 (1953) 195.
14 H. AURICH, *Z. Physiol. Chem.*, 326 (1961) 25.
15 O. HAYAISHI, Y. NISHIZUKA, M. TATIBANA, M. TAKESHITA AND S. KUNO, *J. Biol. Chem.*, 236 (1961) 781.
16 P. R. VAGELOS AND J. M. EARL, *J. Biol. Chem.*, 234 (1959) 2272.
17 H. DEN, W. G. ROBINSON AND M. J. COON, *J. Biol. Chem.*, 234 (1959) 1666.
18 W. K. MAAS, *J. Biol. Chem.*, 198 (1952) 23.
19 W. K. MAAS AND G. D. NOVELLI, *Arch. Biochem. Biophys.*, 43 (1953) 236.
20 G. PFLEIDERER, A. KREILING AND T. WIELAND, *Biochem. Z.*, 333 (1960) 302.
21 T. WIELAND, A. KREILING, W. BUCK AND G. PFLEIDERER, *Biochem. Z.*, 333 (1960) 311.
22 W. K. MAAS, *Abstr. Congr. Intern. Biochem.*, *3rd, Brussels*, (1955) 4.
23 W. K. MAAS, *Federation Proc.*, 15 (1956) 305.
24 W. S. PIERPONT AND D. E. HUGHES, *Abstr. Congr. Intern. Biochem.*, *2nd, Paris*, (1952) 91.
25 G. M. BROWN AND E. E. SNELL, *J. Am. Chem. Soc.*, 75 (1953) 2782.
26 M. B. HOAGLAND AND G. D. NOVELLI, *J. Biol. Chem.*, 207 (1954) 767.
27 Y. ABIKO, T. SUZUKI AND M. SHIMIZU, *J. Biochem. (Tokyo)*, 61 (1967) 309.
28 G. B. WARD, G. M. BROWN AND E. E. SNELL, *J. Biol. Chem.*, 213 (1955) 869.
29 T. E. KING AND V. H. CHELDELIN, *Proc. Soc. Explt. Biol. Med.*, 84 (1953) 591.
30 G. M. BROWN AND E. E. SNELL, *J. Bacteriol.*, 67 (1954) 467.
31 G. M. BROWN, *J. Biol. Chem.*, 226 (1957) 651.
32 R. Q. THOMPSON AND O. D. BIRD, *Science*, 120 (1954) 763.
33 G. M. BROWN, *J. Biol. Chem.*, 234 (1959) 370.
34 Y. ABIKO, *J. Biochem. (Tokyo)*, 61 (1967) 290.
35 Y. ABIKO, *J. Biochem. (Tokyo)*, 61 (1967) 300.
36 Y. ABIKO, M. TOMIKAWA AND M. SHIMIZU, *J. Biochem. (Tokyo)*, 64 (1968) 115.
37 D. CAVALLINI, B. MONDOVI, C. DEMARCO AND G. FERRO-LUZZI, *Enzymologia*, 20 (1959) 359.

Metabolism of Water-Soluble Vitamins

Section e

The Metabolism of Biotin and Analogues

DONALD B. McCORMICK AND LEMUEL D. WRIGHT

Graduate School of Nutrition and Section of Biochemistry and Molecular Biology, Cornell University, Ithaca, N. Y. (U.S.A.)

1. Introduction

Biotin was isolated, characterized and synthesized over 25 years ago. Since biotin is required by animals and some microorganisms in extremely small amounts, is determinable in physiological amounts only by microbiological methods, is synthesized by many microorganisms including those of the intestinal tract so that the production of a deficiency in man and other ani-

biotin

I

mals is complicated, and is covalently bound to apoenzymes rendering resolution of holoenzymes virtually impossible, progress in elucidating the role of biotin in biochemical reactions has been unusually slow. After many false leads, the role of biotin in enzymatic reaction is now becoming clearer.

A number of excellent reviews have appeared covering the chemistry[1-3] and function[4-9] of biotin. Unfortunately, the metabolism of biotin has been less frequently reviewed. After some confusing early reports concerning the naturally-occurring forms of biotin, the subject is now fairly well understood. The biosynthetic pathway of biotin, despite some present enigmas, is on the verge of clarification. Except for a few steps, the degradation of biotin by microorganisms has been essentially delineated. It is appropriate, then, that the present status of the metabolism of biotin be summarized.

2. Biosynthesis

(a) Pimelic acid

Information on the biosynthesis of biotin began to accumulate even before the full structure of the vitamin was elucidated. Du Vigneaud and colleagues reported[10] in 1942 that the requirement of the diphtheria organism, *Corynebacterium diphtheriae*, for pimelic acid[11] could be satisfied by biotin instead. He postulated that pimelic acid is required by the diphtheria organism only to synthesize biotin and that pimelic acid must contribute to the biotin

$$
\begin{array}{c}
\text{COOH} \\
|\\
\text{CH}_2\text{—CH}_2\text{—CH}_2\text{—CH}_2\text{—CH}_2\text{—COOH}
\end{array}
$$

pimelic acid

II

structure. At the same time, Eakin and Eakin[12] reported that the biosynthesis of biotin by *Aspergillus niger* is greatly stimulated by the presence of pimelic acid in the growth medium. Neither biotin nor pimelic acid stimulates the *rate* or *extent* of growth of this mold. The Eakins similarly concluded that pimelic acid is a precursor of biotin. The lower homologues of pimelic acid, succinic, glutaric, and adipic acids, and an isomer, β-methyladipic acid, are inactive in stimulating biotin biosynthesis by *A. niger*, whereas the higher homologues of pimelic acid, suberic and azelaic acids, have activity comparable to pimelic acid. In subsequent studies on biotin biosynthesis by a strain of *A. niger* where the specificity of the dicarboxylic acid was studied, Wright[13] confirmed the activity of azelaic acid (9 carbons) but could show no activity with suberic acid (8 carbons). It was concluded that *A. niger* probably carries out an active β-oxidation and that higher homologues of pimelic acid (7 carbons) containing an odd number of carbons may be expected to show activity, whereas those containing an even number of carbon atoms do not yield pimelic acid on β-oxidation and may be expected to show no activity in stimulating biotin biosynthesis by *A. niger* or other biotin-synthesizing organisms. In an unidentified strain of *Pseudomonas* that utilizes azelaic acid as the sole source of carbon, Bassalik and Wright[14,15] were able to isolate and identify pimelic acid as one metabolic product of azelaic acid. Suberic acid, on the other hand, could not be detected. It is probably safe to assume that the suberic acid used by the Eakins contained pimelic acid and/or azelaic acids as a contaminant.

Although the demonstration of Mueller[11] that the previously unknown growth factor requirement of *C. diphtheriae* may be satisfied with pimelic acid is well founded and confirmed by Du Vigneaud and colleagues[10], the role of pimelic acid as a growth factor seems restricted to that of a precursor of biotin. For example, the biotin requirement of some thermophilic bacteria may be satisfied with pimelic acid[16]. Pimelic acid has been reported to be a vitamin for the larvae of the confused flour beetle, *Tribolium confusum*, but the activity recorded may be due to conversion to biotin by the flora of the intestinal tract[17]. Pimelic acid fails to satisfy the biotin requirement of at least *Lactobacillus casei*[18], *Lactobacillus arabinosus*[19], 13 biotin-requiring fungi[20], yeast[10], or *Rhizobium trifolii*[21]. A thorough search of the literature would probably reveal other species where pimelic acid is inactive in lieu of biotin in promoting growth. Despite the absence of any well-documented growth-stimulating effects of pimelic acid *per se*, pimelic acid is often listed as a component of purified media for the growth of fastidious bacteria. Here it could just as well be eliminated if the medium is adequate with respect to biotin. As far as the present reviewers are aware, the full significance of the urinary excretion of pimelic acid in relatively large amounts by herbivores, as first observed by Mueller[11], remains obscure.

Elford and Wright[22,23] have provided evidence that pimelic acid is incorporated as a unit in the biosynthesis of biotin. *A. niger* was grown with aeration in the presence of [1,7-^{14}C]pimelic acid. The biotin vitamers were concentrated from the medium, and carrier biotin was added to the mixture. The mixture was then treated with hydrogen peroxide to oxidize the biotin vitamers (predominantly biotin *l*-sulfoxide) and carrier biotin to biotin sulfone. The biotin sulfone was subjected to repeated recrystallizations and an aliquot of each crystallization was examined for radioactivity. It was found that the biotin sulfone contained radioactivity and that this radioactivity reached a constant value on recrystallization. Thus, it was confirmed by the isotopic technique that pimelic acid is a precursor of biotin. The isolated and recrystallized radioactive biotin sulfone was then subjected to the Schmidt reaction whereby the carboxyl carbon of the biotin sulfone is removed as carbon dioxide through the action of hydrazoic acid. The remaining portion of the biotin molecule is recovered as "biotin amine". The carbon dioxide evolved in the decarboxylation was found to contain one-half of the radioactivity present in the original biotin sulfone, while the "biotin amine" was found to contain the other half of the original counts. This finding was taken as evidence that both carboxyl carbons of pimelic acid are involved in biotin biosynthesis and that pimelic acid is incorporated as a unit. The positions in the biotin molecule that arise from pimelic acid are apparent from a consideration of the two structures shown above.

Experiments similar in design to those of Elford and Wright involving growth of a microorganism on labeled pimelic acid followed by isolation and counting of biotin vitamers have been carried out by Eisenberg[24,25]. The pimelic acid used was 1,7-^{14}C labeled. The biosynthetic organism was *Phycomyces blakesleeanus*. Following growth, the biotin and vitamers were subjected to several purification steps and then reduced to dethiobiotin with Raney nickel. Carrier dethiobiotin was added and several recrystallizations carried out. The dethiobiotin was found to be radioactive and the radioactivity was constant after the first recrystallization. The constant specific activity of the recrystallized dethiobiotin indicates the incorporation of pimelic acid into either biotin, dethiobiotin, or both.

(b) Pelargonic acid derivatives

As early as 1962, evidence began to accumulate pointing to the existence of an unknown vitamer of biotin in culture filtrates from various microorganisms grown on media suitably supplemented with pimelic acid. For example, Eisenberg[24] reported that the biotin vitamers in the culture filtrate of *Phycomyces blakesleeanus* consisted of true biotin and "biotin sulfoxide" (10–20 %), dethiobiotin (60–70%), biocytin (0–20%), and an unknown vitamer (5–15%). The unknown vitamer was also found to be produced by a biotin-requiring mutant of *Penicillium chrysogenum* and becomes labeled when produced by the mold growing in the presence of ^{14}C]pimelic acid[26]. Yeast cells grown in the presence of this unknown radioactive biotin vitamer incorporate it into biotin[27]. The unknown vitamer does not contain sulfur as it does not become labeled when the vitamer-producing organism, *P. chrysogenum*, is grown on a medium containing ^{35}S]sulphate[28].

Biotin biosynthesis by various molds belonging to *Mucor, Rhizopus, Penicillium, Fusarium, Monascus,* and *Aspergillus,* various bacteria belonging to *Bacillus* and *Pseudomonas,* and actinomyces belonging to *Streptomyces,* was studied by Ogata *et al.*[29]. Large amounts of an unknown biotin vitamer were accumulated by various bacteria. The vitamer was avidin-uncombinable, and suitable experiments ruled out identity with biocytin. A purified sample supports growth of *Saccharomyces cerevisiae* but not *Lactobacillus arabinosus* in lieu of biotin. The electrophoretic mobilities of the unknown vitamer were found to be essentially identical with those of authentic 7-amino-8-keto-pelargonic acid. The vitamer moves more slowly toward the cathode at pH 3.0–4.5 than does 7,8-diaminopelargonic acid. These results indicate the presence of only one free amino group. It was assumed that the vitamer might be an analogue of pelargonic acid. Additional studies on the unknown biotin vitamer were made by Eisenberg and Maseda[26]. The unknown biotin vitamer was found in relatively large amounts in culture filtrates of *P. chrysogenum.* Biosynthesis of the vitamer and dethiobiotin is inhibited by biotin. The unknown vitamer becomes labeled when grown in the presence of radioactive pimelic acid. Some purification of the vitamer was described. Finally, Eisenberg *et al.*[30] have isolated the unknown biotin vitamer in crystalline form from culture filtrates of *P. chrysogenum* and have shown it to be identical with synthetic 7-keto-8-aminopelargonic acid.

$$\begin{array}{c} \text{NH}_2 \quad \text{O} \\ | \qquad \| \\ \text{CH}-\text{C} \\ | \qquad | \\ \text{CH}_3 \quad \text{CH}_2-\text{CH}_2-\text{CH}_2-\text{CH}_2-\text{CH}_2-\text{COOH} \end{array}$$

7-keto-8-aminopelargonic acid

III

The compound is synthesized from pimelyl-CoA and alanine in cell-free extracts of biotin mutants of *Escherichia coli* Y10. Serine is less active than alanine, while cysteine shows apparent inhibition.

An extensive study of biotin biosynthesis has been carried out by a group of Japanese investigators starting in about 1964. Much of their work has been reported in journals not commonly seen by Western investigators. For this reason, their work has not received the consideration that it deserves. The accumulation of biotin vitamers in the culture media of about 700 strains of microorganisms was studied by Ogata *et al.*[31]. The biotin vitamers were determined by microbiological assay with *L. arabinosus* and *S. cerevisiae.* In general, the addition of pimelic acid to the medium was found to enhance the yield of total biotin about 10- to 100-fold, while the cell growth was not affected by pimelic acid. A total biotin content of 200 μg/ml of medium was found in the culture medium of *Bacillus sphaericus* after suitable supplementation with pimelic acid. This is the highest yield of total biotin synthesized yet reported. Biotin biosynthesis in resting cells of *B. sphaericus* was described by Iwahara *et al.*[32]. Total biotin biosynthesis was greatly stimulated by pimelic acid. In the presence of pimelic acid, biotin biosynthesis was greatly enhanced by the presence of various amino acids. Alanine was found to be most effective. When glutamic or aspartic acid was added with pyruvic acid, the same effect as that of alanine was observed.

A paper which appears to have been overlooked until quoted by Eisenberg and Maseda[26] in 1966 is that of Okumura *et al.*[33], which appeared in 1962 written in Japanese. These investigators studied a number of pelargonic acid derivatives as growth factors in lieu of biotin for the biotin auxotrophs *Brevibacterium lactofermentum* 2256 and *Brevibacterium flavum* 2247. Compounds studied included: 7,8-diaminopelargonic acid, 7-keto-8-aminopelargonic acid, 7-amino-8-ketopelargonic acid, 7,8-diketopelargonic acid, 7-amino-8-hydroxypelargonic acid, 8-aminopelargonic acid, 8-ketopelargonic acid, 7-aminopelargonic acid, 7-ketopelargonic acid, 7-nonenoic acid, 8-nonenoic acid, and pelargonic acid. When biotin was assigned a relative activity of 100 in satisfying the biotin requirement of *B. lactofermentum* 2256, several biotin derivatives and those pelargonic acid derivatives found effective in lieu of biotin, exhibited the following activities: *d*-biotin, 100; biotin *d*-sulfoxide, 80; biocytin, 91.5; *dl*-dethiobiotin, 52.5; biotin diaminocarboxylic acid, 0.34; 7,8-diaminopelargonic acid, 3.1; 7-keto-8-aminopelargonic acid, 0.92; 7-amino-8-ketopelargonic acid, 0.14; 7,8-diketopelargonic acid, 0.018; 7-amino-8-hydroxypelargonic acid, 0.0013; and oleic acid, 0.0014. On the basis of these findings, Okumura *et al.*[33] proposed a mechanism for biotin biosynthesis. The inclusion of azelaic acid between oleic and pimelic acid is the suggestion of the present authors. This scheme has some features of that now being advocated by those who hold that the "dethiobiotin" pathway is *the* pathway of biotin biosynthesis (Scheme 1).

Eisenberg and Star[34] have recently described studies on the biosynthesis of 7-keto-8-aminopelargonic acid by cell-free extracts of *E. coli* biotin auxotrophs. Biosynthesis of the pelargonic acid derivative was followed by microbiological assay employing yeast. The enzyme preparation requires the presence of pimelyl-CoA, alanine, and pyridoxal phosphate, although slight activity is found with serine instead of alanine. The presence of cysteine was

Scheme 1

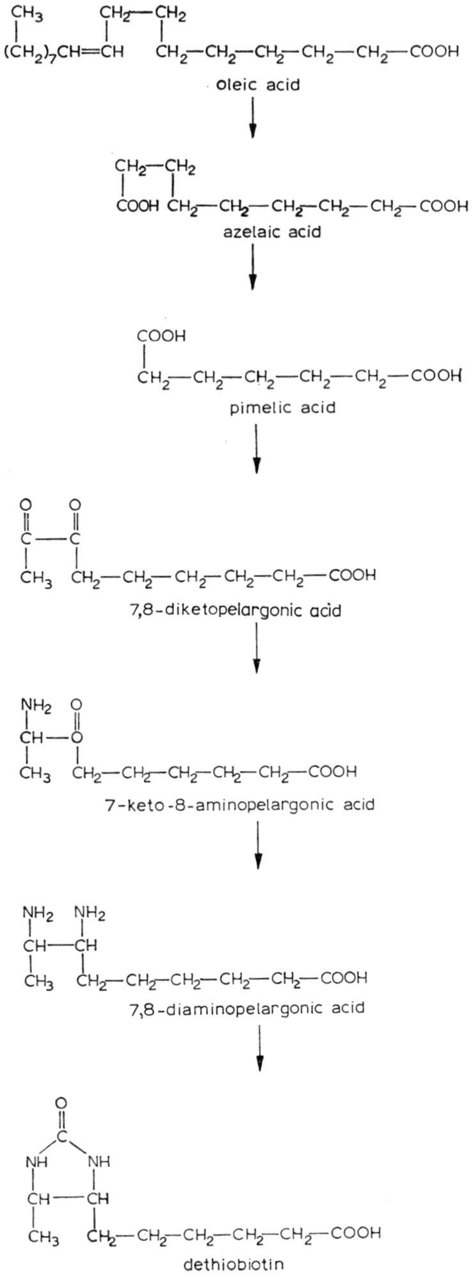

CH₃   CH₂—CH₂

$(CH_2)_7CH=CH$ $CH_2—CH_2—CH_2—CH_2—CH_2—COOH$

oleic acid

↓

CH₂—CH₂

$COOH$ $CH_2—CH_2—CH_2—CH_2—CH_2—COOH$

azelaic acid

↓

$COOH$

$CH_2—CH_2—CH_2—CH_2—CH_2—COOH$

pimelic acid

↓

O   O
‖   ‖
C — C

CH_3 $CH_2—CH_2—CH_2—CH_2—CH_2—COOH$

7,8-diketopelargonic acid

↓

NH₂ O
│   ‖
CH — O
│
CH_3 $CH_2—CH_2—CH_2—CH_2—CH_2—COOH$

7-keto-8-aminopelargonic acid

↓

NH₂ NH₂
│   │
CH — CH
│   │
CH_3 $CH_2—CH_2—CH_2—CH_2—CH_2—COOH$

7,8-diaminopelargonic acid

↓

O
‖
C
NH   NH
│   │
CH — CH
│   │
CH_3 $CH_2—CH_2—CH_2—CH_2—CH_2—COOH$

dethiobiotin

References p. 107

found to be inhibitory for the production of the pelargonic acid derivative. In this connection, attention should be called to the nonenzymatic reaction between pyridoxal phosphate and cysteine leading to the formation of a stable thiazolidine derivative[35] and the loss of pyridoxal phosphate to the system. This possibility as an explanation for the inhibitory effect of cysteine is proposed and discussed in a later section. In addition to the effect of cysteine observed, biotin was found to be inhibitory to the biosynthesis of the pelargonic acid derivatives. A number of *E. coli* auxotrophs were studied and a beginning made towards a knowledge of the genetics of the enzymes involved in biotin biosynthesis.

The biotin requirements of a number of mutants of *E. coli* K12 have been studied by Del Campillo-Campbell *et al.*[36]. On the basis of growth, cross-feeding, and avidin-combinability studies, the mutants fall into 4 nutritional classes: A (bio-0), B (bio-2, bio-17), C (bio-3, bio-12, bio-18, bio-23), and D (bio-19). The simplest scheme compatible with describing the data is:

$$W \xrightarrow{\text{C}} X \xrightarrow{\text{D}} Y \xrightarrow{\text{A}} Z \xrightarrow{\text{B}} \text{biotin}$$

where W, X, Y, and Z are biotin precursors, and X is uncombinable with avidin. The authors conclude that biotin biosynthesis is thus controlled by a closely linked cluster of at least 4 genes mediating different steps in the process.

Pai[37], similarly, has classified a group of about a dozen *E. coli* K12 mutants derived by the nitrosoguanidine method into 4 nutritional groups on the basis of their response to various biotin vitamers, the ability to feed other mutant strains, and the nature of the vitamers excreted into the growth medium. Strains of group I respond only to biotin, but cross-feed members of groups II, III and IV. Strains of group II respond to biotin and dethiobiotin and cross-feed members of groups III and IV. Strains of group III respond to biotin, dethiobiotin, 7,8-diaminopelargonic acid and cross-feed group IV. Lastly, strains of group IV respond to biotin, dethiobiotin, 7,8-diaminopelargonic acid, 7-keto-8-aminopelargonic acid and large amounts of 7-amino-8-ketopelargonic acid and, as expected, do not cross-feed members of any other group. Some of the mutants were examined for biotin vitamer production. Members of group I excreted dethiobiotin. Members of groups II and III excreted 7-keto-8-aminopelargonic acid. Members of group IV failed to excrete biotin vitamers. Pai postulates the biosynthetic pathway of biotin biosynthesis in *E. coli* as follows: Pimelic acid → 7-keto-8-aminopelargonic acid → 7,8-diaminopelargonic acid → dethiobiotin → biotin.

7,8-Diaminopelargonic acid could not be detected in the medium of any strain but the results of the cross-feeding experiments pointed to involvement of the diaminopelargonic acid in biotin biosynthesis.

Krell and Eisenberg[38], in a recent abstract, have submitted preliminary results of a study of dethiobiotin synthesis from 7,8-diaminopelargonic acid by dialyzed, cell-free extracts of E. coli K12. The identity of dethiobiotin was proved by chromatographic and electrophoretic methods. The addition of ATP and HCO_3^- separately to dialyzed extracts resulted in a 2.5- and 6-fold increase, respectively, in the synthesis of dethiobiotin. When added together with Mg^{2+}, a 15-fold increase in yield was obtained. The addition of NH_3 or glutamine produced no further increase. Carbamyl phosphate was only one-sixth as effective as the complete mixture. Radioactive dethiobiotin containing all the radioactivity in the carbonyl group was obtained by the use of radioactive HCO_3^-. The enzyme(s) for the conversion of 7,8-diaminopelargonic acid to dethiobiotin is (are) absent from those mutants that accumulate 7,8-diaminopelargonic acid in the medium. The present enzyme(s) as well as 7-keto-8-aminopelargonic acid synthetase are repressed by biotin. The above studies show that CO_2 rather than a carbamoyl group is the source of the carbonyl group of biotin. In this connection, Pai[39] has also reported in a recent abstract, preliminary results of a study of dethiobiotin biosynthesis from 7,8-diaminopelargonic acid by resting cells of a biotin mutant of E. coli. Dethiobiotin was identified as the product of the reaction on the basis of biological activity as well as chromatographic and electrophoretic characteristics. However, it appears from the abstract that biotin could also be synthesized by the system. The dethiobiotin-synthesizing system is completely repressed by biotin and derepressed when biotin is removed from the growth medium. The derepression is inhibited by chloramphenicol. An excess amount of dethiobiotin also causes repression to some extent, but this occurs only when biotin is allowed to be formed from dethiobiotin by growing cells. Other biotin-mutant strains and the wild-type (K12) of E. coli were also examined for the dethiobiotin-synthesizing system. All strains including the wild-type convert the diaminopelargonic acid to dethiobiotin. An exception to this was a group of mutant strains which was reported previously as being able to grow on dethiobiotin but not on the diaminopelargonic acid. These results suggested that the enzyme system which catalyzes the conversion of the diaminopelargonic acid to dethiobiotin is under physiological control by biotin and that dethiobiotin is a normal precursor of biotin in E. coli.

References p. 107

In retrospect, a compound with the same R_F value as that of 7-keto-8-aminopelargonic acid apparently had been observed as an unknown vitamer of biotin in culture filtrates of *Corynebacterium xerosis* by Genghof[40] in 1956 and in culture filtrates of *E. coli* by Dhyse and Hertz[41] in 1958. Unfortunately, in neither case was the finding exploited to the point of characterizing the unknown compound.

(c) Dethiobiotin

Dethiobiotin (or desthiobiotin) refers to a compound, first described by Melville *et al.*[42], readily derived from biotin by Raney-nickel reduction whereby the sulfur atom of biotin is replaced by two hydrogen atoms. Lilly and Leonian[43] have classified microorganisms with a biotin requirement into 4 groups: (*a*) those whose biotin requirement is satisfied by dethiobiotin; (*b*) those whose biotin requirement is not satisfied by dethiobiotin; (*c*) those whose utilization of biotin is not inhibited by dethiobiotin; and (*d*) those for which dethiobiotin is an antimetabolite of biotin utilization.

The first evidence that dethiobiotin may be converted to biotin by microorganisms was provided by Dittmer *et al.*[44]. These investigators grew *S. cerevisiae* in the presence of dethiobiotin. For this species, dethiobiotin and biotin have equal activity in satisfying the biotin requirement. Following growth, the cells and medium were assayed for biotin with *L. casei*, an organism whose biotin requirement is not satisfied by dethiobiotin. In fact at high levels, dethiobiotin is an antimetabolite of biotin. Their results show that dethiobiotin disappears from the incubating yeast cultures and is replaced by an equivalent amount of a substance possessing growth-promoting activity for *L. casei*. Dittmer *et al.* concluded that "the most logical assumption is that dethiobiotin is transformed to biotin by the growing yeast cell". With this publication began a prolonged and intriguing pursuit as to the role and/or significance of dethiobiotin in the biosynthesis of biotin. Even now, 25 years later, the answer is not yet entirely clear. The conversion of dethiobiotin to biotin by microorganisms seems such a biochemical feat that it is hard to accept this conclusion despite an abundance of evidence in this direction.

Tatum, in a frequently quoted paper[45], provided evidence that an X-ray-produced mutant strain of *P. chrysogenum*, strain 62078, synthesizes dethiobiotin but cannot convert it to biotin, and that as a result the mutant requires biotin for growth. The addition of pimelic acid to the growth medium resulted in a large increase in the biosynthesis of dethiobiotin by the mold as determined by microbiological assay with *Neurospora crassa*. The yield of dethiobiotin was not increased by the presence of cystine either alone or in addition to pimelic acid.

Somewhat direct evidence that dethiobiotin is a precursor of biotin in *A. niger* was obtained by Wright and Driscoll[46]. These investigators grew the mold with aeration on a simple glucose–inorganic salts medium supplemented with 1 mg or 50 μg of *dl*-dethiobiotin per 500 ml, or with unsupplemented medium. It was known that the predominant form of biotin encountered in *A. niger* filtrates when grown with aeration is biotin *l*-sulfoxide[47−49]. The presence of dethiobiotin in the growth medium resulted in the biosynthesis of large amounts of biotin *l*-sulfoxide in the order of 5–10 times that found in the unsupplemented medium. The biotin *l*-sulfoxide was determined by paper chromatography[50] with butanol–acetic acid–water (4:1:5) as solvent where biotin *l*-sulfoxide and dethiobiotin are completely separated (R_F of biotin *l*-sulfoxide is about 0.46; R_F of dethiobiotin is about 0.89). These findings were taken as presumptive evidence that dethiobiotin may be a precursor of biotin in *A. niger*. The reasoning is not entirely sound, since it is quite conceivable that dethiobiotin under the conditions used may spare microbial degradation of biotin synthesized by an alternative pathway, thus indirectly leading to the presence of more biotin *l*-sulfoxide than would have been encountered in the absence of dethiobiotin.

An extended study of the conversion of dethiobiotin to biotin has been made by Iwahara *et al.*[51]. In one phase of this study, about 800 strains of bacteria, yeasts, and molds were examined by differential microbiological assays for true biotin biosynthesis from added dethiobiotin. Some strains of yeasts and bacteria produced comparatively large amounts of true biotin from dethiobiotin. However, most of the bacteria (200 strains) and yeasts (300 strains) tested produced only small amounts of true biotin from dethiobiotin. These data suggest that most bacteria and yeast tested have little activity in the conversion reactions. In contrast with bacteria and yeasts, the molds and *Streptomyces* tested produced, in general, larger amounts of true biotin in their culture filtrates without supplementation with dethiobiotin. However, in the presence of dethiobiotin, biotin biosynthesis by the molds was greatly stimulated. As a consequence of this study, Iwahara *et al.*[51] have classified microorganisms into 4 groups: (*a*) those microorganisms that produce a large amount of dethiobiotin and comparatively large amounts of biotin in their culture filtrates without addition of pimelic acid, (*b*) those microorganisms which are markedly stimulated by the addition of both pimelic acid and dethiobiotin, and only small amounts of biotin are synthesized without the addition of pimelic acid or dethiobiotin, (c) those microorganisms that produce only small amounts of biotin and de- thiobiotin from pimelic acid in their medium but produce comparatively large amounts of biotin from dethiobiotin, and (d) those microorganisms that produce a large amount of dethiobiotin from pimelic acid, but little biotin is produced even in the presence of added dethiobiotin. In these studies of Iwahara et al., the identity of biotin was confirmed by column and paper chromatographic methods. The unequivocal isolation of dethio- biotin from the culture filtrate of B. sphaericus grown on a peptone–soybean meal medium supplemented with pimelic acid has been accomplished by Iwahara et al.[52]. Biotinamide has been isolated by Sekijo et al.[53] from a strain of Rhodotorula flava grown on a rather crude medium containing added biotin. The significance of this compound in biotin biosynthesis and/or degradation is not obvious.

Direct, radioisotopic evidence for the conversion of dethiobiotin to biotin in A. niger has been reported by Tepper et al.[54]. A. niger was grown on a Czapek–Dox medium supplemented with carbonyl- or carboxyl-labeled [^{14}C]dethiobiotin prepared by the desulfuration of [^{14}C]biotin with Raney nickel. Following growth the mycelia were removed, and the biotin and vi- tamers were concentrated from the medium by adsorption and elution from Norit. Carrier biotin was added and the mixture treated with hydrogen peroxide which converts biotin and biotin sulfoxides to biotin sulfone. Biotin sulfone was isolated following anion-exchange chromatography on Dowex-1. The biotin sulfone was subjected to repeated recrystallizations. The biotin sulfone was found to be radioactive, and after the second recrystallization, the activity remained constant with repeated recrystallizations. Similar results were obtained with carbonyl- or carboxyl-labeled dethiobiotin. The conver- sion of labeled dethiobiotin to biotin in these experiments was around 1–2 %. From these experiments, it seems certain that dethiobiotin can be a precursor of biotin, at least in A. niger.

The initial experiments of Tepper et al. were not definitive, because the possibility was not ruled out that the dethiobiotin was degraded microbiolog- ically to small fragments which were then reassembled to biotin. Accord- ingly, essentially the same experiments were repeated with doubly-labeled dethiobiotin[55]. Both carbonyl[^{14}C-U-^{3}H]- and carboxyl[^{14}C-U-^{3}H]- dethiobiotin were used. The doubly-labeled material was prepared by mixing appropriate amounts of the singly-labeled dethiobiotins. The labeled dethiobiotins were prepared by Raney-nickel reduction of labeled biotin. It was found that doubly-labeled dethiobiotin is incorporated into biotin by *A. niger* with a $^3H:^{14}C$ ratio that remains nearly constant. The data provided some evidence that approximately 4 or more hydrogen atoms must be abstracted from dethiobiotin in the overall conversion to biotin. The conclusion appears inescapable, therefore, that in *A. niger* dethiobiotin may be incorporated relatively intact into the biotin molecule.

(d) Sulfur sources

The source of the sulfur in the biosynthesis of biotin from dethiobiotin has not received adequate study. The Eakins reported that cysteine, or cystine, enhance the effect of pimelic acid in stimulating biotin biosynthesis by *A. niger*. This effect of cystine could not be observed by Wright and Cresson[47] nor by Tatum[45]. An extended study of factors influencing biotin biosynthesis by *A. niger* was made by Elford and Wright[56]. Various compounds, either alone or with pimelic acid, were studied with respect to influence on the biosynthesis of biotin by *A. niger* growing in synthetic medium. Of the more than 40 new compounds screened, only lipoic acid and some lipoic acid derivatives showed any effect. Lipoic acid and derivatives did not stimulate biotin biosynthesis in the absence of pimelic acid. When lipoic acid was added to the synthetic medium in conjunction with pimelic acid, a 3- to 5-fold enhancement of the pimelic acid stimulation of biotin biosynthesis occurred. 8-Methyllipoic acid, as well 5-methyllipoic acid, antimetabolites of lipoic acid in pyruvate oxidation, were equally active. Growth of *A. niger* on medium containing [^{35}S]lipoic acid and pimelic acid did not yield radioactive biotin. It was assumed that lipoic acid (and derivatives) enhances the pimelic acid stimulation of biotin biosynthesis by acting as a cofactor in the reduction of sulfate to sulfite as postulated by Fridovich and Handler[57] and by Hilz and Kittler[58]. In the experiments of Elford and Wright[56], as well as in prior experiments of Wright, the sole source of sulfur for the growth of *A. niger* was inorganic sulfate.

Incorporation of sulfur was studied in washed cells of *S. cerevisiae* by Niimura *et al.* In one study[59], it was reported that the addition of several sulfur compounds did not increase the production of biotin, and it was concluded that the sulfur originated from sulfur compounds within the cells. In another study[60], increases in biotin biosynthesis from dethiobiotin by added sulfur compounds was obtained. The most effective sulfur compounds were methionine sulfoxide and methionine, followed by Na_2SO_4, Na_2SO_3, and Na_2S. Cysteine, cystine, and methionine sulfone were less effective. The biosynthesis of biotin by biotin-deficient cells was inhibited by ethionine and reversed only by methionine.

A very significant study of biotin biosynthesis in *Achromobacter* sp. has been made by Lezius *et al.*[61]. This strain of *Achromobacter* was isolated from the bovine rumen. The organism is a rather anomalous one in that it fails to grow on media containing glucose as the main energy source, and it does not synthesize increased amounts of biotin when the medium is supplemented with pimelic acid[62]. Isovaleric acid is utilized as a source of energy, and the experiments of Lezius *et al.* involve the use of a purified isovaleric acid medium. Pimelic acid was not incorporated in the basal medium used, presumably because with this organism biotin biosynthesis is not augmented by the presence of pimelic acid. While the experiments of Lezius *et al.* involve the use of a "peculiar" organism that does not utilize glucose, their conditions for studying biotin biosynthesis may have been more "natural" than those used by most others in that the medium was not fortified with this "unnatural" acid. In the experiments of Lezius *et al.*, the organism was grown with aeration on either [3-^{14}C]cysteine or $^{14}CO_2$. Following growth, cells and medium were separated by centrifugation. Combined biotin was released from the cells by hydrolysis with sulfuric acid. The biotin in each fraction was adsorbed on and eluted from Norit and then adsorbed on and eluted from Dowex-50. Finally, purification of the biotin involved complexing with avidin and separation of the biotin–avidin complex from excess avidin on a column of Sephadex G-25 and release of the biotin from the biotin–avidin complex by heat denaturation of the protein. Biotin isolated following growth on the labeled precursor was degraded by 3 methods. In the first method, the biotin was converted to the methyl ester, this was treated with Raney nickel to obtain dethiobiotin methyl ester, and the latter was treated with chromic acid (Kuhn–Roth oxidation) to yield acetic acid and CO_2. The acetic acid was finally degraded to methylamine and carbon dioxide with the aid of HN_3 (Schmidt–Allan). In the second method, the biotin was converted to the methyl ester, this was treated with hydrazine to form the urethane (Curtius degradation), and the urethane treated with hydrochloric acid to yield the carboxyl carbon of the original biotin as carbon dioxide. In the third method, the biotin was treated with barium hydroxide to yield the carbonyl carbon as carbon dioxide. When the biotin isolated following growth on the appropriate labeled precursor was degraded according to the schemes described, it was found that the radioactive C-atom from $[3\text{-}^{14}C]$-cysteine was incorporated into C-5 of biotin while that of $^{14}CO_2$ appeared in the ureido carbon and the carboxyl group of biotin. These findings led to the hypothesis that biotin is formed from cysteine, pimelyl-CoA and carbamyl phosphate, essentially according to Scheme 2 (after Lezius *et al.*).

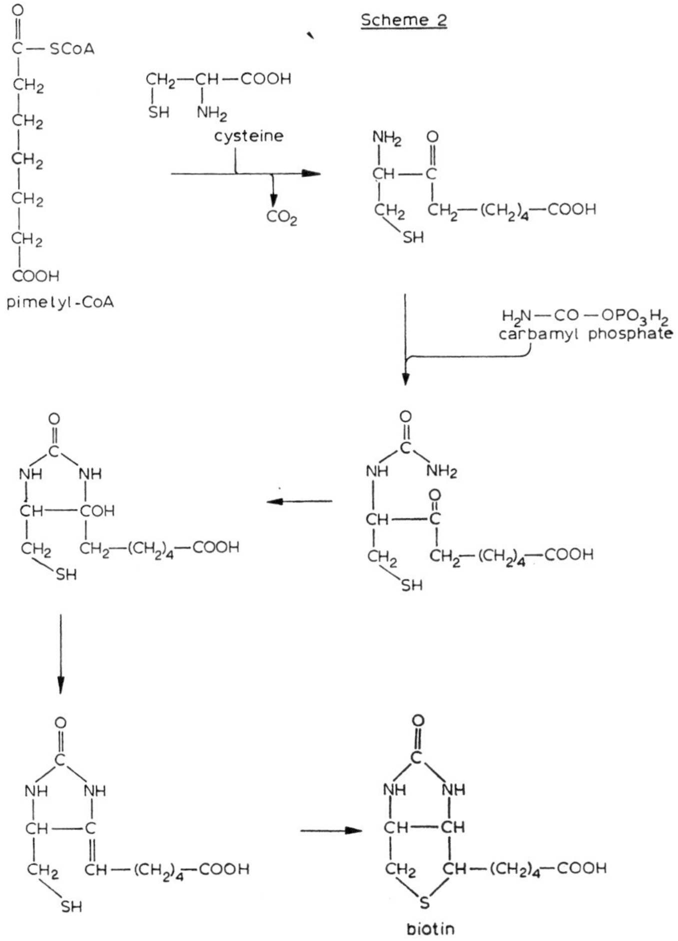

Scheme 2

Some discussion seems imperative with respect to the two essentially conflicting mechanisms of biotin biosynthesis that have been proposed; namely, the pathway proposed by Lynen and co-workers involving the condensation of pimelyl-CoA with cysteine as an early step whereby sulfur is introduced early in the biosynthetic pathway, *versus* the pathway proposed by Ogata and co-workers, Eisenberg, and others who have offered evidence involving the condensation of pimelic acid (as pimelyl-CoA) with alanine to yield pelargonic acid derivatives and dethiobiotin whereby sulfur is introduced late in the biosynthetic pathway. A possible explanation for the discrepancy is that there may be two alternative pathways, one used by some organisms while the other is used by other organisms. If this is the explanation, it would seem that the pelargonic acid–dethiobiotin pathway is the more "popular with microorganisms". In defense of this pathway, present evidence seems irrefutable that dethiobiotin can be converted to biotin by some microorganisms under some conditions. In defense of the Lynen pathway, it should be pointed out that studies tending to substantiate the pelargonic acid–dethiobiotin pathway have generally, if not always, used relatively enormous amounts of pimelic acid or other intermediates to "stimulate" biotin biosynthesis in media either (*a*) supplemented with high levels of alanine, (*b*) supplemented with amino acid mixtures or hydrolyzed casein where the alanine to cysteine ratio is high, or (*c*) supplemented with high levels of carbohydrate leading to precursors of alanine through glycolysis. On the other hand, Lynen and his group studied biotin biosynthesis in an organism where biosynthesis of biotin and/or biotin vitamers is not stimulated by pimelic acid, in a medium where the energy source was furnished by an organic acid rather than by an unphysiological level of carbohydrate. Under these circumstances, the experimental conditions used by Lynen may have been more "natural" than those used by others. It is suggested that possibly the dethiobiotin pathway may be an artifact where the evidence has been obtained by overloading media so as to yield artificial conditions. In further support of the Lynen pathway, is the observation of Eisenberg that when the condensation of pimelyl-CoA with L-alanine is studied in cell-free preparations of *E. coli*, cysteine is found to be inhibitory. As Eisenberg himself has reported, biotin is an inhibitor of biotin biosynthesis. If Eisenberg's cell-free preparations had contained all the enzymes essential to complete biotin biosynthesis it would be expected that the presence of cysteine would inhibit the accumulation of intermediates in the pelargonic acid–dethiobiotin pathway, thus leading to the conclusion that cysteine is not the amino acid with which pimelyl-CoA reacts in the first step of biotin biosynthesis. In this connection, some recent findings of Pai[39] are pertinent. Pai studied the biosynthesis of dethiobiotin (DTB) from 7,8-diaminopelargonic acid by resting cells of *E. coli*. He reports "The DTB-synthesizing system was completely repressed by biotin and derepressed when biotin was removed from the growth medium. ...An excess amount of DTB also caused repression to some extent, *but this occurred only when biotin was allowed to be formed from DTB by growing cells.*" (Italics are by the present authors.) Thus, it would appear that biotin rather than dethiobiotin is the inhibitor of dethiobiotin synthesis and the inhibition observed with cysteine in systems where amino acids are compared as possible precursors of biotin would suggest that cysteine rather than alanine may be the better precursor despite the apparent observation that cysteine is "inhibitory". The solution to the dilemma would appear to necessitate the isolation of the enzyme system involving the condensation of pimelyl-CoA with amino acids and to study the specificity of this reaction uncomplicated by subsequent reactions. This approach involves great difficulties since, as pointed out earlier, cysteine and pyridoxal phosphate react without enzymatic intervention to form a stable thiazolidine derivative, thus depleting the system of two factors suspect of being involved in the condensation. If cysteine is highly active, especially when compared with alanine, this would furnish substantial evidence for the Lynen pathway as the more natural one.

At any rate, it would seem that until more good evidence is at hand, the statement of Eisenberg[27] that "the original hypothesis of Lezius, Ringelmann and Lynen, that cysteine is a precursor in the synthesis of biotin, is no longer tenable in light of the more recent evidence" may be a bit premature.

(e) Miscellaneous factors

Although the effect of the presence of various dicarboxylic acids, primarily pimelic acid, either alone or additionally supplemented, has received most attention, the influence of a few other conditions or supplements has received at least superficial study.

The biosynthesis of biotin and/or biotin vitamers by molds is apparently increased by aeration and/or aerobic conditions[47], but the effect on biotin yield may have been indirect as a consequence of better growth of the organism.

The omission of trace elements was found by Eisenberg to reduce growth and biotin production of *P. blakesleeanus*[63]. The "true" biotin was affected to a greater extent than the "total" biotin. Zinc and iron proved to be the essential trace metals. In the absence of zinc, both the growth and the total biotin production were markedly reduced. The omission of iron affected primarily the biotin production.

Biotin biosynthesis by microorganism is inhibited by the presence of biotin itself. Pai and Lichstein[64-67] studied biotin biosynthesis by an enzyme system from *E. coli* which catalyzes the conversion of dethiobiotin to biotin. The biotin-synthesizing enzyme system is repressed when the medium is supplemented with biotin; complete repression occurs only when the biotin content of the medium is raised above the level found intracellularly in cultures of the organism upon cessation of growth. An excess of dethiobiotin or oxybiotin also causes repression to some extent. Repression by the exogenous biotin vitamers is relieved rapidly when avidin is added to the cultures. Homobiotin is even more effective in relieving the repression. The kinetics of depression follow the pattern characteristic of biosynthetic enzymes that are repressible by their end-products. These findings indicated that the level of the enzyme system varied according to the demand for biotin by the organism.

3. Interconversions

(a) Free and bound forms

Differential microbiological assays of materials which contain compounds having biotin-like activity was an early indication of altered forms (vitamers) of the known vitamin. By the yeast growth assay, biotin vitamers in mammalian urine were classified by Oppel[68] into avidin-combining and non-avidin-combining fractions. A broader survey of materials containing biotin vitamers led to a description by Burk and Winzler[21] of "miotin" as a heat-labile, avidin-uncombinable fraction which is active for *S. cerevisiae* but inactive for *R. trifolii*, "tiotin" as a heat-stable product derived by autoclaving, and "rhiotin" as an avidin-combinable fraction which is active for *Rhizobium* but not yeast. However, Chu and Williams[69] did not detect the presence of an avidin-uncombinable form of biotin in any natural material other than urine and even questioned its existence there. Wright and co-workers[70] re-examined the problem using the versatile *N. crassa* rather than yeast for assay of "total" biotin vitamers and *L. arabinosus* for "free" biotin in human urine. These latter authors reported relatively large amounts of free biotin (~ 20 mμg/ml) and much smaller amounts (~ 2 mμg/ml) of a hydrophilic biotin derivative that was tentatively identified as biocytin sulfoxide on the basis of bioautographic comparison with authentic material which is formed when biocytin is subjected to paper chromatography[50]. McCormick[71] has produced the compound, predominately as d-sulfoxide, by H_2O_2–acetic acid treatment of biocytin.

Biocytin had already been isolated from yeast extract and identified as ε-N-biotinyl-L-lysine[72], and the synthesis was accomplished by Wolf *et al.*[73]. Thoma and Peterson[74] had partially purified an enzyme called biotinidase from hog liver and showed that it would catalyze release of free biotin from biocytin. A similar activity, named biocytinase by Wright *et al.*[75], was found in blood. This enzyme will also catalyze hydrolysis of sulfoxide and sulfone derivatives of biocytin[71]. More recently several groups[76-79] have further purified and characterized these biotin amide amidohydrolases from different sources.

That biotin d-sulfoxide can be converted to biotin in such organisms as *S. cerevisiae* and *L. arabinosus* was recognized by Melville *et al.*[80] who studied the biological properties of the sulfoxides. The facile oxidation of biotin to the d-sulfoxide is also known, and Iwahara *et al.*[81] recently isolated the d-isomer from culture filtrates of a soil pseudomonad grown on the vitamin. The latter organism will also reduce the d-sulfoxide to biotin[82].

Proteolytic release of biocytin and the 1'-N-carboxy derivative from carboxylases is also well known, and trypsin has been used, for example, by Knappe *et al.*[83] in treatment of β-methyl crotonyl carboxylase.

Coon and his co-workers have demonstrated that covalent attachment of biotin to enzymes firstly involves reversible activation to d-biotinyl 5'-adenylate by ATP-dependent enzymes[84] such as found in extracts from a pseudomonad[85] and from liver[86], and secondly as a holocarboxylase synthetase purified from rabbit liver[87]. In proteins, biotin always appears to be amide linked to the ε-amino of L-lysine[88].

The natural interconversions of the vitamin with protein-bound and free forms are as follows:

$$d\text{-biotinyl enzyme} \rightarrow d\text{-biocytin} \rightarrow d\text{-biocytin } d\text{-sulfoxide}$$
$$\uparrow \qquad\qquad\qquad \downarrow \qquad\qquad\qquad \downarrow$$
$$d\text{-biotinyl 5'-adenylate} \rightleftharpoons d\text{-biotin} \rightleftharpoons d\text{-biotin } d\text{-sulfoxide}$$

(b) Thioether and sulfoxide forms

In addition to vitaminic activity[80] and biological interconversion with biotin and catabolites[81,82], biotin d-sulfoxide was found by Melville[89] to predominate from chemical oxidation of biotin. The *d*- and *l*-sulfoxides have been shown by Ruis *et al.*[90] to be chemically interconvertible, and the *l*-isomer is thermodynamically more stable in acid. Biotin *l*-sulfoxide was detected in certain mold filtrates by Wright and Cresson[47] and was isolated[48] and characterized[49] from filtrates of *A. niger*. Brady *et al.*[91] also found the *l*-sulfoxide in culture filtrates of the pseudomonad which degrades the *d*-sulfoxide about as readily as biotin. Both biotin and the *d*-sulfoxide are catabolized faster than the *l*-sulfoxide[92]. Hence, it seems that biotin *d*-sulfoxide is more readily formed and degraded by organisms, wherein the less reactive *l*-sulfoxide accumulates.

The interconversions of biotin and the *d*- and *l*-sulfoxides are as follows:

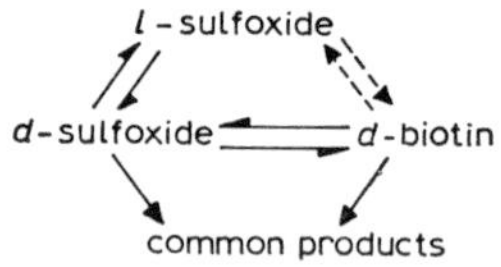

4. Degradations

(a) Side-chain

Evidence for the catabolism of natural *d*-biotin by mammals was supplied by the Fraenkel-Conrats[93] who made a study of the metabolic fate of $[^{14}C]$-biotin, labeled in the ureido carbonyl carbon[94], and the avidin–biotin complex upon parenteral administration to rats. Up to 40% of 10- to 20-μg doses of free biotin administered to fed or fasted rats was excreted within one day, whereas a longer period, 3–5 days, was required to recover similar amounts after administration of the biotin–avidin complex. Most of the residual ^{14}C material was also found to be excreted in one day after biotin administration, but absence of biological activity with *S. cerevisiae* suggested metabolic alteration of this fraction. A small fraction of the administered radiobiotin was found in liver. Only a few percent radioactivity appeared in feces, and little or no $^{14}CO_2$ was expired. Hence, the mechanism of inactivation of the administered biotin in the rat does not appear to involve rupture of the ureido structure of the vitamin. The first clear indication of how the side-chain part of the biotin molecule is degraded in mammals resulted from the work of Baxter and Quastel[95]. These investigators incubated $[^{14}C]$biotin, labeled in the carboxyl carbon[96], with tissue slices, chiefly from renal cortex of the guinea pig. Loss of $^{14}CO_2$ and yeast-growth promoting activity occurred under aerobic conditions. Azide inhibited, and the suppression of oxidizing activity by malonate and fumarate led to the suggestion that there is a removal of one or more two-carbon fragments from the side-chain with subsequent involvement in the citric acid cycle. Fatty acids inhibited biotin β-oxidation in a non-competitive way. Several non-radioactive biotin analogues were tested in the *in vitro* system and found to depress the release of $^{14}CO_2$ from carboxyl-labeled biotin. Schreirer and Pfisterer[97] noted formation of $^{14}CO_2$ from carboxyl-labeled biotin in the rat. Dakshinamurti and Mistry[98] examined the tissue and intracellular distribution of carboxyl-labeled biotin injected at near physiological levels in rats and chicks. Liver was found most effective in concentrating the radioactivity, approximately half of which could be precipitated in protein-bound form from cell supernatant at pH 5.2.

More detailed information on the steps involved in degradation of the biotin side-chain has come from studies using microorganisms. It was shown by Krueger and Peterson[99], among others, that certain biotin-requiring microbes inactivate excess biotin in the medium in which they are grown. Birnbaum and Lichstein[100-103] have reported degradation of biotin and certain analogues in the lactobacilli, but specific metabolites were not identified. Coon and co-workers[85,104] isolated pseudomonads which degrade [2-^{14}C]biotin to $^{14}CO_2$. Cell-free extracts from one organism supplemented with ATP, CoA, and Mg^{2+} produced a significant amount of labeled acetoacetate, suggesting that the side-chain of the vitamin yields an acetate unit. The evidence for activation of the carboxyl function of biotin to form biotinyl 5′-adenylate has already been mentioned, and additional data implicated biotinyl-CoA as the more immediate substrate for β-oxidation[84]. More recently, studies by McCormick, Wright, and their colleagues in a similar bacterium have led to isolation and identification of several metabolites derived from biotin and the *d*-sulfoxide. The soil pseudomonad, which grows on biotin as a sole source of carbon, nitrogen, and sulfur, was found capable in whole- and broken-cell preparations of more rapidly forming $^{14}CO_2$ from carboxyl- than carbonyl-labeled biotin[91,105]. Synthetic [^{14}C]biotinyl-CoA was also shown to be effectively degraded to $^{14}CO_2$ in broken cells. These results confirmed microbial oxidation of the biotin side-chain and additionally indicated that such reactions precede cleavage of the ureido portion of the molecule. Further studies[92,106] with this organism established the cofactor requirements of Mg^{2+}, ATP, CoA and NAD for activation and degradation of the side-chain in washed, broken-cell preparations. Azide is an effective inhibitor, but activity is only slightly decreased by high concentrations of acetate or malonate. Among several ^{14}C analogues tested as substrates[92], the sulfone and diaminocarboxylate from carboxyl-labeled biotin did not give rise to significant $^{14}CO_2$, but the side-chain of carboxyl-labeled dethiobiotin was degraded[107]. As mentioned earlier, however, the d-sulfoxide is even a somewhat better substrate than biotin, whereas the l-sulfoxide is less rapidly metabolized. These findings delimit the specificity of the side-chain oxidizing system, as an intact ureido carbonyl and presence or absence of thioether or sulfoxides, but not sulfone function, are prerequisite to degradation. Additional work from this laboratory[108] has demonstrated that unlabeled α-dehydrobiotin competitively decreases release of $^{14}CO_2$ from carboxyl-labeled biotin, as the former substrate must be an early intermediate of the β-oxidation of the biotin side-chain. α-Dehydrobiotin had already been isolated from Streptomyces lydicus by Hanka et al.[109]. Direct proof for the pathway of side-chain oxidation resulted from isolation and identification of bisnorbiotin, α-dehydrobisnorbiotin and tetranorbiotin from carbonyl-labeled [^{14}C]biotin and also the tetranor-catabolite from tritium-labeled biotin[81]. Certain of these compounds and the sulfoxide derivatives have been isolated from culture filtrates of the pseudomonad grown on carbonyl-labeled biotin d-sulfoxide as well[82]. Both bisnorbiotin and the sulfoxide, probably d-, were also isolated from culture filtrates of Endomycopsis[110]. The intermediate β-hydroxy and β-keto compounds normally do not accumulate in the media.

The interconnected pathways by which the side-chains of biotin and the d-sulfoxide are β-oxidized are indicated by the compounds given in Scheme 3, most of which are known to accumulate.

Among analogues tested for their effect in decreasing formation of $^{14}CO_2$ from carboxyl-labeled biotin, Baxter and Quastel[95] had noted that unlabeled dl-dethiobiotin was active in the guinea-pig kidney slice. Birnbaum and Lichstein had indicated metabolism of dl-dethiobiotin in L. plantarum[102] and L. casei[103] occurs. As already mentioned, the pseudomonad is able to degrade the side-chain of the natural analogue, d-dethiobiotin[107]. Optically inactive bisnordethiobiotin was isolated by Iwahara et al.[111] who grew Aspergillus oryzae on dl-dethiobiotin. Both principal catabolites which accumulate from β-oxidation of the side-chains of carbonyl-labeled d- and dl-dethiobiotins in growing cultures of A. niger were isolated and characterized by Li et al.[112,113] as the bisnor and tetranor derivatives. From the

Scheme 3

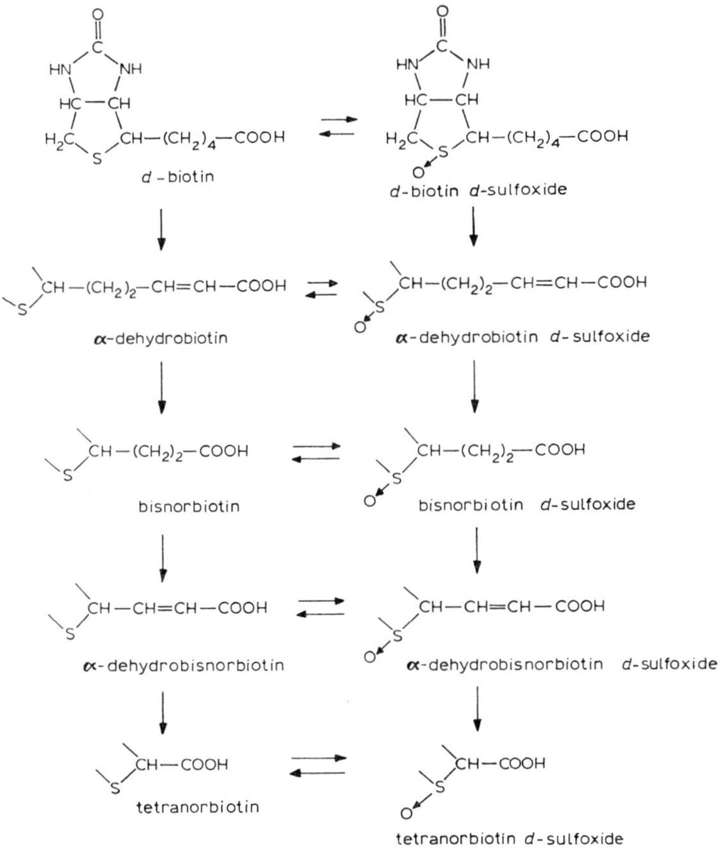

Scheme 4

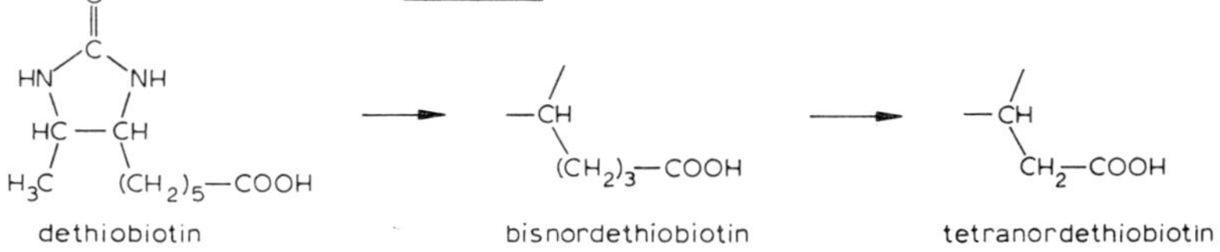

foregoing, it is obvious that both d- and l-forms of dethiobiotin are metabolized and compete with biotin as substrate.

The natural catabolites from β-oxidation of the side-chain of dethiobiotin are given in Scheme 4.

References p. 107

The suppression of $^{14}CO_2$ formation from carboxyl-labeled biotin with guinea-pig kidney slices to which unlabeled dl-homobiotin or d-norbiotin was added had been reported[95]. These synthetic analogues were also found to exhibit the same behavior with particulate preparations of the pseudomonad[92]. More recently, Ruis et al.[107,114] isolated and characterized the two principal catabolites which were formed by incubation of carbonyl-labeled d-homobiotin with broken-cell preparations of the pseudomonad. The first catabolite which accumulates in considerable amounts, d-norbiotin, was also shown to be subsequently converted to the other, d-trisnorbiotin.

The catabolites from β-oxidation of the side-chain of synthetic d-homobiotin are given in Scheme 5.

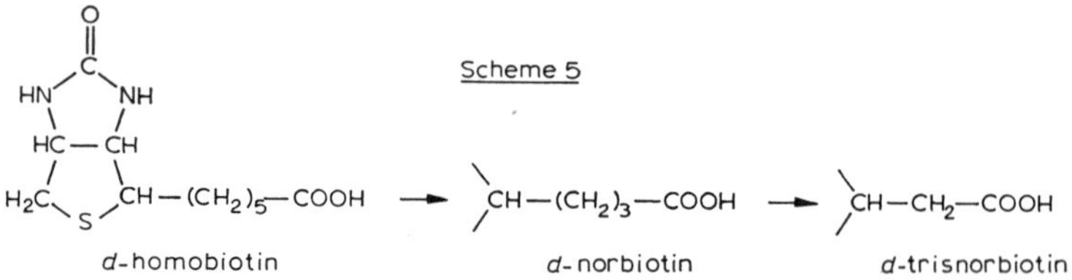

Scheme 5

d-homobiotin d-norbiotin d-trisnorbiotin

Another synthetic analogue of biotin which has been investigated by several groups is d-biotinol, the alcohol formed by chemical reduction of biotin methyl ester. Drekter et al.[115] reported that biotinol does not replace d-biotin in the nutrition of L. arabinosus, L. casei, or S. cerevisiae but can be converted to biotin efficiently in both rat and human. As with certain other analogues already described, the presence of biotinol during incubation of carboxyl-labeled biotin with guinea-pig kidney slices depressed formation[95] of $^{14}CO_2$. The particulate system from the pseudomonad, like other microorganisms, does not significantly metabolize d-biotinol[107], but this analogue is an effective competitive inhibitor against catabolism of biotin[108].

(b) Bicyclic ring system

The recent work by McCormick, Wright, and their co-workers has led to partial elucidation of the sequence of catabolic events occurring with ultimate cleavage of the biotin ring structure in the biotin-degrading pseudomonad. Compounds, the side-chains of which cannot be degraded to d-tetranorbiotin or its sulfoxide, e.g. d-homo- and d-norbiotins which are β-oxidized only as far as the 2-carboxymethyl side-chain of d-trisnorbiotin and dethiobiotin which lacks the thioether sulfur in a tetrahydrothiophene portion, are not subject to loss of the carbonyl function in the ureido portion of these molecules, as evidenced by failure to yield $^{14}CO_2$ upon incubation of carbonyl-labeled homo and nor derivatives[92] or the dethio compound[107] with particulate preparations of the pseudomonad. Also in growing *A. niger*, much less label is lost from ureido carbonyl than carboxyl function of *dl*-[^{14}C]dethiobiotin[113]. It has already been pointed out that *d*-biotin sulfone is not subject to β-oxidative cleavage of the side-chain[92]. Moreover, the carbonyl-labeled compound does not[92] yield $^{14}CO_2$. All of these results, the more rapid production of $^{14}CO_2$ from carboxyl- than carbonyl-labeled biotin[91, 105], the observed accumulation of side-chain degraded catabolites from biotin[81] and from the *d*-sulfoxide[82], especially the relatively large amounts of *d*-tetranorbiotin formed[81], emphasize that β-oxidation of the side-chain must precede cleavage of the ring system and that a thioether or sulfoxy function is obligatory to ultimate ring cleavage. In this latter regard, it was also found[92] that carbonyl-labeled oxybiotin does not yield $^{14}CO_2$. As the *d*-sulfoxide is a somewhat better substrate than *d*-biotin[81], it is probable that *d*-tetranorbiotin *d*-sulfoxide is the proximal substrate for cleavage of the tetrahydrothiophene portion of the molecule. This intermediate catabolite may be hydrolyzed to the ring-opened product. The earlier finding that L-cysteine, among numerous amino acids tested, is exceptionally well metabolized by the bacterium is in concert with the expectation that catabolites in common with sulfur-containing ones derived from biotin are actively degraded. [^{14}C]Urea with complete retention of radioactivity has been isolated from culture filtrates of the pseudomonad grown on carbonyl-labeled *d*-biotin[81]. This proves that oxidation at both carbons bridging the imidazolidine to tetrahydrothiophene portions must occur in the terminal steps. [^{14}C]Uracil was also found in the culture filtrate[81]. This metabolite was less radioactive than calculated if complete retention of the original ureido carbonyl function of the [^{14}C]biotin occurred, but the specific radioactivity of the pyrimidine was greater than could be accounted for by resynthesis from CO_2 and NH_3 formed in the overall oxidation of the vitamin. Formation of 5′-uridylate *via* the known biosynthetic pathway from L-ureidosuccinate to dihydroorotate to orotate to 5′-orotidylate was recently shown operable in this microbe which also contains active phosphatase and ribosidase to hydrolyze, respectively, the UMP to uridine and finally uracil[116].

That controlling factors are in operation in the pseudomonad is clear from the repression in synthesis of biotin-catabolizing enzymes reported by Baumann and Lichstein to occur when glucose is supplied to a growing culture[117].

The possible steps by which the bicyclic ring system of *d*-tetranorbiotin is degraded are given in Scheme 6.

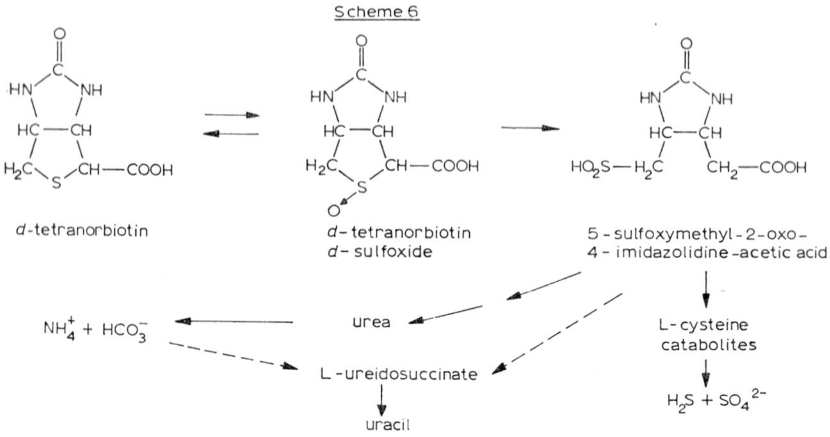

Scheme 6

More work is in progress in this and other laboratories to elucidate completely the individual intermediates between early and late ring-opened products.

Overall, the degradation of protein-bound biotin can be summarized by the numbered sequence of major steps as indicated in formula IV.

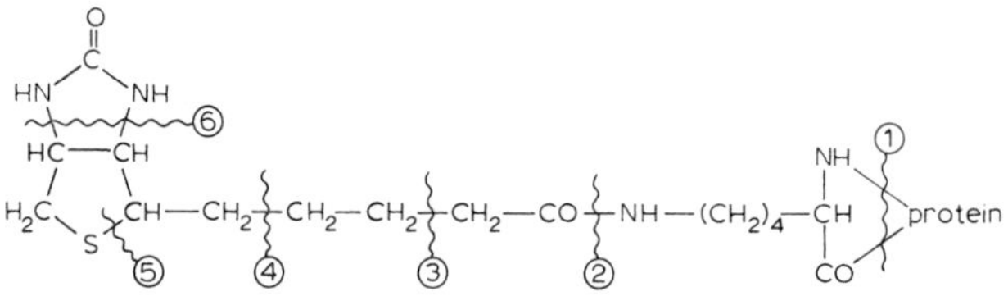

IV

METABOLISM OF BIOTIN AND ANALOGUES 107

REFERENCES

1 K. Hoffmann, in F. F. Nord and C. H. Werkman (Eds.), *Advances in Enzymology,* Vol. III, Interscience, New York, 1943, p. 289.
2 D. B. Melville, in R. S. Harris and K. V. Thimann (Eds.), *Vitamins and Hormones,* Vol. II, Academic Press, New York, 1944, p. 29.
3 L. H. Sternbach, in M. Florkin and E. H. Stotz (Eds.), *Comprehensive Biochemistry,* Vol. 11, Elsevier, Amsterdam, 1963, p. 66.
4 F. Lynen, J. Knappe, E. Lorch, G. Jütting and E. Ringelmann, *Angew. Chem.,* 71 (1959) 481.
5 F. Lynen, J. Knappe, E. Lorch, G. Jütting and E. Ringelmann, *Angew. Chem.,* 73 (1961) 513.
6 S. Ochoa and Y. Kaziro, *Federation Proc.,* 20 (1961) 982.
7 F. Lynen, J. Knappe and E. Lorch, *Proc. Vth Intern. Congr. Biochem.,* 4 (1961) 225.
8 F. Lynen, *Bull. Soc. Chim. Biol.,* 46 (1964) 1775.
9 S. P. Mistry and K. Dakshinamurti, in R. S. Harris, I. G. Wool and J. A. Loraine (Eds.), *Vitamins and Hormones,* Vol. XXII, Academic Press, New York, 1964, p. 1.
10 V. du Vigneaud, K. Dittmer, E. Hauge and B. Long, *Science,* 96 (1942) 186.
11 J. H. Mueller, *J. Biol. Chem.,* 119 (1937) 121.
12 R. E. Eakin and E. A. Eakin, *Science,* 96 (1942) 187.
13 L. D. Wright, in E. E. Snell (Ed.), *The National Vitamin Foundation Symposium on Vitamin Metabolism,* New York, 1956, p. 104.
14 L. J. Bassalik and L. D. Wright *J. Gen. Microbiol.,* 36 (1964) 405.
15 L. J. Bassalik and L. D. Wright, *Nature,* 204 (1964) 501.
16 L. L. Campbell Jr. and G. B. Williams, *J. Bacteriol.,* 65 (1953) 146.
17 H. Rosenthal and C. A. Brog, *Z. Vitaminforsch.,* 17 (1946) 27.
18 L. D. Wright, *Proc. Soc. Exptl. Biol. Med.,* 51 (1942) 27.
19 L. D. Wright and H. R. Skeggs, *Proc. Soc. Exptl. Biol. Med.,* 56 (1944) 95.
20 W. J. Robbins and R. Ma, *Science,* 96 (1942) 406.
21 D. Burk and R. J. Winzler, *Science,* 97 (1943) 57.
22 H. L. Elford and L. D. Wright, *Federation Proc.,* 21 (1962) 467.
23 H. L. Elford and L. D. Wright, *Biochem. Biophys. Res. Commun.,* 10 (1963) 373.
24 M. A. Eisenberg, *Federation Proc.,* 21 (1962) 467.
25 M. A. Eisenberg, *Biochem. Biophys. Res. Commun.,* 8 (1962) 437.
26 M. A. Eisenberg and R. Maseda, *Biochem. J.,* 101 (1966) 601.
27 M. A. Eisenberg, *Biochem. J.,* 101 (1966) 598.
28 M. A. Eisenberg, *Biochem. J.,* 98 (1965) 156.
29 K. Ogata, T. Tochikura, S. Iwahara, K. Ikushima, S. Takasawa, M. Kikuchi and A. Nishimura, *Agr. Biol. Chem. (Tokyo),* 29 (1965) 895.
30 M. A. Eisenberg, R. Maseda and C. Star, *Federation Proc.,* 27 (1968) 762.
31 K. Ogata, T. Tochikura, S. Iwahara, S. Takasawa, K. Ikushima, A. Nishimura and M. Kikuchi, *Agr. Biol. Chem. (Tokyo),* 29 (1965) 889.
32 S. Iwahara, T. Tochikura and K. Ogata, *Agr. Biol. Chem. (Tokyo),* 29 (1965) 262.
33 S. Okumura, R. Tsugawa, T. Tsunoda and S. Motozaki, *J. Agr. Chem. Soc. Japan,* 36 (1962) 204.
34 M. A. Eisenberg and C. W. Star. *J. Bacteriol.,* 96 (1968) 1291.
35 M. V. Buell and R. E. Hansen, *J. Am. Chem. Soc.,* 82 (1960) 6042.
36 A. del Campillo-Campbell, G. Kayajanian, A. Campbell and S. Adhya, *J. Bacteriol.,* 94 (1967) 2065.
37 C. H. Pai, *Can. J. Microbiol.,* 15 (1969) 21.
38 K. Krell and M. A. Eisenberg, *Federation Proc.,* 28 (1969) 352.

39 C. H. PAI, *Bacteriological Proceedings, Abstracts of the 69th Annual Meeting*, 1969, p. 130.
40 D. S. GENGHOF, *Arch. Biochem. Biophys.*, 62 (1956) 63.
41 F. G. DHYSE AND R. HERTZ, *Arch. Biochem. Biophys.*, 74 (1958) 7.
42 D. B. MELVILLE, K. DITTMER, G. B. BROWN AND V. DU VIGNEAUD, *Science*, 98 (1943) 497.
43 V. G. LILLY AND L. H. LEONIAN, *Science*, 98 (1944) 205.
44 K. DITTMER, D. B. MELVILLE AND V. DU VIGNEAUD, *Science*, 99 (1944) 203.
45 E. L. TATUM, *J. Biol. Chem.*, 160 (1945) 455.
46 L. D. WRIGHT AND C. A. DRISCOLL. *J. Am. Chem. Soc.*, 76 (1954) 4999.
47 L. D. WRIGHT AND E. L. CRESSON, *J. Am. Chem. Soc.*, 76 (1954) 4156.
48 L. D. WRIGHT, E. L. CRESSON, J. VALIANT, D. E. WOLF AND K. FOLKERS, *J. Am. Chem. Soc.*, 76 (1954) 4160.
49 L. D. WRIGHT, E. L. CRESSON, J. VALIANT, D. E. WOLF AND K. FOLKERS, *J. Am. Chem. Soc.*, 76 (1954) 4163.
50 L. D. WRIGHT, E. L. CRESSON AND C. A. DRISCOLL, *Proc. Soc. Exptl. Biol. Med.*, 86 (1954) 480.
51 S. IWAHARA, S. TAKASAWA, T. TOCHIKURA AND K. OGATA, *Agr. Biol. Chem. (Tokyo)*, 30 (1966) 385.
52 S. IWAHARA, Y. EMOTO, T. TOCHIKURA AND K. OGATA, *Agr. Biol. Chem. (Tokyo)*, 30 (1966) 64.
53 C. SEKIJO, T. TSUBOI, Y. YOSHIMURA AND O. SHOJI, *Agr. Biol. Chem. (Tokyo)*, 32 (1968) 1181.
54 J. P. TEPPER, D. B. MCCORMICK AND L. D. WRIGHT, *J. Biol. Chem.*, 241 (1966) 5734.
55 H.-C. LI, D. B. MCCORMICK AND L. D.WRIGHT, *J. Biol. Chem.*, 243 (1968) 6442.
56 H. L. ELFORD AND L. D. WRIGHT, *Arch. Biochem. Biophys.*, 123 (1968) 145.
57 I. FRIDOVICH AND P. J. HANDLER, *J. Biol. Chem.*, 223 (1956) 321.
58 H. HILZ AND M. KITTLER, *Biochem. Biophys. Res. Commun.*, 3 (1960) 140.
59 T. NIIMURA, T. SUZUKI AND Y. SAHASHI, *Bitamin*, 26 (1962) 38.
60 T. NIIMURA, T. SUZUKI AND Y. SAHASHI, *Bitamin*, 27 (1963) 355.
61 A. LEZIUS, E. RINGELMAN AND F. LYNEN, *Biochem. Z.*, 336 (1963) 510.
62 L. D. WRIGHT AND F. LYNEN, unpublished.
63 M. A. EISENBERG, *J. Bacteriol.*, 86 (1963) 673.
64 C. H. PAI AND H. C. LICHSTEIN, *Biochim. Biophys. Acta*, 100 (1965) 28.
65 C. H. PAI AND H. C. LICHSTEIN, *Biochim. Biophys. Acta*, 100 (1965) 36.
66 C. H. PAI AND H. C. LICHSTEIN, *Biochim. Biophys. Acta*, 100 (1965) 43.
67 C. H. PAI AND H. C. LICHSTEIN, *Arch. Biochem. Biophys.*, 114 (1966) 138.
68 T. OPPEL, *Am. J. Med. Sci.*, 204 (1942) 869.
69 E. J.-H. CHU AND R. J. WILLIAMS *J. Am. Chem. Soc.*, 66 (1944) 1678.
70 L. D. WRIGHT, E. L. CRESSON AND C. A. DRISCOLL, *Proc. Soc. Exptl. Biol. Med.*, 91 (1956) 248.
71 D. B. MCCORMICK, *Proc. Soc. Exptl. Biol. Med.*, 132 (1969) 502.
72 L. D. WRIGHT, E. L. CRESSON, H. R. SKEGGS, R. L. PECK, D. E. WOLF, T. R. WOOD, J. VALIANT AND K. FOLKERS, *Science*, 114 (1951) 635.
73 D. E. WOLF, J. VALIANT, R. L. PECK AND K. FOLKERS, *J. Am. Chem. Soc.*, 74 (1952) 2002.
74 R. W. THOMA AND W. H. PETERSON, *J. Biol. Chem.*, 210 (1954) 569.
75 L. D. WRIGHT, C. A. DRISCOLL AND W. P. BOGER, *Proc. Soc. Exptl. Biol. Med.*, 86 (1954) 335.
76 J. KNAPPE, W. BRUMMER AND K. BIEDERBICK, *Biochem. Z.*, 338 (1963) 599.

77 M. KOIVUSALO, C. ELORRIAGA, Y. KAZIRO AND S. OCHOA, *J. Biol. Chem.*, 238 (1963) 1038.
78 M. KOIVUSALO AND J. PISPA, *Acta Physiol. Scand.*, 58 (1963) 13.
79 J. PISPA, *Ann. Med. Exptl. Biol. Fenniae (Helsinki)*, Suppl. 5 (1965) 39.
80 D. B. MELVILLE, D. S. GENGHOF AND J. M. LEE, *J. Biol. Chem.*, 208 (1954) 503.
81 S. IWAHARA, D. B. MCCORMICK, L. D. WRIGHT AND H.-C. LI, *J. Biol. Chem.*, 244 (1969) 1393.
82 W. B. IM, J. A. ROTH, D. B. MCCORMICK AND L. D. WRIGHT, *J. Biol. Chem.*, in the press (1970).
83 J. KNAPPE, K. BIEDERBICK AND W. BRUMMER, *Angew. Chem.*, 74 (1962) 432.
84 J. E. CHRISTNER, M. J. SCHLESINGER AND M. J. COON, *J. Biol. Chem.*, 239 (1964) 3997.
85 M. J. COON, J. L. FOOTE AND J. E. CHRISTNER, *Proc. Vth Internat. Congr. Biochem.*, 7 (1963) 248.
86 J. L. FOOTE, J. E. CHRISTNER AND M. J. COON, *Federation Proc.*, 21 (1962) 239.
87 L. SIEGEL, J. L. FOOTE, J. E. CHRISTNER AND M. J. COON, *Biochem. Biophys. Res. Commun.*, 13 (1963) 307.
88 M. D. LANE AND F. LYNEN, *Proc. Natl. Acad. Sci. (U.S.)*, 49 (1963) 379.
89 D. B. MELVILLE, *J. Biol. Chem.*, 208 (1954) 495.
90 H. RUIS, D. B. MCCORMICK AND L. D. WRIGHT, *J. Org. Chem.*, 32 (1967) 2010.
91 R. N. BRADY, L.-F. LI, D. B. MCCORMICK AND L. D. WRIGHT, *Federation Proc.*, 24 (1965) 625.
92 R. N. BRADY, H. RUIS, D. B. MCCORMICK AND L. D. WRIGHT, *J. Biol. Chem.*, 241 (1966) 4717.
93 J. FRAENKEL-CONRAT AND H. FRAENKEL-CONRAT, *Biochim. Biophys. Acta*, 8 (1952) 66.
94 D. B. MELVILLE, J. G. PIERCE AND C. W. H. PARTRIDGE, *J. Biol. Chem.*, 180 (1949) 299.
95 R. M. BAXTER AND J. H. QUASTEL, *J. Biol. Chem.*, 201 (1953) 751.
96 S. B. BAKER, D. E. DOUGLAS AND A. E. SEATH, *Nucl. Sci. Abstr.*, 5 (1951) 802, abstract 5158.
97 VON K. SCHREIRER AND R. PFISTERER, *Intern. Z. Vitaminforsch.*, 29 (1959) 217.
98 K. DAKSHINAMURTI AND S. P. MISTRY, *J. Biol. Chem.*, 238 (1963) 294.
99 K. K. KRUEGER AND W. H. PETERSON, *J. Bacteriol.*, 55 (1948) 693.
100 J. BIRNBAUM AND H. C. LICHSTEIN, *J. Bacteriol.*, 89 (1965) 1035.
101 J. BIRNBAUM AND H. C. LICHSTEIN, *J. Bacteriol.*, 92 (1965) 913.
102 J. BIRNBAUM AND H. C. LICHSTEIN, *J. Bacteriol.*, 92 (1965) 920.
103 J. BIRNBAUM AND H. C. LICHSTEIN, *J. Bacteriol.*, 92 (1965) 925.
104 J. E. CHRISTNER, J. L. FOOTE AND M. J. COON, *Federation Proc.*, 20 (1961) 271.
105 R. N. BRADY, L.-F. LI, D. B. MCCORMICK AND L. D. WRIGHT, *Biochem. Biophys. Res. Commun.*, 19 (1965) 777.
106 R. N. BRADY, H. RUIS, D. B. MCCORMICK AND L. D. WRIGHT, *Federation Proc.*, 25 (1966) 218.
107 H. RUIS, R. N. BRADY, D. B. MCCORMICK AND L. D. WRIGHT, *J. Biol. Chem.*, 243 (1968) 547.
108 J. A. ROTH, S. IWAHARA, R. N. BRADY, D. B. MCCORMICK AND L. D. WRIGHT, *Federation Proc.*, 27 (1968) 256.
109 L. J. HANKA, M. E. BERGY AND R. B. KELLY, *Science*, 154 (1966) 1667.
110 H.-C. YANG, M. KUSUMOTO, S. IWAHARA, T. TOCHIKURA AND K. OGATA, *Agr. Biol. Chem. (Tokyo)*, 32 (1968) 399.
111 S. IWAHARA, S. TAKASAWA, T. TOCHIKURA AND K. OGATA, *Agr. Biol. Chem. (Tokyo)*, 30 (1966) 1069.

112 H.-C. LI, D. B. MCCORMICK AND L. D. WRIGHT, *Federation Proc.*, 27 (1968) 763.
113 H.-C. LI, D. B. MCCORMICK AND L. D. WRIGHT, *J. Biol. Chem.*, 243 (1968) 4391.
114 H. RUIS, R. N. BRADY, D. B. MCCORMICK AND L. D. WRIGHT, *Federation Proc.*, 26 (1967) 344.
115 L. DREKTER, J. SCHEINDER, E. DE RITTER AND S. H. RUBIN, *Proc. Soc. Exptl. Biol. Med.*, 78 (1951) 381.
116 J. A. ROTH, S. IWAHARA, D. B. MCCORMICK AND I. D. WRIGHT, *Federation Proc.*, 28 (1969) 560.
117 R. J. BAUMANN AND H. C. LICHSTEIN, *Abstr. Bacteriol. Proc., Annual Meeting* (1969) 130.

Metabolism of Water-Soluble Vitamins

Section f

The Biosynthesis of Folic Acid and 6-Substituted Pteridine Derivatives

T. SHIOTA

Department of Microbiology and Division of Nutrition, University of Alabama, Birmingham, Ala. (U.S.A.)

1. Introduction

The intent of this chapter is to review the current status of pteridine biosynthesis in relation to folic acid compounds and other naturally occurring 6-substituted pteridines. Pteridines with substituents on positions 7 and 8 will not be reviewed. The ring system of naturally occurring pteridine derivatives with the structure 2-amino-4-hydroxypteridine will be referred to by the trivial name "pterin" [XXVIII] as recommended by Pfleiderer[1] and Kaufman[2]. Over the past years, worthy reviews have appeared dealing with the early discoveries and chemistry of pteridines[1-8], the biology of pteridines in insects[9], the nutrition, biochemistry and chemistry of folic acid[5,6,10-12] and the biosynthesis of pteridines, riboflavin and folic acid[3,6,7,10,13,14]. Interested readers are encouraged to refer to these works.

Roman numerals which are followed by a subscript $_a$ denote the 7,8-dihydro structure of the pteridine ring. Roman numerals without subscript denote the aromatic form of the pteridine ring.

2. The biosynthesis of folate compounds

(a) Involvement of purines

The pattern of incorporation of radioactive precursors into pteridines by the larvae of butterfly[15,16] and by *Drosophila melanogaster*[17] suggested that pteridines and purines shared common precursors. Similar conclusions were reached with *Ashbya gossypii* and *Eremothecium ashbyii*[18,19] which incorporated radioactive precursors into riboflavin, which contains a pteridine structure. These results suggested that purines may be precursors of pteridines and were supported by the report of Zegler-Günder *et al.*[20] who showed that [2-^{14}C]guanine was incorporated into a pteridine by *Xenopus* larvae.

McNutt, in a series of reports, demonstrated the incorporation of radioactivity of [U-^{14}C]adenine into riboflavin and little incorporation from [8-^{14}C]adenine[21] by *E. ashbyii*. Most of the radioactivity found in riboflavin was in the pyrimidine ring of the vitamin[22]. By the use of [^{15}N]adenine and [U-^{14}C]xanthine, it was demonstrated that the purine, with the exception of the carbon atom 8, was incorporated into the pteridine ring of riboflavin[23]. With regard to the role of purines as precursors of folic acid, Vieira and Shaw[24] reported that pteroyltriglutamate isolated from cultures of *Corynebacterium* sp. was radioactive when the organism was grown in the presence of [2-^{14}C]adenine and not radioactive when grown in the presence of [8-^{14}C]-adenine. The involvement of a purine except its carbon atom 8 requires a ring-opening reaction as a step prior to pteridine formation. Weygand *et al.*[25] using butterfly larvae and Brenner-Holzach and Leuthardt[17,26] with *D. melanogaster* followed the fate of several radioactive precursors into various pteridines. From these studies, they concluded that guanine or guanylic acid was the precursor, and proposed that carbon atoms 1 and 2 of ribose were incorporated as carbon atoms 7 and 6 of the pteridine, respectively.

Consequently, these results permitted the formulation of a biosynthetic pathway[25,27] which, with minor modifications, has been incorporated into a proposed unified pathway (Fig. 1). A scission of the imidazole bond of a guanine nucleotide (guanosine triphosphate, GTP, [I]) ultimately results in the elimination of carbon atom 8 and the formation of 2,5-diamino-6-(5′-triphosphoribosyl)-amino-4-hydroxypyrimidine [III]. The ribose moiety then undergoes an Amadori-type rearrangement resulting in a deoxypentulose derivative [IV] and contributes its carbon atoms 1′ and 2′ to form the pteridine ring [V$_a$]. Hence, the ribose carbon atoms 1′ and 2′ are incorporated as

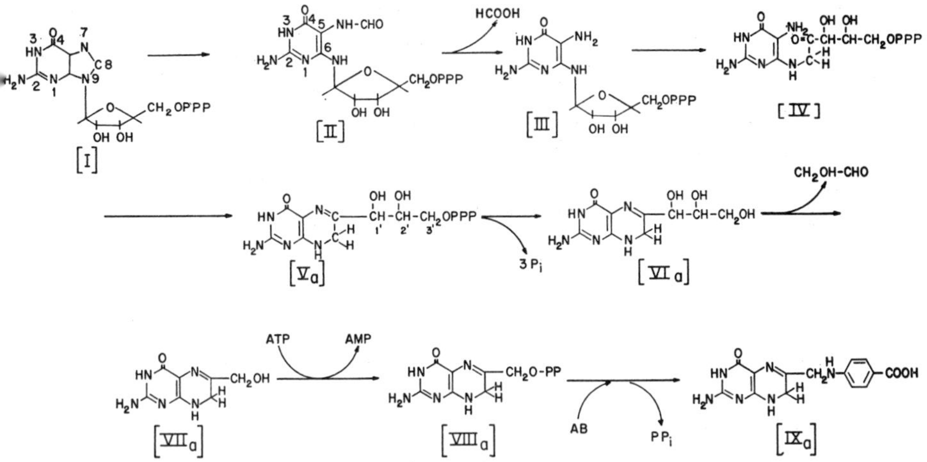

Fig. 1. A proposed unified biosynthetic pathway for dihydropteroate. AB=p-aminobenzoic acid.

carbon atoms 7 and 6 of the pterin, and the remaining ribose carbons constitute the side-chain at position 6.

The participation of ribose was further borne out by the work of Krumdieck et al.[27] which indicated that carbon atoms 6, 7 and 1' of the 6-substituted pteridines and carbon atoms 6, 7 and 9 of folic acid (or pteroic acid [IX]) originate from carbon atoms 2', 1' and 3' respectively of the ribosyl moiety of GMP. Additional information on the origin of the pterin nucleus was provided by Levy[28]. The incubation of the skin of the tadpole, Rana catesbiana, with [2-^{14}C]guanine incorporated the radioactivity into the corresponding position of 6-L-erythro-dihydroxypropylpterin (biopterin, [X]) and 2-amino-4-hydroxypteridine-6-carboxylic acid (carboxypterin). Furthermore, by using [2,4-^{14}C]guanine, the results indicated that the intact pyrimidine ring of guanine was incorporated into the corresponding position of the pterin. As expected, experiments with [8-^{14}C]guanine indicated that carbon atom 8 was not incorporated into pteridines.

(b) GTP as the pteridine precursor

Reynolds and Brown[29,30] reported that cell-free extracts of Escherichia coli converted a guanine nucleotide to the pterin moiety of folic acid. When the E. coli extracts were incubated with [U-^{14}C]guanine or guanine plus [1-^{14}C]-

ribose, under conditions where dihydrofolate is synthesized, the dihydrofolate was found to be radioactive. However, when [8-^{14}C]guanosine or guanine
plus [5-^{14}C]ribose 5-phosphate was used, the dihydrofolate which was synthesized was not radioactive. In order to accommodate the production of dihydrofolate, Reynolds and Brown[30] proposed that the three-carbon side-
chain of the 6-substituted pteridine [VI$_a$] undergoes an aldolase-type cleavage between carbon atoms 1' and 2' (Fig. 1). Similar results were soon reported whereby enzyme preparations from *Lactobacillus plantarum*[31], *Salmonella typhimurium*[32] and *Brassica pekinensis*[33,34] catalyzed the conversion
of guanine nucleotides to the pteridine moiety of dihydrofolate or dihydropteroate and from *Pseudomonas* sp.[35] to pteridines. Carbon atom 8 from
GTP or GDP which is not involved in pteridine biosynthesis was identified
as formic acid by Shiota and Palumbo[31] (Fig. 1). Subsequently, Dalal and
Gots[32] showed that among the various guanine compounds, GTP was the
most efficient substrate for pteridine biosynthesis. As other enzyme preparations were purified, the superior efficacy of GTP as the substrate was shown
for pteridine and formate production[36,41].

(c) Pyrimidine derivatives as intermediates

Little is known about the intermediary reactions of GTP [I] in Fig. 1; that is,
the ring-opening reactions leading to the pyrimidine derivative, the removal
of the formyl group, the Amadori rearrangement of the ribosyl moiety and
finally, ring closure resulting in 2-amino-4-hydroxy-6-(D-*erythro*-3'-triphos-
photrihydroxypropyl)-7,8-dihydropteridine (D-*erythro*-dihydroneopterin triphosphate, [V$_a$]). There are other reactions in biochemistry worthy of mention which undergo similar ring-opening of the imidazole ring of purines, and
the Amadori-type rearrangements of a ribosyl moiety. Examples of the former are riboflavin[23], toxoflavin[36,38] and tubercidin[42] biosynthesis, and of the
latter, tryptophan[43] and histidine biosynthesis[44]. Enzymes are probably involved in each of these reactions. The ring closure which is also part of these
reactions may not be enzymatic.

Two hydrolytic steps of the ureido carbon of GTP are required to produce
the C$_1$ unit at the oxidation state of formic acid and the pyrimidine derivative
2,5-diamino-6-(5'-triphosphoribosyl)-amino-4-hydroxypyrimidine [III].
Levenberg and Kaczmarek[37] reported the stoichiometric cleavage of GTP
by an enzyme preparation of *Pseudomonas cocovenenans* to formic acid and
ring-opened 5,6-diaminopyrimidine residues. These pyrimidine derivatives

were identified and quantitated after chemical condensation with glyoxal to form the corresponding pteridines. Furthermore, in these reaction mixtures, a compound containing a labile formyl group was ascertained suggesting a ring-opened 5- or 6-formamidopyrimidine derivative. Accordingly, they[37] speculated that the expulsion of carbon atom 8 involved two steps; the initial step being scission of the imidazole bond resulting in a N-formylpyrimidine intermediate followed by a subsequent step of deformylation.

Shiota et al.[41] discovered that GTP undergoes an interesting, but poorly understood, non-enzymatic reaction with 2-mercaptoethanol, Fe^{2+} and O_2. Among the several products formed is a ring-opened compound, 5- or 6-form-amido-2-amino-6-(5'-triphosphoribosyl)-amino-4-hydroxypyrimidine which was as active as GTP as a substrate in the production of formic acid, dihydro-folate and a pteridine by extracts of L. plantarum. Recently, by a controlled degradation procedure, the formyl substituent of the ring-opened compound was assigned to position N-5 ([II], Fig. 1), thus defining the structure as 2-amino-5-formamido-6-(5'-triphosphoribosyl)-amino-4-hydroxypyrim-idine[45]. The authors proposed the trivial name formasine triphosphate (FTP) for this compound and formadine for the free base.

Recently, Brown[46] reported a unique approach to demonstrate a ring-opened intermediate. His group prepared 7-methyl-GTP and it was found to be enzymatically converted to a ring-opened, Amadori-rearranged product and formic acid. The N-methyl substituent on position 5 of the pyrimidine derivative prevented ring closure to a pteridine.

These results lend support for the existence of pyrimidine derivatives of GTP as intermediates. But as will be pointed out under dihydroneopterin triphosphate synthesis, their direct detection may prove to be difficult since these intermediates may be enzyme-bound and are extremely unstable.

There have been several reports concerning the ability of 2,5,6-triamino-4-hydroxypyrimidine (TOP) to serve as a pteridine precursor based primarily on feeding TOP to intact organisms. Accordingly, TOP has been found to spare the folic acid requirement of Tetrahymena pyriformis[47], replace the bio-pterin requirement by Crithidia fasciculata[48] and to be converted to xantho-pterin by Pierid, caterpillars[15] and to pteridines by Corynebacterium sp.[49]. Furthermore, the latter organism was able to incorporate the radioactivity of [2-^{14}C]guanine into TOP[49]. These results suggest that growing organisms can transform TOP to a more direct precursor, such as a ribosylated TOP prior to its conversion to a 6-substituted 2-amino-4-hydroxypteridine. Wey-gand et al.[50] in a later experiment failed to find any incorporation of [2-^{14}C]-

TOP into folic acid by bacteria. Burg and Brown[51] proposed the explanation that since dicarbonyl compounds readily condense with TOP, the media used to support the growth of whole organisms may have contained ingredients which may have condensed with TOP non-enzymatically to produce pteridines. In a cell-free *E. coli* system, guanosine (with ATP) is a precursor of the pteridine moiety of dihydrofolate[30]. This reaction was found to be inhibited by TOP. Because of the similarity of TOP to several of the structurally related intermediates such as FTP ([II], Fig. 1) and its deformylated derivative ([III], Fig. 1), the authors[30] suggested that the site of inhibition may lie with the reactions of these pyrimidine compounds. Guanine nucleotides which were found to be precursors of dihydrofolate compounds could not be replaced by TOP[30,31,33] or TOP supplemented with 1-pyrophosphoribosyl-5-phosphate[30], ribose 5-phosphate[31] or ribose[33].

(d) Dihydroneopterin triphosphate synthesis

Goto and Forrest[52] extracted a pteridine from *E. coli* which was identified as 2-amino-4-hydroxy-6-(trihydroxypropyl)-pteridine monophosphate (neopterin monophosphate). The isomeric form of the trihydroxypropyl side-chain was not established. A year later, Rembold and Buschmann[53] isolated and identified the D-*erythro* isomer of neopterin[VI] from bee larvae and honey. The existence of such a pteridine in nature supported the working hypothesis of a neopterin-like compound as a possible intermediate in the biosynthesis of folate ([V$_a$ and VI$_a$] in Fig. 1) from GTP.

The first experimental evidence which showed that neopterin is produced from GTP was provided by Guroff and Strenkowski[40]. These investigators, in studying the enzymatic synthesis of the pteridine cofactor for phenylalanine hydroxylase, identified a small amount of neopterin, xanthopterin and a cyclic phosphate derivative of neopterin from a reaction mixture containing an extract of *Pseudomonas* sp. and GTP. The production of neopterin from GTP was also demonstrated by Jones and Brown[54] with an extract of *E. coli*. The isomeric form of neopterin was identified as *erythro*-neopterin by phosphocellulose column chromatography of the reaction mixture previously treated with alkaline phosphatase. Attempts to show whether this compound was L- or D-*erythro*-neopterin were unsuccessful because of limitations of the procedures available. Recently, Burg and Brown[51] performed additional experiments with an *E. coli* enzyme preparation which was highly purified with respect to its ability to produce radioactive formate from [8-^{14}C]GTP. These authors proposed the trivial name GTP cyclohydrolase for this enzyme. By incubating [U-^{14}C]GTP and GTP cyclohydrolase, the authors identified a radioactive fluorescent product as neopterin and the residual unreacted GTP. They stated that the expected dihydroneopterin would have been oxidized to neopterin due to its lability to oxidation. Other experiments which were performed in order to determine the number of phosphate residues per mole of neopterin indicated that the product from GTP was neopterin triphosphate ([V], Fig. 1). Shiota[55], using a purified *L. plantarum* extract found that [U-^{14}C]GTP is converted stoichiometrically to formic acid and dihydroneopterintriphosphate. Rather than GTP cyclohydrolase, Shiota[55] designated this enzyme dihydroneopterin triphosphate synthetase based on the ability of the enzyme, or enzyme complex, to transform either GTP or FTP to a compound with a new bond between the 2'-carbon of the ribosyl moiety and the nitrogen on position 5 of the pyrimidine derivative of GTP.

Although Burg and Brown[51] demonstrated the synthesis of neopterin triphosphate from GTP by a purified extract of *E. coli*, the isomeric form was not established. The assumption was, however, that D-*erythro*-dihydroneopterin triphosphate was the product formed because of the configuration of the D-ribose moiety of GTP, and the earlier finding of *erythro*-neopterin referred to above[53,54].

The structure of the reduced pyrazine ring of dihydroneopterin produced from GTP has not been directly established. However, the bathochromic shift exhibited by the dihydroneopterin isolated from a reaction mixture of GTP and the *L. plantarum* enzyme from pH 7 to acid medium[55] suggests a 7,8-dihydro structure according to Pfleiderer and Zondler[56], Nagai[57] and Fukushima and Akino[58].

In each of the studies cited above, the use of paper chromatography[40], paper electrophoresis[51] or DEAE-cellulose chromatography[55] to separate GTP from its products revealed only the unreacted GTP and either pteridines[40], neopterin triphosphate[51] or dihydroneopterin triphosphate[55]. In no instance was a pyrimidine intermediate observed. However, in the work of Levenberg and Kaczmarek[37] who used an enzyme preparation which synthesized the pteridine, toxoflavin, 4,5-diaminopyrimidine derivatives were reported to be found. The ring-opened pyrimidine intermediate reported by Brown[46] was an *N*-methylpyrimidine derivative arising from an unnatural substrate, 7-methyl GTP. Hence it appears that GTP cyclohydrolase from *E. coli*[51] and dihydroneopterin triphosphate synthetase from *L. plantarum*[55] catalyze the conversion of GTP to dihydroneopterin triphosphate and formic acid without the involvement of a free ring-opened pyrimidine intermediate. Burg and Brown[51] suggested that GTP cyclohydrolase, which was purified some 700-fold, may be a single protein. The dihydroneopterin triphosphate synthetase[55] from *L. plantarum*, in contrast, exists as multiple forms. Nevertheless, like the *E. coli* enzyme, each synthetase catalyzed the complete set of reactions from GTP to dihydroneopterin triphosphate.

(e) Dihydroneopterin triphosphate utilization

For several years, 2-amino-4-hydroxy-6-hydroxymethyldihydropteridine (hydroxymethyldihydropterin) has been known to function as a precursor of dihydropteroate or dihydrofolate[59,62]. Its occurrence as an intermediate was established by Krumdieck et al.[63]. By tracer experiments, 2-amino-4-hydroxy-6-hydroxymethylpteridine (hydroxymethylpterin) was shown to be derived from [2-14C]guanine in a *Corynebacterium* sp. Direct experimentations on the production of hydroxymethylpterin were performed by Reynolds and Brown[30]. These authors reported that a 2-amino-4-hydroxy-6-trihydroxy-propyldihydropterin, whose side-chain structure was not specified, was active as a precursor of dihydrofolate. This finding indicated that the trihydroxy-propyl side-chain must undergo a reaction resulting in the loss of two of its carbons.

Although the expected configuration of the trihydroxypropyl side-chain in position 6 of the pteridine arising from GTP is the D-erythro-trihydroxy-propyl isomer, four stereoisomeric forms of the 6-trihydroxypropyl side-chain are possible. Accordingly, Jones et al.[64] prepared D-erythro-neopterin [VI], L-erythro-neopterin [XX], D-threo-neopterin [XIX] and L-threo-neopterin [XVIII] by condensing D-ribose, L-arabinose, D-xylose and L-xylose, respectively with TOP[65]. After reducing these 6-trihydroxypropylpterin isomers to their dihydro forms and testing their efficacy as precursors of dihydropteroate, it was found that D-erythro-dihydroneopterin [VI$_a$] was the most efficient; next came L-threo-dihydroneopterin [XVIII$_a$] and last, D-threo-dihydroneopterin [XIX$_a$] and L-erythro-dihydroneopterin [XX$_a$] which were equivalent (Table I). In addition to these isomers of neopterin, Mitsuda et al.[33], prepared three 6-substituted butylpterin derivatives by condensing TOP with L-rhamnose, D-glucose and D-galactose to obtain 2-amino-4-hydroxy-6-(L-arabo-1',2',3'-trihydroxy-4'-deoxybutyl)-pteridine, 2-amino-4-hydroxy-6-(D-arabo-1',2',3',4'-tetrahydroxybutyl)-pteridine and 2-amino-4-hydroxy-6-(D-lyxo-1',2',3',4'-tetrahydroxybutyl)-pteridine, respectively. Each of these

TABLE I

RELATIVE ACTIVITIES OF 2-AMINO-4-HYDROXY-6-POLYHYDROXYALKYLDIHYDROPTERIN TO PRODUCE HYDROXYMETHYLDIHYDROPTERIN[a]

6-Substituted polyhydroxyalkylpterin	Polyhydroxyalkyl substituent on position 6	Relative activity[b]	Reference
D-erythro-Dihydroneopterin	OH OH –C – C – CH₂OH H H	100	33,31,64
L-threo-Dihydroneopterin	OH H –C – C – CH₂OH H OH	70	33,64
D-threo-Dihydroneopterin	H OH –C – C – CH₂OH OH H	40	33,64
L-erythro-Dihydroneopterin	H H –C – C – CH₂OH OH OH	30	33,31,64
D-erythro-Neopterin		0	46
D-erythro-Tetrahydroneopterin		0	46
D-erythro-Dihydroneopterin-P	OH OH –C – C – CH₂–O–PO₃H₂ H H	0	46
D-erythro-Dihydroneopterin-PPP	OH OH –C – C – CH₂–O–P₃O₉H₄ H H	0	46
Dihydrobiopterin	H H –C – C – CH₃ OH OH	0	30,31,34,54
L-arabo-1',2',3'-Trihydroxy-4'-deoxybutyldihydropterin[c]	OH H H –C – C – C – CH₃ H OH OH	117	33
D-arabo-1',2',3',4'-Tetrahydroxybutyldihydropterin[c]	H OH OH –C – C – C – CH₂OH OH H H	70	33,54
D-lyxo-1',2',3',4'-Tetrahydroxybutyldihydropterin[c]	H H OH –C – C – C – CH₂OH OH OH H	30	33
D-1',2'-Dihydroxyethyldihydropterin[c]	OH –C – CH₂OH H	20	54

The production of hydroxymethyldihydropterin which Jones and Brown[54] showed as the pteridine product of dihydroneopterin aldolase was measured indirectly as a precursor of dihydropteroate or dihydrofolate by a microbial assay for folate-like compound.

Relative activity is an arbitrary ranking of the efficacy of various 6-polyhydroxyalkyldihydropterins, using D-erythro-dihydroneopterin as 100, to produce hydroxymethyldihydropterin.

2-amino-4-hydroxypterin derivatives.

References p. 147

pteridines was reduced to its dihydro form and tested for precursor activity of dihydropteroate. The results summarized in Table I indicate that D-*erythro*-dihydroneopterin and the 6-L-*arabo*-1',2',3'-trihydroxy-4'-deoxybutyldihydropterin were efficiently converted to the pterin moiety of dihydropteroate by cell-free extracts of *B. pekinensis*. Jones and Brown[54] similarly tested various polyalkyl dihydropterins and found that D-*arabo*-tetrahydroxybutyldihydropterin and D-dihydroxyethyldihydropterin, though active, were utilized less efficiently than D-*erythro*-dihydroneopterin.

Examination of the structures of the 6-substituted polyhydroxyalkylpterins tested (Table I) reveals that the enzyme responsible for the cleavage of the 1',2'-carbon bond of the side-chain appears to have a broad specificity, but favors the configuration of the 1'-hydroxyl group contained in D-*erythro*-, L-*threo*- and L-*arabo*-polyhydroxyalkylpterin.

Reynolds and Brown[30] suggested that an aldolase-type reaction may be responsible for the cleavage of the 1',2'-carbon bond of the side-chain of dihydroneopterin [VI$_a$] resulting in hydroxymethyldihydropterin ([VII] Fig. 1) and glycolaldehyde. Jones and Brown[54] who studied various polyalkyldihydropterins as precursors to dihydropteroate with extracts of *E. coli* found that D-*erythro*-dihydroneopterin triphosphate which is the immediate pteridine product from GTP must lose its phosphate residues prior to cleavage to hydroxymethyldihydropterin. More recently, Brown[46] reported that this enzyme from *E. coli* has been purified 500-fold, and that D-*erythro*-dihydroneopterin was converted to hydroxymethyldihydropterin and glycolaldehyde. Brown[46] proposed the trivial name dihydroneopterin aldolase for this enzyme. The enzyme requires the dihydro-dephosphorylated form of D-*erythro*-neopterin and will not utilize D-*erythro*-neopterin, D-*erythro*-tetrahydroneopterin or D-*erythro*-dihydroneopterin mono- or triphosphate[46]. L-*erythro*-Dihydroneopterin which was shown to be a poor substrate[33,64] inhibited the utilization of the D-*erythro*-dihydroneopterin[31,34].

Another mechanism proposed for hydroxymethyldihydropterin [VII$_a$] formation involves the eventual removal of the three-carbon side-chain of D-*erythro*-dihydroneopterin [VI$_a$] resulting in pterin [XXVIII]. This could occur by either the removal of the three-carbon side-chain or the successive removal of a two-, and then a one-carbon unit. The addition of a C$_1$ unit to pterin [XXVIII] would result in hydroxymethylpterin [VII$_a$]. This proposed mechanism, based on chemical studies[66,67] finds support in the reports of the formation of isoxanthopterin[68] [XXIX] and pterin[69] [XXVIII] and hydroxymethylpterin [VII] by *D. melanogaster*. The reports of incorporation of radioactivity from [U-^{14}C]-threonine into the side-chain of biopterin [X] by *Anacystis nidulans*[70] and from [10-^{14}C]pterin [XXVIII] into sepiapterin [XI$_a$] by *D. melanogaster*[68] as well as the enzymatic synthesis of folic acid from xanthopterin [XXXI] and *p*-aminobenzoylglutamic acid (ABG) by cell-free extracts of *Mycobacterium avium*[71] suggest the addition of nucleophiles to position 6 of a pteridine.

The interpretation of the data of biopterin synthesis by *A. nidulans*[70] was questioned by Krumdieck *et al.*[27] and Watt[72] because of the extremely low incorporation of radioactivity into biopterin. The report that xanthopterin[71] is a precursor of the pteridine nucleus of folic acid is subject to critical review. Crude cell-free extracts of *M. avium* were employed to demonstrate folate synthesis from xanthopterin and ABG. However, an experiment demonstrating the requirement for xanthopterin was not presented, and it is highly possible that the crude extract employed does not require xanthopterin but utilizes an endogenous 6-substituted pterin such as hydroxymethyldihydropterin. The possibility that an endogenous precursor pteridine was present in the extract was suggested by their finding that a resin-treated enzyme required a boiled crude extract. Crude extracts of *L. plantarum*[59] and *E. coli*[60] have been demonstrated to contain a precursor pteridine of folate compounds. Direct testing for the addition of various nucleophiles to position 6 of 2-amino-4-hydroxydihydropterin with enzyme preparations derived from *E. coli*[30] and from green plants[34] failed to show any addition. This controversy dealing with the side-chain of position 6 of pterins will be further discussed on the topic of non-folate-related pteridines (Section 4b, p. 131).

(f) Hydroxymethyldihydropterin utilization

Early studies on folate biosynthesis were principally concerned with the immediate reactions leading to folate-like compounds by bacteria. Resting-cell suspensions were found to be capable of synthesizing folate-like compounds from *p*-aminobenzoic acid (AB), indicating that these cells contained an endogenous precursor pteridine[73,74]. In these experiments, 2-amino-4-hydroxy-6-formylpteridine (formylpterin) which was tested as the precursor pteridine was without effect[74]. Subsequently, growing bacterial cells were found to incorporate [^{14}C]xanthopterin into folic acid[75] and to utilize formylpterin[76]. Cell-free extracts of *M. avium* were reputed to synthesize folate from xanthopterin, ABG or AB and glutamic acid[71]. These initial studies demonstrated that bacteria were capable of synthesizing folate-like compounds, but revealed little about the nature of the precursor pteridine.

Appropriately, studies on the nature of the precursor pteridines of folate biosynthesis indicated that the reduced formylpterin and hydroxymethyl-pterin were effective precursors when crude extracts of *L. plantarum*[59] or *E. coli*[60] were incubated with ATP, $MgCl_2$ and AB or ABG. The products which were initially believed to be the tetrahydropteroate and tetrahydrofolate[59] were later shown to be dihydropteroate and dihydrofolate[60-62,77]. The active reduced precursor pteridine was shown to be hydroxymethyldihydro-pterin[60-62,77]. The dihydro structure of hydroxymethylpterin produced by dithionite reduction was shown to be the 7,8-form[58,78], thus completing the structural assignment of the active precursor pteridine as hydroxymethyl-7,8-dihydropterin ([VII$_a$], Fig. 1). Similar reactions were found in *S. typhimu-rium*[32], pneumococci[79,80], *Staphylococcus epidermidis*[81,82], soybeans[83,84], spinach leaves[84], cabbage leaves[33,34] and pea seedlings[85,86].

(g) Hydroxymethyldihydropterin-PP as an intermediate

In each of the systems described above, using hydroxymethyldihydropterin, the requirement for ATP indicated a phosphorylated pteridine intermediate. Jaenicke and Chan[77] proposed the structure 2-amino-4-hydroxy-6-hydroxy-methylpyrophosphoryl-7,8-dihydropteridine (hydroxymethyldihydropterin-PP, [VIII$_a$] Fig. 1) for this intermediate. Accordingly, Shiota *et al.*[87] prepared mono- and diphosphate derivatives of hydroxymethylpterin. Their results showed that only hydroxymethyldihydropterin diphosphate was active in condensing with AB or ABG. Subsequent studies[88] regarding structure gave proof for 2-amino-4-hydroxy-6-hydroxymethylpyrophosphoryl pteridine. Other enzyme systems using hydroxymethyldihydropterin and ATP for di-hydropteroate or dihydrofolate synthesis were shown to utilize hydroxy-methyldihydropterin-PP[33,34,62,80,82-85,89]. Stoichiometry of the reaction indicated that for each mole of hydroxymethyldihydropterin-PP used, a mole of dihydropteroate[84,89] or dihydrofolate[86,88] and PP_i[86,88,89] were formed. Based on the observation that hydroxymethyldihydropterin-P and ATP failed to be utilized for dihydropteroate or dihydrofolate synthesis, pyrophospho-rylation rather than successive phosphorylation by ATP was suggested as the mode of phosphorylation[87]. The formulation of these reactions is:

Hydroxymethyldihydropterin + ATP →   (1)

        hydroxymethyldihydropterin-PP + AMP

Hydroxymethyldihydropterin-PP + AB or ABG →   (2)

dihydropteroate or dihydrofolate + PP$_i$

Until recently, the enzymes which catalyze reactions (1) and (2) had not been separated and studied thoroughly. Furthermore, the isolation and identification of the phosphorylated product of reaction (1) had not been accomplished. Weisman and Brown[62] reported the separation of two enzyme fractions from *E. coli*. Fraction A accumulated a product from hydroxymethyldihydropterin and ATP which, in the presence of AB and Fraction B, was converted to dihydropteroate. This product was believed to be hydroxymethyldihydropterin-PP. Jones and Williams[82] identified a product from a reaction mixture containing a cell-free extract of *S. epidermidis*, hydroxymethyldihydropterin and ATP as hydroxymethyldihydropterin-PP by paper chromatography. The latter work was confirmed by Mitsuda and Suzuki[90] and Mitsuda *et al.*[91] who demonstrated the conversion of [10-^{14}C]hydroxymethyldihydropterin to radioactive hydroxymethylpterin-PP by *E. coli* extracts.

Recently, Richey and Brown[89] succeeded in a clear separation of the two enzymes catalyzing reactions (1) and (2) from *E. coli*. Thorough studies using a 400-fold purified enzyme, hydroxymethyldihydropterin and ATP revealed that a product was formed which behaved like hydroxymethylpterin-PP in several chromatographic solvents. A double-labeling experiment using [β,γ-^{32}P]ATP and [^{14}C]hydroxymethyldihydropterin indicated the transfer of two phosphate residues to hydroxymethyldihydropterin. A stoichiometric experiment comparing AMP and hydroxymethyldihydropterin-PP ([VIII$_a$], Fig. 1) production suggested pyrophosphorylation. The authors proposed the name hydroxymethyldihydropterin pyrophosphokinase for this enzyme. The second enzyme was shown to catalyze the stoichiometric formation of dihydropteroate ([IX$_a$], Fig. 1) and PP$_i$. Richey and Brown[89] proposed that this enzyme be called dihydropteroate synthetase. Similar results were obtained with two enzyme fractions from *L. plantarum*[92]. One fraction catalyzed the pyrophosphorylation by ATP to produce hydroxymethyldihydropterin-PP (hydroxymethyldihydropterin pyrophosphokinase) and the second fraction converted hydroxymethyldihydropterin-PP to dihydrofolate or dihydropteroate (dihydropteroate synthetase).

(h) Role of p-aminobenzoate and p-aminobenzoylglutamate

In all the systems described, hydroxymethyldihydropterin-PP condenses with either AB or ABG with varying efficacy for either dihydropteroate or dihydrofolate synthesis[60,80,84,85,88]. This variance has evoked discussions[62,80,88,89] as to whether there are two enzymes, one specific for hydroxymethyldihydropterin-PP and AB and the other for hydroxymethyldihydropterin-PP and ABG or a single enzyme with broad specificity which is altered by the purification procedure favoring either AB or ABG. Recently[92] it was found that the dihydropteroate synthetase activity for AB and ABG was not separable by polyacrylamide gel electrophoresis. These results suggested a single enzyme. That a single enzyme is responsible for either AB or ABG was also concluded by Brown[46]. In all probability, the cosubstrate for dihydropteroate synthetase is AB. The synthesis of AB from shikimic acid[93], shikimic acid 5'-phosphate[94] and chorismic acid[95] is well established while the biosynthesis of ABG, reported by Katsunuma et al.[71], has not been verified to date. These authors reported that in M. avium, ABG is synthesized from AB and glutamic acid and that the reaction required ATP and coenzyme A. Others have sought ABG synthesis without success[59,60,85].

Although the hydroxymethyldihydropterin pyrophosphokinase and dihydropteroate synthetase have been resolved into two separable enzymes from E. coli[89] and L. plantarum[92], the system found in pea seedlings[96] is reported to be an enzyme complex which catalyzes these two reactions without the involvement of a free hydroxymethyldihydropterin-PP. In this system, either hydroxymethyldihydropterin or its pyrophosphorylated derivative is active. Somewhat similarly, in L. plantarum[92] which contains a separable pyrophosphokinase and synthetase, an additional form, an enzyme complex, may exist. Unlike the pea-seedling, the enzyme complex from L. plantarum can catalyze dihydropteroate or dihydrofolate synthesis from hydroxymethyldihydropterin, ATP and AB or ABG, but cannot use the pyrophosphorylated pteridine derivative. Additional work is necessary in order to determine the importance and physiological role of the enzyme complex in folate biosynthesis.

Another set of reactions involving hydroxymethyldihydropterin-PP is the synthesis of rhizopterin (N^{10}-formylpteroic acid). Preliminary studies[81,97,98] indicated that S. epidermidis synthesized a phosphate-containing pteroic acid and rhizopterin-like materials. Subsequently, Jones and Williams[82] reported that cell-free extracts of S. epidermidis were found to catalyze formyldihydropteroate synthesis in a reaction mixture containing hydroxymethyldihydropterin-PP, AB, ADP and a C_1 donor. The formulation of the reactions was postulated to take place as follows:

Hydroxymethyldihydropterin-PP + AB →   (3)
                dihydropteroatephosphate + P_i

Dihydropteroatephosphate + C_1 donor + ADP →   (4)
                dihydrorhizopterin + ATP

The nature of dihydropteroate phosphate and the C_1 donor remains unknown and the importance of these reactions is yet to be determined. These sets of reactions thus far are unique for *S. epidermidis* since they have not been reported for other dihydrofolate- or dihydropteroate-synthesizing systems.

(i) Dihydrofolate synthesis

Brown et al.[60] provided evidence indicating that *E. coli* extracts catalyzed the synthesis of dihydrofolate from dihydropteroate and L-glutamate. These findings were extended by Griffin and Brown[99] who demonstrated that neither pteroate nor tetrahydropteroate but only dihydropteroate was active for dihydrofolate synthesis. This reaction was found to require ATP, divalent and monovalent cations. The enzyme that catalyzed the addition of L-glutamate to dihydropteroate to form dihydrofolate is widely distributed among microorganisms, including *M. avium*[99]. Contrarily, Katsunuma et al.[71] had previously reported that *M. avium* catalyzed the addition of L-glutamate to AB to form ABG. This pathway of ABG synthesis has not as yet been duplicated in any other system to date. However, the pathway of Griffin and Brown[99] has now been reported to occur in partially purified extracts from pea seedlings[84,96].

3. Pteroylpolyglutamates

(a) Occurrence

Pteroylpolyglutamates is a term applied to describe pteroylglutamate derivatives with additional L-glutamic acid residues in γ-peptide linkages. These compounds are also referred to as the bound forms of folate, folate conjugates, polyglutamates of folate derivatives, N^5-formyl, N^{10}-formyl or N^5-methyl pteroylpolyglutamate and N^5-substituted polyglutamyl pteroic acid derivatives. These compounds are often found as the di- or tetrahydro-forms. They are found widely distributed in nature: in yeast, pteroylheptaglutamate[100], N^{10}-formyl, N^5-formyl and N^5-methyl pteroylpolyglutamates (probably as the tetrahydro- forms)[101]; in liver, N^5-formyl pteroylpolygluta-mate[102,103], N^5-methyl pteroylpolyglutamate[103], their tetrahydro- deriva-tives and N^5-formyl tetrahydropteroyltriglutamate[104]; in plants, N^{10}-for-myl pteroyltriglutamate[105]; in whole blood, pteroyldiglutamate, N^5-for-myl tetrahydropteroyl di- and triglutamates[106], and in the red cells, primarily N^5-methyl tetrahydropteroylpolyglutamate[107]; in bacteria, N^5-formyl tetra-hydropteroyltriglutamate[108,109], and in *Clostridium cylindrosporum*, pteroyl mono-, di- and heptaglutamate derivatives containing additional amino acids, serine, glycine and alanine, and some with pentose and phosphates[110].

(b) Biosynthesis

Little is known about the biosynthesis of the pteroylpolyglutamate derivatives. One possibility of pteroylpolyglutamate formation is by way of condensing *p*-aminobenzoylpolyglutamate to an appropriate pteridine. However, the reasons presented in sections 2h and 2i argue against this. In contrast, one can view the elongation of the pteroic acid molecule by either the successive ad-dition of L-glutamic acid residues or by the addition of a homo-γ-glutamyl peptide to pteroic acid or folic acid (or their reduced forms). The report by Sirotnak et al.[111] can be interpreted in favor of the latter view. A comparison was made of the cellular content of pteroylpolyglutamates in *Diplococcus pneumoniae* grown in the presence of excess dihydrofolate with and without sulfanilamide as an inhibitor. In the presence of dihydrofolate and the drug, little synthesis of pteroylpolyglutamate took place, whereas, in the absence of the inhibitor, considerable amounts of pteroylpolyglutamates were found.

The mechanism involving successive additions of L-glutamic acid residues to pteroic acid and folic acid is supported by the work of Griffin and Brown[99]. Extracts of *E. coli* catalyzed the synthesis of dihydrofolate by the addition of L-glutamate to dihydropteroate. The enzyme required ATP, divalent and mon-ovalent cations and was inactive for pteroate or tetrahydropteroate. These extracts were also found to convert tetrahydropteroylglutamate to tetra-hydropteroyl-γ-diglutamate and tetrahydropteroyl-γ,γ-triglutamate. The reactions required L-glutamate and ATP. Neither pteroylglutamate nor di-hydropteroylglutamate could substitute for tetrahydropteroylglutamate.

Little is known about the origin of the C_1 unit of N^5-formyl, N^{10}-formyl and N^5-methyl tetrahydropteroylpolyglutamate. Since the coenzyme func-tion of tetrahydropteroyltriglutamate has been shown for several reactions, there is little doubt that the addition of C_1 and its transformation occur at the

polyglutamate level of tetrahydrofolate. Wood and Wise[112] reported that
cell suspensions of *Streptococcus faecalis* synthesized N^5-formyl tetrahydropteroylpolyglutamates. The reaction was dependent upon the addition of
either folic acid or N^5-formyl tetrahydrofolate and formic acid was required
for each case. L-Glutamic acid or L-glutamine was shown to be required when
cells depleted of endogenous amino acids were used. With cell-free extracts of
S. faecalis, Hakala and Welch[109] were able to demonstrate the synthesis of
N^5-formyl tetrahydropteroyltriglutamate. The latter compound was synthesized from pteroyltriglutamate, but not from pteroylglutamate or pteroyldiglutamate.

It is interesting to point out that tetrahydropteroyltriglutamate has been
shown to be an irreplaceable requirement for some of the enzyme systems in the
de novo generation of one-carbon fragments. Thus, the conversion of serine to
glycine and formic acid in *Clostridium sticklandii*[113] and *Pseudomonas* AM-
1[114], as well as in *S. faecalis* which synthesized N^5-formyl tetrahydropteroyltriglutamate[109] shows an absolute requirement for the tetrahydropteroyltriglutamate. Furthermore, the cobalamin-independent pathway of methionine
biosynthesis[115] also showed an absolute dependence upon a triglutamyl derivative. In addition, other enzymes involved in the *de novo* synthesis of one-
carbon fragments such as formyl tetrahydrofolate synthetase[116] and serine
hydroxymethylase of mammalian origin[117] have been shown to function better
with tetrahydropteroyltriglutamate than with the monoglutamate derivative. Finally, the tetrahydropteroyltriglutamate can be replaced by the monoglutamate derivative with no loss of activity in the cobalamin-dependent sys-

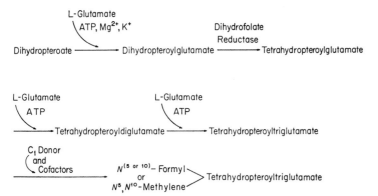

Fig. 2. A proposed biosynthetic pathway for *N*-formyl tetrahydropteroyltriglutamate and
N^5,N^{10}-methylene tetrahydropteroyltriglutamate.

tem of methionine biosynthesis[118-120] and in formiminoglycine formimino transferase[121]. These reactions for N-formyl tetrahydropteroyltriglutamate[60,84,96,99,109,112,117] are summarized in Fig. 2.

(c) Conjugase(s)

Conjugases are γ-glutamic acid carboxypeptidases[122,123] which liberate L-glutamic acid from pteroylpoly-γ-L-glutamyl–L-glutamate derivatives. These enzymes are widely distributed and are found in rat liver[124], hog kidney[125], chicken pancreas[126], tissues from other animals[104,127,128], bacteria[109, 129,130] and plants[131]. There are two types of conjugases reported in the literature, one which hydrolyzes pteroylpolyglutamate derivatives to pteroylglutamate derivatives and another to pteroyldiglutamate derivatives. The former is exemplified by the conjugase found in hog kidney[132,133] and the latter by that found in chicken pancreas[122,123]. The conjugases from *Physalia physalis* (Portuguese man-of-war)[104], *Bacillus subtilus*[109] and green leaves of soybeans[131] are apparently of the hog-kidney type. The enzyme from chicken pancreas has been partially purified[126] and is reported to have a requirement for calcium with an optimum temperature and pH of 32° and 7.8 respectively. The enzyme from hog kidney has not been extensively purified[128] and is reported to have an optimum temperature and pH of 45° and 4.5. The conjugase from chicken pancreas[122] preferentially hydrolyzed the C-terminal glutamic acid, which is linked through its amino group to the γ-carboxyl of the preceding glutamic acid. Subsequently, Dabrowska *et al.*[123] showed that the chicken-pancreas conjugase required a γ-polyglutamate derivative with a minimum of three glutamate residues since it failed to hydrolyze pteroyl-γ-diglutamate. The enzyme which was previously characterized as a carboxypeptidase[100] was established to be a γ-glutamic acid carboxypeptidase[122,123]. These conjugases were shown also to hydrolyze N^5-formyl, N^{10}-formyl and N^{10}-methyl tetrahydropteroylpolyglutamates[101-107,133-135].

Two types of *Flavobacterium*, isolated from soil by an enrichment technique, have been reported to contain γ-polyglutamic acid peptidases. Volcani and Margalith[129] reported a peptidase from *Flavobacterium polyglutamicum* which specifically hydrolyzed the γ-glutamyl bond of γ-polyglutamic acid and pteroyl-γ-polyglutamate but failed to split pteroyl-γ-diglutamate. Pratt *et al.*[130] also found in another *Flavobacterium* sp. a peptidase which was specific for the γ-glutamyl bond of not only γ-glutamylglutamate and pteroyl-

γ,γ-triglutamate, but also for pteroyl-γ-diglutamate. These investigators reported the separation of two additional enzymes, a peptidase which was specific for α-glutamylglutamate and folate amidase. This latter enzyme which hydrolyzed pteroylglutamate to pteroate and glutamate has a relatively strict specificity for an unmodified pteroyl residue as compared to the enzyme found in a pseudomonad reported by Levy and Goldman[136].

The γ-polyglutamylglutamate peptidases from various animals have not been studied in detail and yet have been used to hydrolyze pteroylpolyglutamate derivatives in order to assay for total folates. The procedures described for this purpose have limitations caused by the presence of inhibitors of conjugases[122,128,137] or by the varied responses by the assay organisms to-

TABLE II

DEAE-CELLULOSE CHROMATOGRAPHIC AND MICROBIOLOGICAL
CHARACTERIZATION OF FOLATE COMPOUNDS

Compound	Peak tube number	Activity of test organism[a]			Reference
		PC	SF	LC	
N^{10}-Formyl H_4pteroylglutamate	8	+	+	+	134
N^{10}-Formyl H_2pteroylglutamate	9	−	+	+	103,134
N^{10}-Formyl pteroylglutamate	10	−	+	+	134
N^5-Formyl H_4pteroylglutamate	12	+	+	+	103,134
N^5-Methyl H_4pteroylglutamate	13	−	−	+	103,134
N^{10}-Formyl H_4pteroyltriglutamate	14	−	−	+	135,138
H_4Pteroylglutamate	15	+	+	+	134
N^5-Methyl H_4pteroyldiglutamate	16	−	−	+	104
N^{10}-Formyl H_4pteroylpolyglutamate	18	−	−	−	135
N^5-Formyl H_4pteroyltriglutamate	22	±	−	+	135,138
H_2Pteroylglutamate	26	−	+	+	134
Pteroylglutamate	28	−	+	+	103,134
N^5-Formyl H_4pteroylpolyglutamate	28–30	−	−	−	104,135
Pteroyltriglutamate	38	−	−	+	134
N^5-Methyl pteroylpolyglutamate	40,53,83	−	−	−	104,107
N^5,N^{10}-Methenyl H_4pteroylglutamate		−	+	+	141
Pteroic acid		−	+	−	141
N^{10}-Formyl pteroic acid		−	+	−	141
N^5-Formyl pteroic acid		−	+	−	141

[a] PC, *Pediococcus cerevisiae*; SF, *Streptococcus faecalis*; LC, *Lactobacillus casei*.

References p. 147

ward the multiplicity of folate derivatives[109,133,138] or by the varied rate or completion of hydrolysis of the pteroylpolyglutamate derivatives by different preparations of conjugases[127,139].

(d) Assays

For the microbiological assays of folate compounds, three microorganisms have been employed, namely, *S. faecalis*, *Lactobacillus casei* and *Pediococcus cerevisiae*, formerly *Leuconostoc citrovorum*[140]. Often these three organisms have been used in conjunction with conjugase-treated and untreated material. This differential assay permits the recognition of the forms of some of the folate derivatives because of the varied response of the assay organisms to some of the folate compounds. Silverman *et al.*[134] extended the procedure by first subjecting the material to an analytical DEAE-cellulose chromatography so that the various forms of folate compounds were sequentially eluted. The fractions were then assayed by the differential procedure before and after conjugase treatment. By this method, the characterization of some of the folate derivatives is possible by their peak elution and the responses of the assay organisms. The peak elution and the responses by the three assay organisms for the various folate compounds which are summarized in Table II indicate a refinement of the procedure is desirable in order to obtain better resolution of these compounds.

4. Biosynthesis of unconjugated pteridines

(a) Occurrence of pteridines with a side-chain in position 6

Pteridines with a number of three-carbon side-chains in position 6 have been reported to occur in nature. In *Drosophila*[142] and in human urine[143], a factor designated biopterin was found to be a growth factor for *C. fasciculata*. Its structure was established to be 2-amino-4-hydroxy-6-(L-*erythro*-1',2'-dihydroxy-3'-deoxypropyl)-pteridine[144,145] [X]. Since then, biopterin has been found in royal jelly[65] and amphibia[28,146]. A related pteridine, sepiapterin [XI$_a$], has been found in *D. melanogaster*[147] and in amphibia[146,148] and the structure, 2-amino-4-hydroxy-6-lactyl-7,8-dihydropterin proposed by Nawa[149] and Forrest and Nawa[150] has now been accepted. The term drosopterins is used to designate a mixture of at least three red pteridines found in *Drosophila*[147] and lizards[151,152]. Their structures still remain in doubt.

Viscontini[153] proposed a structure for drosopterin [XII], isodrosopterin [XIII] and neodrosopterin [XIV] which are geometrical and optical isomers. Alternatively, Pfleiderer[154] recently suggested that drosopterins are dimers linked through position 7 [XV]. The structure [XV] is one of several dimeric forms of drosopterins proposed by Pfleiderer[154]. Additional pteridines with a three-carbon side-chain in position 6 have been reported. They are: D- or L-*threo*-neopterin[40] [XIX] or [XVIII], an *erythro*-neopterin triphosphate[51], and neopterin monophosphate[52] in *E. coli*; a 2,4-dihydroxypolyalkylpteridine[155], in *Pseudomonas fluorescens*; a neopterin[40] in a pseudomonad and photosynthetic bacteria[156]; a *threo*-neopterin[157] in male red ants; L-*threo*-neopterin in pseudomonads[158,159] and in *Serratia indica*[160], and L-*threo*-biopterin in *T. pyriformis*[161].

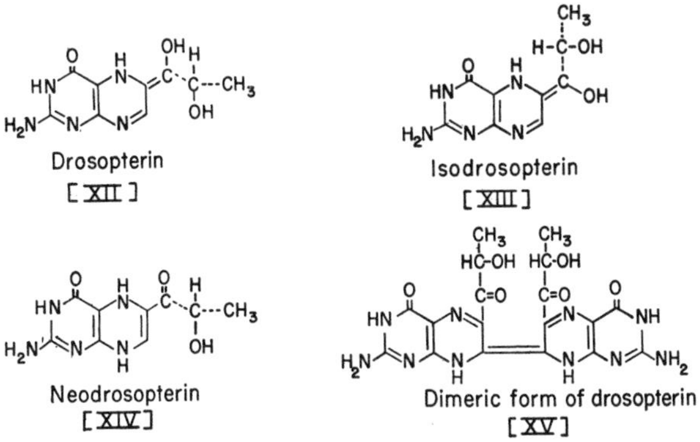

Fig. 3.

(b) Origin of the side-chain of biopterin and sepiapterin

Presently, there are three hypotheses offered for the origin of the side-chain of biopterin-like compounds from neopterin[161] (Fig. 5). Hypothesis *one* maintains that the side-chain of neopterin is retained and undergoes alterations. Hypothesis *two* requires the removal of carbon atoms 2' and 3' of D-*erythro*-neopterin followed by the addition of a two-carbon unit. Hypothesis *three* indicates that all three carbon atoms of the side-chain are replaced by a new three-carbon unit.

The hypothesis proposing that the side-chain of biopterin: L-*erythro*-1',2'-dihydroxy-3'-deoxypropyl group arises by the alteration of the side-chain of D-*erythro*-neopterin is well supported by several different reports. Brenner-

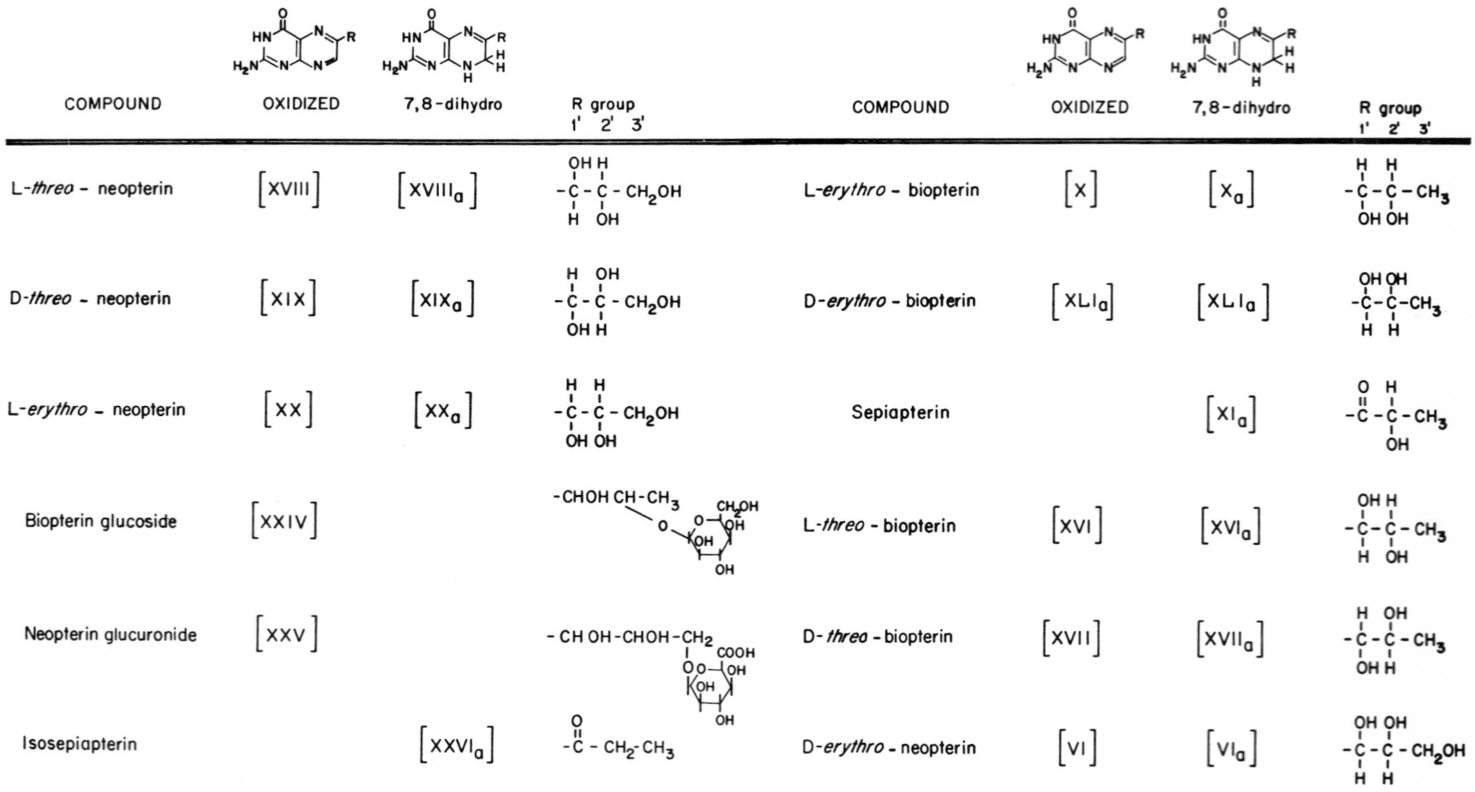

COMPOUND	OXIDIZED	7,8-dihydro	R group 1' 2' 3'
L-*threo* – neopterin	[XVIII]	[XVIIIa]	OH H -C-C-CH$_2$OH H OH
D-*threo* – neopterin	[XIX]	[XIXa]	H OH -C-C-CH$_2$OH OH H
L-*erythro* – neopterin	[XX]	[XXa]	H H -C-C-CH$_2$OH OH OH
Biopterin glucoside	[XXIV]		-CHOH CH-CH$_3$
Neopterin glucuronide	[XXV]		-CH OH-CHOH-CH$_2$
Isosepiapterin		[XXVIa]	O -C-CH$_2$-CH$_3$

COMPOUND	OXIDIZED	7,8-dihydro	R group 1' 2' 3'
L-*erythro* – biopterin	[X]	[Xa]	H H -C-C-CH$_3$ OH OH
D-*erythro* – biopterin	[XLI]	[XLIa]	OH OH -C-C-CH$_3$ H H
Sepiapterin		[XIa]	O H -C-C-CH$_3$ OH
L-*threo* – biopterin	[XVI]	[XVIa]	OH H -C-C-CH$_3$ H OH
D-*threo* – biopterin	[XVII]	[XVIIa]	H OH -C-C-CH$_3$ OH H
D-*erythro* – neopterin	[VI]	[VIa]	OH OH -C-C-CH$_2$OH H H

Fig. 4.

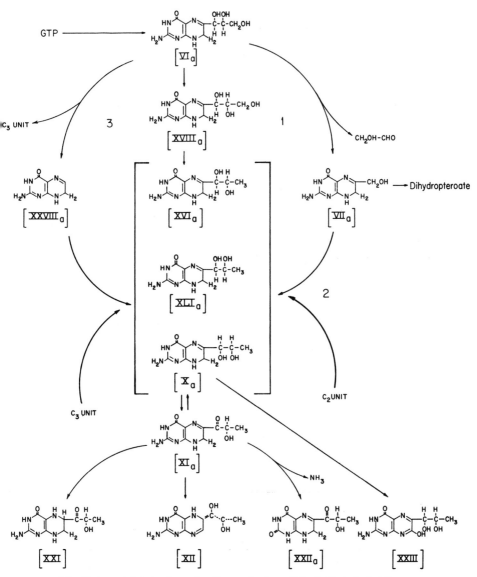

Fig. 5. A postulated scheme for the production of 6-substituted pteridines.

Holzach and Leuthardt[26,162] found that carbon atoms 5 and 6 of glucose were incorporated into the side-chain of biopterin and drosopterins. Similar findings with biopterin-like compounds reported by Watt[72] and Kidder and

Dewey[161] were consistent with the view that the side-chains of sepiapterin[72] [XI$_a$] and ciliapterin[161] [XVI] originate from the ribose moiety of guanosine and GMP respectively. Watt[72] isolated [^{14}C]sepiapterin from wings of *Colias eurytheme* incubated with [U-^{14}C]guanosine, and Kidder and Dewey [161], [^{14}C]ciliapterin from *T. pyriformis* cultures incubated with [U-^{14}C]GMP. For each of these pteridines, an experiment was performed in order to determine the origin of the side-chain. The pteridines were oxidized to carboxy-pterin by alkaline permanganate and the specific radioactivities of [^{14}C]-sepiapterin and [^{14}C]ciliapterin compared with that of the corresponding [^{14}C]carboxypterin. Watt[72] and Kidder and Dewey[161] were thus able to test the hypotheses of the origin of the side-chain. Their results which are summarized in Table III are very impressive and the retention of 7/9 of the original specific radioactivities in [^{14}C]carboxypterin proves that the three carbons of the side-chain of either sepiapterin or ciliapterin originate from the ribose moiety of guanosine and GMP respectively. Had the specific radioactivities

TABLE III

Origin of the Side Chain

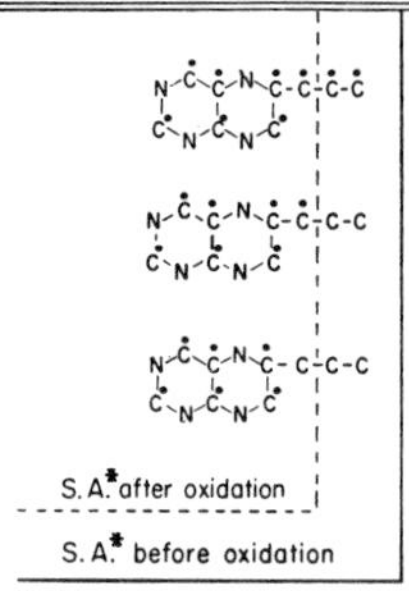

Hypothesis	Proposal of the Labeling Pattern	Alkaline KM$_n$O$_4$ Oxidation	% of S.A.* Remaining After Oxidation
1	N~C~C~N~C-C:C-C / C~N~C~C	⟶	77.8
2	N~C~C~N~C-C:C-C / C~N~C~C	⟶	100.0
3	N~C~C~N~C- C:C-C / C~N~C~C	⟶	100.0

S.A.* after oxidation
_____
S.A.* before oxidation

• C^{14}
* Specific radioactivity

Compound	SPECIFIC ACTIVITY		
	BEFORE OXIDATION	AFTER OXIDATION	% REMAINING
[^{14}C]Sepiapterin **	11.47×10^3	8.47×10^3	73.8
[^{14}C]Ciliapterin ***	25.9×10^5	21.0×10^5	81.1

** W.B. Watt, *J. Biol. Chem.*, 242 (1967) 565

*** G.W. Kidder and V.C. Dewey, *J. Biol. Chem.*, 243 (1968) 826

of [^{14}C]carboxypterin remained unchanged, the decarboxylation of carboxy-pterin by a procedure developed by Shaw et al.[163] would have determined the validity of hypotheses *two* or *three*.

The second hypothesis (Fig. 5) for the origin of the side-chain of biopterin-like compounds proposes the cleavage of the three-carbon chain of a 6-sub-stituted pteridine to a pterin with a one-carbon unit in position 6. To the latter, a new two-carbon unit would be added to produce a biopterin-like compound. Kidder et al.[164] by a carefully controlled experiment discovered that *Crithidia*, which was grown with [2-^{14}C]folic acid produced labeled bio-pterin [X] and labeled hydroxymethylpterin [VII]. The authors suggested that *Crithidia* cleaved folic acid and from the resulting hydroxymethylpterin, bio-pterin was synthesized. Recently, Rembold[165] stated that when [6,7,1',2', 3'-^{14}C]D-*erythro*-tetrahydroneopterin [XXVII] was injected intraperitoneally into a rat, [6,7,1'-^{14}C]D-*erythro*-biopterin [XLI] was excreted. This result in-dicated that carbon atoms 2' and 3' of 6-(D-*erythro*-1',2',3'-trihydroxypropyl)-pterin were replaced by a 2-carbon unit and the new 6-substituted side-chain of pterin bears the structure D-*erythro*-1',2'-dihydroxy-3'-deoxypropyl.

The third hypothesis, concerning the origin of the side-chain of biopterin (Fig. 5), involves the addition of a new three-carbon unit to position 6 of pterin. The non-enzymatic addition of nucleophiles to carbon 6 of pterin prompted Forrest and Nawa[67] and Wood et al.[66] to propose that biopterin-like compounds could arise by an analogous enzymatic reaction. Goto et al.[166] accordingly, reported that radioactive drosopterins were detected by paper chromatography when larvae of *Drosophila* were fed radioactive re-duced pterin. This result suggested that biopterin synthesis would occur by the addition of a three-carbon unit to pterin. Additionally, Maclean et al.[70] claimed that the alga, *A. nidulans*, when grown with [U-^{14}C]DL-threonine produced labeled biopterin. This work drew criticism from Krumdieck et al.[27] and Watt[72] who found contradictory evidence which supports the pro-posal for the retention of the 6-substituted side-chain originating from the D-ribose moiety of guanine nucleoside or nucleotide. The incorporation of label from [U-^{14}C]DL-threonine into biopterin[70] is so low that it is reasonable to suspect that it arises after randomization of the label of the supplemented [U-^{14}C]DL-threonine. For similar reasons, the low incorporation of radio-active reduced pterin, hydroxymethylpterin and neopterin into sepiapterin[68] and biopterin[69] lack the necessary strength to support the interpretation given, *viz.*, that these compounds are direct precursors of biopterin. In a recent report by Sugiura and Goto[167], low levels of incorporation of [8a-^{14}C]-

dihydroneopterin and [8a-^{14}C]dihydropterin into biopterin were again shown. However, the data presented regarding the testing for efficacy of incorporation of the label from various neopterin derivatives suggested that neopterin 3'-phosphate is superior to other precursors tested. Similar to the report by Watt[72], Sugiura and Goto[167] also failed to show the incorporation of any radioactivity from labeled L-threonine, L-serine, L-glutamic acid, pyruvic acid and D-ribose into biopterin.

The data presented by the Forrest's[70,166] and Goto's[68,69,167] groups are difficult to accept in terms of precursor–product relationships. It would be of great interest if these groups could perform an experiment similar to that of Watt[72] or Kidder and Dewey[161] in order to determine the origin of the side-chain of biopterin in their experimental systems.

With regard to the use of labeled compounds with intact animals, Krumdieck et al.[27] and Kidder and Dewey[161] recognized that the ribose moiety of [U-^{14}C]guanine nucleosides or nucleotides could equilibrate with the free ribose pool, by way of the action of nucleosidases (-tidases), followed by ribosylation of the guanine to regenerate a guanine nucleotide bearing radioactive label principally in the ring system. Obviously, in the presence of these reactions, the use of [U-^{14}C]guanosine or [U-^{14}C]guanine nucleotide would result in diminished or no incorporation of radioactivity into the side-chain of neopterin.

(c) *The reactions concerning the side-chain of neopterin, biopterin and sepiapterin*

Virtually nothing is known about the details of the individual reactions involved in the conversion of pteridines substituted in position 6 with a 1',2',3'-trihydroxypropyl residue to pteridines with 1',2'-dihydroxy-3'-deoxypropyl or lactyl side-chains. Nevertheless, there is some evidence which may be of general interest (Fig. 5).

Fig. 1 which summarizes the pathway of dihydropteroate [IX$_a$] synthesis places D-*erythro*-dihydroneopterin triphosphate [V$_a$] as the first pteridine produced from GTP[51,54,55] [I]. Accordingly, the pterin nucleus contained in D-*erythro*-dihydroneopterin triphosphate must be the origin for the pterin moiety of folate-like compounds[24,27,29–31] and of all non-folate pteridine derivatives. The latter statement is based on isotope-tracer studies which conclude that pterin[69] [XXVIII], xanthopterin[25,72] [XXXI], isoxanthopterin[26,69] [XXIX], leucopterin[25,72] [XXXII], lumazine[69] [XXXIII], erythropterin[72], biopterin[28,69,162] [X], sepiapterin[72] [XI$_a$] and drosopterin[26] [XII] are all derived from GTP. Hence, the precursor D-*erythro*-dihydroneopterin [VI$_a$] must undergo a variety of changes that include modification and removal of the side-chain and additions of nucleophiles to positions 6, 7 and 8. In the case of lumazine, the pteridine must be deaminated.

The metabolic origin of a *threo*-neopterin[54,152] or specifically L-*threo*-neopterin[158,160] can be reasonably envisioned to be derived from D-*erythro*-neopterin. Jones and Brown[54] provided information of this interesting possibility. They showed that an extract of *E. coli* catalyzed the enzymatic conversion of GTP to *erythro*-dihydroneopterin and *threo*-dihydroneopterin and suggested that an epimerase was most likely involved. The occurrence of a cyclic phosphate of neopterin[40,159] suggested to Guroff and Rhoads[159] that the cyclic phosphate is formed between the 2'- and 3'-hydroxyl groups of D-*erythro*-neopterin and that the hydrolysis of the cyclic phosphate results in an inversion of the 2'-hydroxyl group. In accordance with the central role assigned to D-*erythro*-dihydroneopterin, it would be converted by dihydroneopterin aldolase (see section 2 e, p. 118) into hydroxymethyldihydropterin and by an inversion (a dihydroneopterin epimerase) into *threo*-neopterin. Although L-*threo*-dihydroneopterin is a substrate for dihydroneopterin aldolase, it is not as efficient as D-*erythro*-dihydroneopterin[33,64]. Therefore, it seems improbable that the epimerization around carbon atom 2' of D-*erythro*-dihydroneopterin provides the cell with an additional pathway for hydroxymethyldihydropterin synthesis. In the absence of any experimental data, the fate of L-*threo*-dihydroneopterin is unknown. However, it is tempting to speculate that this compound is the first intermediate in the conversion of D-*erythro*-dihydroneopterin to biopterin, sepiapterin and the drosopterins as proposed in hypothesis *one*, Fig. 5. Hence, by epimerization of carbon atoms 1' and 2' and reduction of carbon atom 3' of neopterin, the side-chain of biopterin would be originated. On the other hand, no enzyme reactions have been reported which could support either hypothesis *two*[164,181] (Fig. 5) suggesting the addition of a two-carbon unit to a pteridine having a one-carbon substituent on position 6, *viz.*, the condensation of acetaldehyde to the one-carbon unit of hydroxymethylpterin, or hypothesis *three*[59,68,69,166,167] which requires the addition of a three-carbon unit to pterin.

The reactions which are best described for the alteration of the three-carbon side-chain of 6-substituted pteridines are those involving L-*erythro*-dihydrobiopterin and sepiapterin. These important reactions were initially observed by Taira[168] who found that a yellow pigment (sepiapterin) from

Drosophila was reduced by a NADPH-dependent dihydrofolate reductase preparation from chicken liver. A preparation of dihydrofolate reductase from *Drosophila* also catalyzed the NADPH-dependent reduction of sepiapterin. The reduced product was then converted non-enzymatically in the absence of air to biopterin[169]. Matsubara *et al.*[170] noted that an enzyme preparation from silk worms which reduced folate and sepiapterin, lost the ability to reduce folate upon storage. This observation, which suggested two distinct enzymes, was substantiated by the independent rise and fall of dihydrofolate reductase and sepiapterin reductase activity during the development of the silk worm larvae. Additional evidence for two enzymes was presented by Matsubara and Akino[171] when they discovered that dihydrofolate reductase and sepiapterin reductase were separable by fractionating extracts of chicken liver. Furthermore, aminopterin at a concentration which completely inhibited dihydrofolate reductase activity caused only about 10% reduction in sepiapterin reductase activity.

About this time, Kaufman had identified the phenylalanine hydroxylation cofactor in rat liver as dihydrobiopterin[172]. Although the products of sepiapterin reduction by sepiapterin reductase and dihydrofolate reductase were not known, Matsubara *et al.*[173], from various lines of evidence, recognized that sepiapterin and biopterin had a 7,8-dihydro structure and a preparation of dihydrofolate reductase reduced both pteridines to the tetrahydro level[174].

However, since a highly purified preparation of dihydrofolate reductase was catalytically inactive for sepiapterin, they proposed the existence of another enzyme which reduced sepiapterin to a product which in turn was a substrate for dihydrofolate reductase. Accordingly, Matsubara *et al.*[173] tested this hypothesis in a series of experiments. They found that the product of sepiapterin reductase was a reduced biopterin which was oxidizable to biopterin, that the keto function of sepiapterin disappeared stoichiometrically with reduction, that the ultraviolet absorption spectrum of the reduced product was similar to dihydrobiopterin, that the reduced product was a substrate for the highly purified dihydrofolate reductase, and that the latter product had cofactor activity in the phenylalanine hydroxylation reaction. From these results, the authors formulated the reaction catalyzed by sepiapterin reductase.

$$\text{Sepiapterin} + \text{NADPH} + \text{H}^+ \rightleftharpoons \text{7,8-dihydrobiopterin} + \text{NADP}^+$$

Recently, Nagai[57] confirmed this by isolating the reaction product of sepiapterin reductase prepared from rat liver and identifying it as 7,8-dihydrobiopterin.

Sepiapterin reductase also catalyzes the reverse reaction in the presence of NADP[+], namely the oxidation of dihydrobiopterin[175]. The reductase also oxidizes L-*erythro*-dihydroneopterin but not D-*erythro*-dihydroneopterin[175]. In the same report by Katoh *et al.*[175], purified dihydrofolate reductase from chicken liver which was said not to reduce sepiapterin[173], was found to do so in acidic pH to form the tetrahydro form of sepiapterin [XXI]. Sepiapterin is a substrate for still another enzyme, a deaminase found in the silkworm, *Bombyx mori*[176,177]. The deaminated product was identified as 2,4-dihydroxy-6-lactyl-7,8-dihydropteridine [XXII$_a$]. Properties of sepiapterin deaminase indicate that this enzyme is distinct from guanase (EC 3.5.4.3).

There is very little information regarding the formation of other 6-propyl-pteridine derivatives found in nature. A pteridine which is closely related to biopterin, ichtyopterin (fluorescyanine) found in the skin of fish, *Cyprinid*, was reported by Tschesche and Glaser[178] to be 6-dihydroxypropylisoxanthopterin [XXIII]. Hama *et al.*[179] subsequently isolated two unknown pteridines from fish which were enzymatically converted to 6-dihydroxypropylisoxanthopterin. These two unknown pteridines were recently characterized by Ohta and Goto[180] as 6-(1'-hydroxy-2'-acetoxypropyl)-isoxanthopterin and 6-(1'-acetoxy-2'-hydroxypropyl)-isoxanthopterin. Hence, it would appear that Hama *et al.*[179] had observed that the acetylated dihydroxypropyliso-xanthopterin derivatives were deacetylated enzymatically to 6-dihydroxy-propylisoxanthopterin. Drosopterins, which are structurally related to sepiapterin with respect to the side-chain, may arise from the biopterin-sepiapterin pool. This is suggested by the enlargement of the biopterin-sepiapterin pool in mutants of *Drosophila* showing reduced production of drosopterins[147]. In Fig. 5, a postulated scheme for the production of 6-substituted pterins is presented.

Another group of naturally occurring 6-substituted pteridines are the pteridine glycosides. From *A. nidulans*, Forrest *et al.*[181] isolated a 6-dihydroxypropylpterin with a glucose, attached through an α-glucosidic linkage, to either carbon atom 1' or 2' [XXIV]. Examination of a number of blue-green algae[182], by paper chromatography of acetic acid–manganese dioxide treated cell extracts, revealed a predominant fluorescent material differing from specie to specie of algae. By acid hydrolysis followed by paper chromatography of the fluorescent material, pteridines and sugars were tentatively identified. The pteridine reportedly found in all algae, except in one case, was believed to be 6-dihydroxypropylpterin. The one exceptional case had a pteridine glucoside of hydroxymethylpterin. The sugars found were 6-deoxyglucose, galactose, glucose, xylouronic acid or xylose (ribose). Suzuki and Goto[183] found in *Azotobacter agilis* a pteridine glycoside and proposed for it the structure of neopterin-3'-β-D-glucuronide [XXV].

The complexity of these pteridine glycosides is indicated by those having in position 6, substituents of various chain lengths. As an example, Goto *et al.*[184] have reported a pteridine glycoside from *Mycobacterium smegmatis* which was active for *C. fasciculata*, as 2'-β-glucuronide of 2-amino-4-hydroxy-6-(1',2',3',4',5'-pentahydroxypentyl)-pteridine.

Another interesting pteridine glycoside, though not a pteridine with a three-carbon side-chain substituent in position 6, is asperopterin. This compound isolated by Kaneko[185] from cultures of *Aspergillus oryzae* is a derivative of isoxanthopterin with probably a methyl group in position N[8] and a hydroxymethyl group in position 6, linked to the reducing end of D-ribose[186]. The metabolic pathway for the synthesis of the pteridine glycosides is not known, but it is probable that each of the aglycones is the immediate precursor of the corresponding pteridine glycoside.

Two pteridines which have been studied and are probably not of metabolic consequence are isosepiapterin [XXVI] and L-*erythro*-neopterin [XX]. The former, which has been reported in *D. melanogaster*[187] and *A. nidulans*[4] is probably a degradation product[188]. L-*erythro*-Neopterin which is as effective as biopterin as a growth factor for *C. fasciculata*[65] has not been found in nature.

(d) Pterin and hydroxypterins

Pterin [XXVIII], isoxanthopterin (7-hydroxypterin, [XXIX]), xanthopterin (6-hydroxypterin, [XXXI]), and leucopterin (6,7-dihydroxypterin, [XXXII]) are widely found in nature. Pterin has been reported in insects[189,192], bacteria[193] and amphibia[146,194]; xanthopterin in butterflies[195] and urine[196]; isoxanthopterin in insects[191,192,197], amphibia[146,194] and urine[198]; and leucopterin in insects[199-202] and rat liver[203]. Early reports on the occurrence of these and other pteridines contained little information on the derivation of these compounds and only the more recent ones revealed data on the process of simple pteridine formation.

Pterin and the hydroxypterins probably arise from the non-enzymatic degradation of 6-substituted pterins. With respect to the non-enzymatic reactions, folic acid yields ABG[204] and a pteridine[205] or formylpterin[206] upon exposure to sunlight or ultraviolet light; dihydrofolate produces ABG and formyl 6,7-dihydropterin by a phosphate-catalyzed reaction[207,208]; tetrahydrofolate[209,210], tetrahydrosepiapterin[211] and tetrahydrobiopterin[203] yield an unidentified pteridine, carboxypterin and 7,8-dihydropterin, respectively, when exposed to air. The formyl pterin produced by the reactions in-

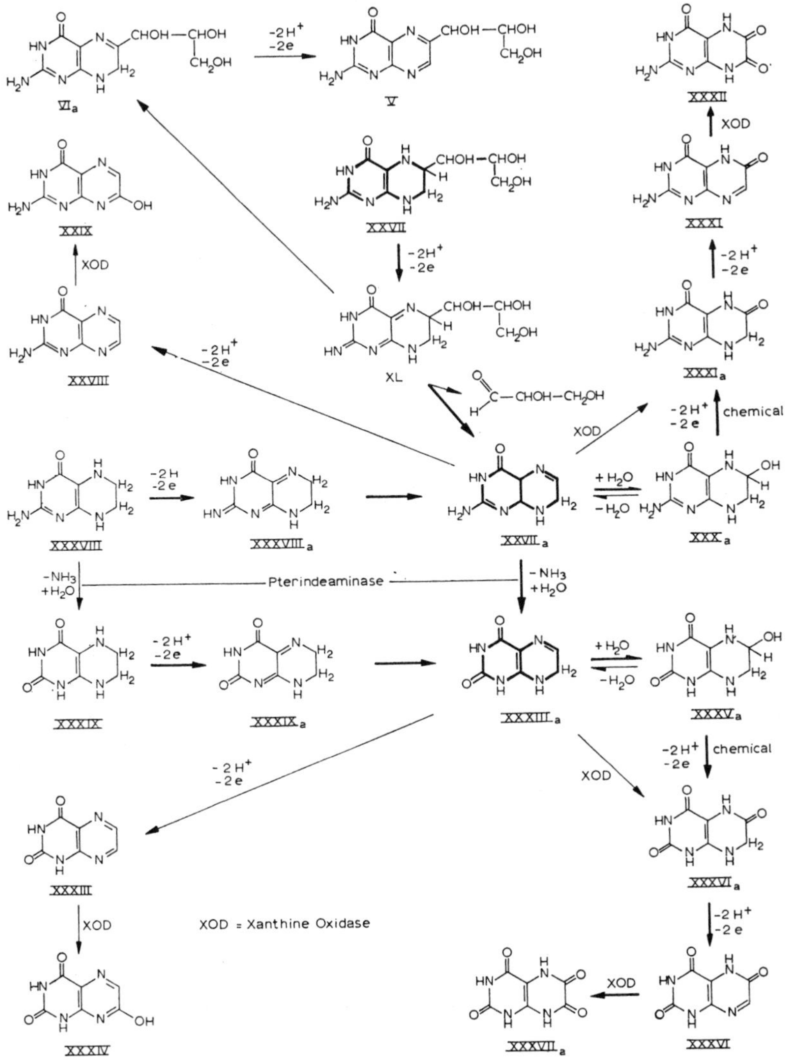

Fig. 6. Proposed catabolic pathway for hydroxypteridines. Permission granted by H. Rembold[203].

References p. 147

dicated above could be oxidized by xanthine oxidase (EC 1.2.3.2) to carboxy-
pterin[206,212] and later decarboxylated by ultraviolet light to pterin[163,206].
In the manner described above, pterin and 7,8-dihydropterin are available as
substrates for xanthine oxidase for the production of hydroxypterins.

The ubiquitousness of xanthine oxidase is indicated by its presence in insects[213-215], milk[213,216,217] and liver[203,216]. The similarity of the heterocyclic systems of purines and pteridines suggested that xanthine oxidase
could attack pteridines. In effect, the enzyme was found to catalyze the insertion of an oxygen function into position 7 of pterin [XXVIII] and 6-hy-
droxypterin [XXXI] resulting in 7-hydroxypterin[206,211,213,215,218,219]
[XXIX] and 6,7-dihydroxypterin[215,218-220] [XXXII], respectively. The enzymatic insertion of an oxygen function to position 6 of pterin [XXVIII] to
yield 6-hydroxypterin [XXXI] has not been demonstrated. These results are
summarized in Fig. 6.

Bergmann and Kwietny[217,221], who studied the mode of action of xanthine
oxidase with 2-, 4- and 7-monohydroxypteridines and 2,4-dihydroxypteridine
(lumazine, [XXXIII]), found that each of the pteridines studied eventually was
oxidized to 2,4,7-trihydroxypteridine, analogous to the oxidation of 2-
amino-4-hydroxypteridine to 2-amino-4,7-dihydroxypteridine (isoxanthopterin). Furthermore, the oxidation of 2,4,6-trihydroxypteridine to 2,4,6,7-
tetrahydroxypteridine is analogous to the insertion of an oxygen function at
position 7 of xanthopterin.

Recently, Rembold and Gutensohn[216] reported a significant finding bearing
on 6-hydroxypterin formations. When tetrahydrobiopterin, tetrahydrofolic
acid or tetrahydropterin was incubated with xanthine oxidase from rat liver,
6-hydroxypterin was found. They believed that the tetrahydropteridine derivatives were degraded non-enzymatically to 7,8-dihydropterin which in turn
was oxidized enzymatically to xanthopterin. With a rat-liver homogenate,
tetrahydrobiopterin was catabolyzed to 6-hydroxylumazine. This reaction
required the cleavage of the side-chain, a deamination of the amino function
in position 2 and an insertion of an oxygen function in position 6 of the
pteridine. From these studies, Rembold and Gutensohn[216] proposed that the
hydrated form of a $C = N$ double bond is required for the dehydrogenation by
xanthine oxidase. Accordingly, with all oxidized pteridines, the 7,8-double
bond must become hydrated faster than the 5,6-double bond for dehydrogenation to 7-hydroxypterin (Fig. 6).

In the case of the 6-hydroxylumazine [XXXVI] or 6-hydroxypterin[XXXI]
formation, a hydration of the 5,6-double bond [XXX$_a$] or [XXXV$_a$] is requir-

ed by xanthine oxidase. A pteridine structure fulfilling this requirement would be the 7,8-dihydropteridine [XXVIII$_a$] or [XXXIII$_a$]. This proposal is illustrated in Fig. 6.

From rat-liver homogenates, Rembold[203] described two enzymes involved in the catabolism of tetrahydroneopterin. One was a specific pterin deaminase (EC 3.5.4.11) which replaces the amino group with an oxygen function in position 2 and xanthine oxidase which inserts an oxygen function in positions 6 and 7. In Fig. 6, Rembold summarizes the non-enzymatic and enzymatic steps from tetrahydroneopterin [XXVII] and tetrahydropterin [XXXVIII] to dihydropterin [XXVIII$_a$], pterin [XXVIII], quinoid forms of dihydropterin [XXXVIII$_a$] and dihydrolumazine [XXXIX$_a$], 7-hydroxypterin [XXIX], 6-hydroxypterin [XXXI], 6,7-dihydroxypterin [XXXII], dihydrolumazine [XXXIII$_a$], 6-hydroxylumazine [XXXVI], 6,7-dihydroxylumazine [XXXVII] and lumazine [XXXIII] and 7-hydroxylumazine [XXXIV].

5. Coenzyme function

(a) Hydroxylation

The biochemical functions of folate coenzymes have been thoroughly reviewed elsewhere[12,222-225] and accordingly will not be discussed here. The cofactor role of unconjugated pteridines so widely anticipated to occur in a variety of biochemical reactions is well documented for only the hydroxylation reactions. The reactions involving the tetrahydropteridine dependent hydroxylation have been reviewed by Kaufman[2,226] and will be summarized only briefly. Kaufman[227] discovered in 1958 that a cofactor in rat liver, which functioned in the enzymatic hydroxylation of phenylalanine to tyrosine, had certain chemical properties suggestive of a compound with a pteridine ring. In that communication[227], the natural cofactor could be replaced by tetrahydrofolate. Kinetic data indicated, however, that tetrahydrofolate did not function catalytically as did the natural cofactor. Other pteridines were also found to replace the natural cofactor[228,229] which was finally isolated from rat liver by Kaufman[172] and structurally identified as dihydrobiopterin. From these studies, the phenylalanine-hydroxylation process was formulated[2,174,226] to occur as depicted in Fig. 7. The rat-liver fraction, which contains phenylalanine hydroxylase (EC 1.14.3.1), catalyzes the tetrahydrobiopterin-dependent hydroxylation of phenylalanine to tyrosine. In this process, the tetrahydropteridine is oxidized to the quinoid form of dihydropteridine. The quinoid form of dihydropteridine is reduced back to the tetrahydro form by dihydropteridine reductase (sheep-liver enzyme) using NADPH as coenzyme. In order for 7,8-dihydrobiopterin to enter the reaction, it must first be reduced to tetrahydrobiopterin by dihydrofolate reductase (EC 1.5.1.4)[172,226].

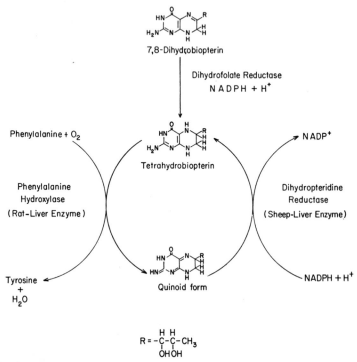

Fig. 7. The scheme for the process of the hydroxylation of phenylalanine[2,174,226].

Other tetrahydropteridine-dependent oxygenase reactions are tryptophan hydroxylation[230,231], tyrosine hydroxylation[232,233], cinnamic acid hydroxylation[234] and the oxidation of long-chain alkyl ethers of glycerol to fatty acids and glycerol[235].

The sparing effect exerted on the neopterin requirement of *C. fasciculata* by a saponified lipid fraction of *Crithidia* cells, and by a mixture of oleic acid and linoleic acid was observed by Kidder and Dewey[236]. In a subsequent report, Dewey and Kidder[237] extended this observation to include the sparing effect of arachidonic acid and linoleic acid. These results suggested to the authors the possible involvement of unconjugated pteridines in the production of unsaturated fatty acids from saturated fatty acids, and the omega oxidation of long-chain saturated fatty acids. Pteridines may also be involved in the dehydrogenation of the carotenoid pigments phytoene to lycopene[238]. This was indicated by the accumulation of saturated forms of carotenoids in cells of *R. rubrum* by a pteridine inhibitor, 4-phenoxy-2,6-diaminopyridine.

(b) Possible involvement in the photosynthetic process

Evidence suggesting a role of pteridines in the photosynthetic process came from Fuller *et al.*[239] who found that the pteridine content of *Euglena* and *Rhodospirillum rubrum* increased 20-fold and 5-fold, respectively, when the condition of growth was changed from the absence to the presence of light. Similar findings were reported by Maclean *et al.*[240] who found an increased pteridine level in light-grown cells of photosynthetic bacteria. These findings, perhaps circumstantial, were followed by others, supporting a function of pteridines in photosynthesis.

A factor designated phosphodoxin, which was isolated from spinach chloroplasts by Black *et al.*[241], stimulated photosynthetic phosphorylation. The ultraviolet absorption spectrum of the factor suggested to Maclean *et al.*[242] its similarity to 6-substituted pteridines. Since Forrest *et al.*[181] had earlier reported that a blue-green alga, *A. nidulans*, contained a biopterin glucoside, Maclean *et al.*[242] concluded that pteridines may be involved in photosynthetic phosphorylation. A number of pteridines were tested and tetrahydrofolate, tetrahydrobiopterin glucoside and the synthetic pteridines 6,7-dimethyltetrahydropterin and 6,7-dimethyltetrahydrolumazine were found to be active in stimulating photosynthetic phosphorylation. Comparable activities were found for 6,7-dimethyldihydropterin and biopterin but low activities for folic acid and lumazine. Additionally, these workers isolated from spinach a pteridine-containing fraction and its mobility on paper chromatograms was alluded to be similar to the mobility of phosphodoxin.

In a subsequent work, Maclean *et al.*[243] reported the parallel stimulation of cytochrome *c* photooxidation and photophosphorylation by pteridines. The effect of pteridines on photophosphorylation, which was also shown to occur efficiently with riboflavin 5'-phosphate and phenazine methosulfate, and the effect of pteridines on cytochrome *c* photooxidation, which was most efficiently promoted by riboflavin 5'-phosphate, suggested to the authors the function of pteridines in mediating the electron flow from system 1 to molecular oxygen. The varying activities of the aromatic, the dihydro and the tetrahydro forms of the pteridines in photophosphorylation indicated that the active pteridines function as the quinoid form. These authors have also proposed that pteridines may be involved as intermediary phosphopteridines in the photophosphorylation process.

With respect to the possible electron-transport role of reduced pteridines in the photosynthetic process, Fuller and Nugent[244] recently reported the photoreduction of dihydrobiopterin by chromatophore preparations of *R. rubrum* or *Chromatium* strain D. Furthermore, they found that the synthetic compound, 6,7-dimethyltetrahydropterin reduced ferredoxin non-enzymatically and produced a spectral shift of the pteridine-free chromatophore or chloroplast preparation. This spectral shift, which is similar to that found with illuminated, unwashed chromatophore preparations, suggested to the authors that these pteridines interact with reaction-center chlorophyll or bacteriochlorophyll. In addition, the authors believed that since pteridines are known to have potentials of -0.7 V, their hypothetical role as primary photochemical electron acceptors for system 1 of higher plants and for the single-primary photochemical act of bacterial photosynthesis is supported. Fuller and Nugent envisaged the process as the capture of a single electron, produced by the excited state of chlorophyll, by a semiquinone form of the reduced pteridine.

ACKNOWLEDGEMENT

This work was supported in part by U.S. Public Health Service, Research Grants, CA-10399 from the National Cancer Institute.

ADDENDUM

After the submission of the proofs to the publisher, Dr. H. Rembold stated that the information concerning the conversion of D-*erythro*-tetrahydroneopterin to D-*erythro*-biopterin, requires further verification inasmuch as the report was preliminary in nature[165].

REFERENCES

1   W. PFLEIDERER, *Angew. Chem. (Intern. Ed.)*, 3 (1964) 114.
2   S. KAUFMAN, *Ann. Rev. Biochem.*, 36 (1967) 171.
3   *Ciba Foundation Symposium on Chemistry and Biology of Pteridines*, Little, Brown, Boston, 1954.
4   H. S. FORREST, in M. FLORKIN AND H. S. MASON (Eds.), *Comparative Biochemistry*, Vol. IV, Academic Press, New York, 1962, p. 615.
5   W. SHIVE, in M. FLORKIN AND E. H. STOTZ (Eds.), *Comprehensive Biochemistry*, Vol. 11, Elsevier, Amsterdam, 1963, p. 82.
6   W. PFLEIDERER AND E. C. TAYLOR (Eds.), *Pteridine Chemistry*, Macmillan, Londen, 1964.
7   G. W. E. PLAUT, *Ann. Rev. Biochem.*, 30 (1961) 409.
8   M. GATES, *Chem. Rev.*, 41 (1947) 63.
9   I. ZIEGLER AND R. HARMSEN, *Advan. Insect Physiol.*, 6 (1969) 139.
10 T. H. JUKES AND H. P. BROQUIST, in D. M. GREENBERG (Ed.), *Metabolic Pathways*, Vol. II, Academic Press, New York, 1961, p. 713.
11 E. L. R. STOKSTAD AND J. KOCH, *Physiol. Rev.*, 47 (1967) 83.
12 J. C. RABINOWITZ, in P. D. BOYER, H. LARDY AND K. MYRBÄCK (Eds.), *The Enzymes*, Vol. 2, Part A, Academic Press, New York, 1960, p. 185.
13 G. M. BROWN, *Physiol. Rev.*, 40 (1960) 331.
14 G. M. BROWN AND J. J. REYNOLDS, *Ann. Rev. Biochem.*, 32 (1963) 419.
15 F. WEYGAND AND M. WALDSCHMIDT, *Angew. Chem.*, 67 (1955) 328.
16 F. WEYGAND, H. J. SCHLIEP, H. SIMON AND G. DAHMS, *Angew. Chem.*, 71 (1959) 522.
17 O. BRENNER-HOLZACH AND F. LEUTHARDT, *Helv. Chim. Acta*, 42 (1959) 2254.
18 G. W. E. PLAUT AND J. J. BETHEIL, *Ann. Rev. Biochem.*, 25 (1956) 463.
19 G. W. E. PLAUT, *Nutrition Symposium Ser.*, No. 13 (1956) 20.
20 I. ZIEGLER-GÜNDER, H. SIMON AND A. WALKER, *Z. Naturforsch.*, 11b (1956) 82.
21 W. S. MCNUTT, *J. Biol. Chem.*, 210 (1954) 511.
22 W. S. MCNUTT, *J. Biol. Chem.*, 219 (1955) 365.
23 W. S. MCNUTT, *J. Am. Chem. Soc.*, 83 (1961) 2303.
24 E. VIEIRA AND E. SHAW, *J. Biol. Chem.*, 236 (1961) 2507.
25 F. WEYGAND, H. SIMON, G. DAHMS, M. WALDSCHMIDT, H. J. SCHLIEP AND H. WALKER, *Angew. Chem.*, 73 (1961) 402.
26 O. BRENNER-HOLZACH AND F. LEUTHARDT, *Helv. Chim. Acta*, 44 (1961) 1480.
27 C. L. KRUMDIECK, E. SHAW AND C. M. BAUGH, *J. Biol. Chem.*, 241 (1966) 383.
28 C. C. LEVY, *J. Biol. Chem.*, 239 (1964) 560.
29 J. J. REYNOLDS AND G. M. BROWN, *J. Biol. Chem.*, 237 (1962) P. C. 2713.
30 J. J. REYNOLDS AND G. M. BROWN, *J. Biol. Chem.*, 239 (1964) 317.
31 T. SHIOTA AND M. P. PALUMBO, *J. Biol. Chem.*, 240 (1965) 4449.
32 F. R. DALAL AND J. S. GOTS, *Biochem. Biophys. Res. Commun.*, 20 (1965) 509.
33 H. MITSUDA, Y. SUZUKI, K. TADERA AND F. KAWAI, *J. Vitaminol. (Kyoto)*, 12 (1966) 192.
34 H. MITSUDA AND Y. SUZUKI, *J. Vitaminol. (Kyoto)*, 14 (1968) 106.
35 G. GUROFF, *Biochim. Biophys. Acta*, 90 (1964) 623.
36 B. LEVENBERG, *Federation Proc.*, 24 (1965) 669.
37 B. LEVENBERG AND D. K. KACZMARECK, *Biochim. Biophys. Acta*, 117 (1966) 272.
38 B. LEVENBERG AND S. N. LINTON, *J. Biol. Chem.*, 241 (1966) 846.
39 A. W. BURG AND G. M. BROWN, *Biochim. Biophys. Acta*, 117 (1966) 275.
40 G. GUROFF AND C. A. STRENKOWSKI, *J. Biol. Chem.*, 241 (1966) 2220.
41 T. SHIOTA, M. P. PALUMBO AND L. TSAI, *J. Biol. Chem.*, 242 (1967) 1961.

42 B. SULADONICK AND M. E. SMULSON, *J. Biol. Chem.*, 242 (1967) 2872.
43 O. H. SMITH AND C. YANOFSKY, *J. Biol. Chem.*, 235 (1960) 2051.
44 D. W. E. SMITH AND B. N. AMES, *J. Biol. Chem.*, 239 (1964) 1848.
45 T. SHIOTA, C. M. BAUGH AND J. MYRICK, *Biochim. Biophys. Acta*, 192 (1969) 205.
46 G. M. BROWN, *Abstr., IVth International Congress on Pteridines*, Toba, Japan, 1969, p. 34.
47 G. W. KIDDER AND V. C. DEWEY, *Biochem. Biophys. Res. Commun.*, 5 (1961) 324.
48 V. C. DEWEY, G. W. KIDDER AND F. P. BUTLER, *Biochem. Biophys. Res. Commun.*, 1 (1959) 25.
49 C. M. BAUGH AND E. SHAW, *Biochem. Biophys. Res. Commun.*, 10 (1963) 28.
50 F. WEYGAND, A. WACKER, A. TREBST AND O. P. SWOBODA, *Z. Naturforsch.*, 11b (1956) 689.
51 A. W. BURG AND G. M. BROWN, *J. Biol. Chem.*, 243 (1968) 2349.
52 M. GOTO AND H. S. FORREST, *Biochem. Biophys. Res. Commun.*, 6 (1961) 180.
53 H. REMBOLD AND L. BUSCHMANN, *Chem. Ber.*, 96 (1963) 1406.
54 T. H. D. JONES AND G. M. BROWN, *J. Biol. Chem.*, 242 (1967) 3989.
55 T. SHIOTA, *Abstr., IVth International Congress on Pteridines*, Toba, Japan, 1969, p. 36.
56 W. PFLEIDERER AND H. ZONDLER, *Chem. Ber.*, 99 (1966) 3008.
57 M. (MATSUBARA) NAGAI, *Arch. Biochem. Biophys.*, 126 (1968) 426.
58 T. FUKUSHIMA AND M. AKINO, *Arch. Biochem. Biophys.*, 128 (1968) 1.
59 T. SHIOTA, *Arch. Biochem. Biophys.*, 80 (1959) 155.
60 G. M. BROWN, R. A. WEISMAN AND D. A. MOLNAR, *J. Biol. Chem.*, 236 (1961) 2534.
61 T. SHIOTA AND M. N. DISRAELY, *Biochim. Biophys. Acta*, 52 (1961) 467.
62 R. A. WEISMAN AND G. M. BROWN, *J. Biol. Chem.*, 239 (1964) 326.
63 C. L. KRUMDIECK, C. M. BAUGH AND E. SHAW, *Biochim. Biophys. Acta*, 90 (1964) 573.
64 T. H. D. JONES, J. J. REYNOLDS AND G. M. BROWN, *Biochem. Biophys. Res. Commun.*, 17 (1964) 486.
65 H. REMBOLD AND H. METZGER, *Chem. Ber.*, 96 (1963) 1395.
66 H. C. S. WOOD, T. ROWAN AND A. STUART, in W. PFLEIDERER AND E. C. TAYLOR (Eds.), *Pteridine Chemistry, Third International Symposium, Stuttgart, 1962*, Pergamon, Oxford, 1964, p. 129.
67 H. S. FORREST AND S. NAWA, in W. PFLEIDERER AND E. C. TAYLOR (Eds.), *Pteridine Chemistry, Third International Symposium, Stuttgart, 1962*, Pergamon, Oxford, 1964, p. 281.
68 T. OKADA AND M. GOTO, *J. Biochem. (Tokyo)*, 58 (1965) 458.
69 K. SUGIURA AND M. GOTO, *Biochem. Biophys. Res. Commun.*, 28 (1967) 687.
70 F. I. MACLEAN, H. S. FORREST AND J. MYERS, *Biochem. Biophys. Res. Commun.*, 18 (1965) 623.
71 N. KATSUNUMA, T. SHODA AND H. NODA, *J. Vitaminol. (Kyoto)*, 3 (1957) 77.
72 B. WATT, *J. Biol. Chem.*, 242 (1967) 565.
73 R. H. NIMMO-SMITH, J. LASCELLES AND D. D. WOODS, *Brit. J. Exptl. Pathol.*, 29 (1948) 264.
74 J. LASCELLES AND D. D. WOODS, *Brit. J. Exptl. Pathol.*, 33 (1952) 288.
75 F. KORTE, H. G. SCHICKE AND H. WEITKAMP, *Angew. Chem.*, 69 (1957) 96.
76 F. KORTE, H. U. ALDAG, H. BANNUSCHER, H. BARKEMEYER, G. LUDWIG, G. SYNNATSCHKE AND H. WEITKAMP, *Abstr., IVth International Congress of Biochemistry, Vienna, 1958*, Pergamon, Oxford, 1958, p. 93.
77 L. JAENICKE AND PH. C. CHAN, *Angew. Chem.*, 72 (1960) 752.
78 H. MITSUDA, Y. SUZUKI AND F. KAWAI, *J. Vitaminol. (Kyoto)*, 12 (1966) 205.
79 B. WOLF AND R. D. HOTCHKISS, *Biochemistry*, 2 (1963) 145.
80 P. J. ORTIZ AND R. D. HOTCHKISS, *Biochemistry*, 5 (1966) 67.

81 F. D. WILLIAMS AND B. W. KOFT, *Can. J. Microbiol.*, 12 (1966) 565.
82 L. P. JONES AND F. D. WILLIAMS, *Can. J. Microbiol.*, 14 (1968) 933.
83 H. MITSUDA, F. KAWAI AND Y. SUZUKI, *Bitamin*, 28 (1963) 453.
84 H. MITSUDA, Y. SUZUKI, K. TADERA AND F. KAWAI, *J. Vitaminol. (Kyoto)*, 11 (1965) 122.
85 K. IWAI, O. OKINAKA AND N. SUZUKI, *J. Vitaminol. (Kyoto)*, 14 (1968) 160.
86 K. IWAI AND O. OKINAKA, *J. Vitaminol. (Kyoto)*, 14 (1968) 170.
87 T. SHIOTA, M. N. DISRAELY AND M. P. MCCANN, *Biochem. Biophys. Res. Commun.*, 7 (1962) 194.
88 T. SHIOTA, M. N. DISRAELY AND M. P. MCCANN, *J. Biol. Chem.*, 239 (1964) 2259.
89 D. P. RICHEY AND G. M. BROWN, *J. Biol. Chem.*, 244 (1969) 1582.
90 H. MITSUDA AND Y. SUZUKI, *Biochem. Biophys. Res. Commun.*, 36 (1969) 1.
91 H. MITSUDA, Y. SUZUKI, F. KAWAI AND K. YASUMOTO, *Abstr., IVth International Congress on Pteridines*, Toba, Japan, 1969, p. 42.
92 T. SHIOTA, C. M. BAUGH, R. JACKSON AND R. A. DILLARD, *Biochemistry*, 8 (1969) 5022.
93 B. D. DAVIS, *J. Biol. Chem.*, 191 (1951) 315.
94 B. WEISS AND P. R. SRINIVASAN, *Proc. Natl. Acad. Sci. (U.S.)*, 45 (1959) 1941.
95 S. HENDLER AND P. R. SRINIVASAN, *Biochim. Biophys. Acta*, 141 (1967) 656.
96 K. IWAI, O. OKINAKA, M. IKEDA AND N. SUZUKI, *Abstr., IVth International Congress on Pteridines*, Toba, Japan, 1969, p. 40.
97 A. J. MEROLA AND B. W. KOFT, *Bacteriol. Proc.*, (1959) 122.
98 S. BOCCHIERI AND B. W. KOFT, *Bacteriol. Proc.*, (1965) 74.
99 M. J. GRIFFIN AND G. M. BROWN, *J. Biol. Chem.*, 239 (1964) 310.
100 J. J. PFIFFNER, D. G. CALKINS, E. S. BLOOM AND B. L. O'DELL, *J. Am. Chem. Soc.*, 68 (1946) 1392.
101 M. E. SCHERTEL, J. W. BOEHNE AND D. A. LIBBY, *J. Biol. Chem.*, 240 (1965) 3154.
102 V. M. DOCTOR AND J. R. COUCH, *J. Biol. Chem.*, 200 (1953) 223.
103 O. D. BIRD, V. M. MCGLOHAN AND J. W. VAITKUS, *Anal. Biochem.*, 12 (1965) 18.
104 J. M. NORONHA AND M. SILVERMAN, *J. Biol. Chem.*, 237 (1962) 3299.
105 K. IWAI, S. NAKAGAWA AND O. OKINAKA, *Mem. Res. Inst. Food Sci., Kyoto Univ.*, No. 19 (1959) 17.
106 E. USDIN, P. M. PHILLIPS AND G. TOENNIES, *J. Biol. Chem.*, 221 (1956) 865.
107 J. M. NORONHA AND V. S. ABOOBAKER, *Arch. Biochem. Biophys.*, 101 (1963) 445.
108 S. F. ZAKRZEWSKI AND C. A. NICHOL, *Federation Proc.*, 14 (1955) 311.
109 M. T. HAKALA AND A. D. WELCH, *J. Bacteriol.*, 73 (1957) 35.
110 B. E. WRIGHT, *J. Am. Chem. Soc.*, 77 (1955) 3930.
111 F. M. SIROTNAK, G. J. DONATI AND D. J. HUTCHISON, *J. Bacteriol.*, 85 (1963) 658.
112 R. C. WOOD AND M. F. WISE, *Texas Repts. Biol. Med.*, 23 (1965) 512.
113 B. E. WRIGHT, *J. Biol. Chem.*, 219 (1956) 873.
114 P. J. LARGE AND J. R. QUAYLE, *Biochem. J.*, 87 (1963) 386.
115 J. R. GUEST AND K. M. JONES, *Biochem. J.*, 75 (1960) 12p.
116 R. H. HIMES AND J. C. RABINOWITZ, *J. Biol. Chem.*, 237 (1962) 2903.
117 R. L. BLAKLEY, *Biochem. J.*, 65 (1957) 342.
118 J. R. GUEST, M. FOSTER AND D. D. WOODS, *Biochem. J.*, 92 (1964) 488.
119 J. R. GUEST, S. FRIEDMAN, M. A. FOSTER, G. TEJERINA AND D. D. WOODS, *Biochem. J.*, 92 (1964) 497.
120 S. TAKEYAMA, F. T. HATCH AND J. M. BUCHANAN, *J. Biol. Chem.*, 236 (1961) 476.
121 J. C. RABINOWITZ AND R. H. HIMES, *Federation Proc.*, 19 (1960) 963.
122 A. KAZENKO AND M. LASKOWSKI, *J. Biol. Chem.*, 173 (1948) 217.
123 W. DABROWSKA, A. KAZENKO AND M. LASKOWSKI, *Science*, 110 (1949) 95.

124 V. Mims, J. R. Totter and P. L. Day, *J. Biol. Chem.*, 155 (1944) 401.
125 O. D. Bird, S. B. Binkley, E. S. Bloom, A. D. Emmett and J. J. Pfiffner, *J. Biol. Chem.*, 157 (1945) 413.
126 V. Mims and M. Laskowski, *J. Biol. Chem.*, 160 (1945) 493.
127 M. Laskowski, V. Mims and P. L. Day, *J. Biol. Chem.*, 157 (1945) 731.
128 O. D. Bird, M. Robbins, J. M. Vandenbelt and J. J. Pfiffner, *J. Biol. Chem.*, 163 (1946) 649.
129 B. E. Volcani and P. Margalith, *J. Bacteriol.*, 74 (1957) 646.
130 A. G. Pratt, E. J. Crawford and M. Friedkin, *J. Biol. Chem.*, 243 (1968) 6367.
131 K. Iwai, *Mem. Res. Inst. Food Sci., Kyoto Univ.*, No. 13 (1957) 1.
132 V. G. Allfrey and C. G. King, *J. Biol. Chem.*, 182 (1950) 367.
133 A. Bolinder, E. Widoff and L. E. Ericson, *Arkiv Kemi*, 6 (1953) 487.
134 M. Silverman, L. W. Law and B. Kaufman, *J. Biol. Chem.*, 236 (1961) 2530.
135 J. B. Wittenberg, J. M. Noronha and M. Silverman, *Biochem. J.*, 85 (1962) 9.
136 C. C. Levy and P. Goldman, *J. Biol. Chem.*, 242 (1967) 2933.
137 V. Mims, M. E. Swendseid and O. D. Bird, *J. Biol. Chem.*, 170 (1947) 367.
138 M. Silverman and B. E. Wright, *J. Bacteriol.*, 72 (1956) 373.
139 A. Sreenivasan, A. E. Harper and C. A. Elvehjem, *J. Biol. Chem.*, 177 (1949) 117.
140 E. A. Felton and C. F. Niven Jr., *J. Bacteriol.*, 65 (1953) 482.
141 E. L. R. Stokstadt and J. Koch, *Physiol. Rev.*, 47 (1967) 83.
142 H. S. Forrest and H. K. Mitchell, *J. Am. Chem. Soc.*, 77 (1955) 4865.
143 H. P. Broquist and A. M. Albrecht, *Proc. Soc. Exptl. Biol. Med.*, 89 (1955) 178.
144 E. L. Patterson, M. H. von Saltza and E. L. R. Stokstadt, *J. Am. Chem. Soc.*, 78 (1956) 5871.
145 E. L. Patterson, R. Milstrey and E. L. R. Stokstadt, *J. Am. Chem. Soc.*, 76 (1956) 5868.
146 T. Hama and M. Obika, *Nature*, 187 (1960) 326.
147 J. L. Hubby and L. H. Throckmorton, *Proc. Natl. Acad. Sci. (U.S.)*, 46 (1960) 65.
148 J. T. Bagnara and M. Obika, *Comp. Biochem. Physiol.*, 15 (1965) 33.
149 S. Nawa, *Bull. Chem. Soc. Japan*, 33 (1960) 1555.
150 H. S. Forrest and S. Nawa, *Nature*, 196 (1962) 372.
151 E. Ortiz, L. H. Throckmorton and H. G. Williams-Ashman, *Nature*, 196 (1962) 595.
152 E. Ortiz, E. Bächli, D. Price and H. G. Williams-Ashman, *Physiol. Zool.*, 36 (1963) 97.
153 M. Viscontini, in W. Pfleiderer and E. C. Taylor (Eds.), *Pteridine Chemistry*, MacMillan, London, 1964, p. 267.
154 W. Pfleiderer, *Abstr., IVth International Congress on Pteridines*, Toba, Japan, 1969, p. 36.
155 A. M. Chakrabarty and S. C. Roy, *Biochem. J.*, 93 (1964) 144.
156 K. Kobayashi and H. S. Forrest, *Biochim. Biophys. Acta*, 141 (1967) 642.
157 G. H. Schmidt and M. Viscontini, *Helv. Chim. Acta*, 49 (1966) 344.
158 M. Viscontini and R. Bühler-Moor, *Helv. Chim. Acta*, 51 (1968) 1548.
159 G. Guroff and C. A. Rhoads, *J. Biol. Chem.*, 244 (1969) 142.
160 K. Iwai, M. Kobashi and H. Yokomizo, *Abstr., IVth International Congress on Pteridines*, Toba, Japan, 1969, p. 70.
161 G. W. Kidder and V. C. Dewey, *J. Biol. Chem.*, 243 (1968) 826.
162 O. Brenner-Holzach and F. Leuthardt, *Z. Physiol. Chem.*, 348 (1967) 605.
163 E. Shaw, C. M. Baugh and C. L. Krumdieck, *J. Biol. Chem.*, 241 (1966) 379.
164 G. W. Kidder, V. C. Dewey and H. Rembold, *Arch. Mikrobiol.*, 59 (1967) 180.

165 H. REMBOLD, *Abstr.*, *IVth International Congress on Pteridines*, Toba, Japan, 1969, Discussion and Personal Communications.
166 M. GOTO, T. OKADA AND H. S. FORREST, *J. Biochem. (Tokyo)*, 56 (1964) 379.
167 K. SUGIURA AND M. GOTO, *J. Biochem. (Tokyo)*, 64 (1968) 657.
168 T. TAIRA, *Nature*, 189 (1961) 231.
169 T. TAIRA, *Japan. J. Genet.*, 36 (1961) 244.
170 M. MATSUBARA, M. TSUSUE AND M. AKINO, *Nature*, 199 (1963) 908.
171 M. MATSUBARA AND M. AKINO, *Experientia*, 20 (1964) 574.
172 S. KAUFMAN, *Proc. Natl. Acad. Sci. (U.S.)*, 50 (1963) 1085.
173 M. MATSUBARA, S. KATOH, M. AKINO AND S. KAUFMAN, *Biochim. Biophys. Acta*, 122 (1966) 202.
174 S. KAUFMAN, *J. Biol. Chem.*, 239 (1964) 332.
175 S. KATOH, M. NAGAI, Y. NAGAI, T. FUKUSHIMA AND M. AKINO, *Abstr.*, *IVth International Congress on Pteridines*, Toba, Japan, 1969, p. 32.
176 M. TSUSUE, *Experientia*, 23 (1967) 116.
177 M. GOTO, M. KONISHI, K. SUGIURA AND M. TSUSUE, *Bull. Chem. Soc. Japan*, 39 (1966) 929.
178 R. TSCHESCHE AND A. GLASER, *Chem. Ber.*, 91 (1958) 2081.
179 T. HAMA, J. MATSUMOTO AND Y. MORI, *Proc. Japan. Acad.*, 36 (1960) 346.
180 K. OHTA AND M. GOTO, *J. Biochem. (Tokyo)*, 63 (1968) 127.
181 H. S. FORREST, C. VAN BAALEN AND J. MYERS, *Arch. Biochem. Biophys.*, 78 (1958) 95.
182 D. L. HATFIELD, C. VAN BAALEN AND H. S. FORREST, *Plant Physiol.*, 36 (1961) 240.
183 A. SUZUKI AND M. GOTO, *J. Biochem. (Tokyo)*, 63 (1968) 798.
184 M. GOTO, K. KOBAYASHI, H. SATO AND F. KORTE, *Ann. Chem.*, 689 (1965) 221.
185 Y. KANEKO, *Agr. Biol. Chem. (Tokyo)*, 29 (1965) 965.
186 Y. KANEKO AND M. SANADA, *J. Ferment. Technol.*, 47 (1969) 8.
187 H. S. FORREST AND H. K. MITCHELL, *J. Am. Chem. Soc.*, 76 (1954) 5656.
188 S. KATOH AND M. AKINO, *Experientia*, 22 (1966) 793.
189 H. S. FORREST AND H. K. MITCHELL, *J. Am. Chem. Soc.*, 77 (1955) 4865.
190 M. VISCONTINI, M. SCHOELLER, E. LOESER, P. KARRER AND E. HADORN, *Helv. Chim. Acta*, 38 (1955) 397.
191 H. REMBOLD AND L. BUSCHMANN, *Ann. Chem.*, 662 (1963) 1406.
192 H. REMBOLD AND L. BUSCHMANN, *Ann. Chem.*, 662 (1963) 72.
193 H. S. FORREST, C. VAN BAALEN AND J. MYERS, *Science*, 125 (1957) 699.
194 H. L. STACKHOUSE, *Comp. Biochem. Physiol.*, 17 (1966) 219.
195 H. WIELAND AND C. SCHÖPF, *Chem. Ber.*, 58 (1925) 2178.
196 W. KOSCHARA, *Z. Physiol. Chem.*, 240 (1936) 127.
197 H. WIELAND, H. METZGER, C. SCHÖPF AND M. BÜLOW, *Ann. Chem.*, 507 (1933) 226.
198 J. A. BLAIR, *Biochem. J.*, 68 (1958) 385.
199 F. G. HOPKINS, *Phil. Trans. Roy. Soc. (London)*, *Ser.*, B186 (1893) 661.
200 C. SCHÖPF AND E. BECKER, *Ann. Chem.*, 507 (1933) 266.
201 C. SCHÖPF AND H. WEILAND, *Chem. Ber.*, 59 (1926) 2067.
202 R. PURRMANN AND M. MASS, *Ann. Chem.*, 556 (1944) 186.
203 H. REMBOLD, *Abstr.*, *IVth International Congress on Pteridines*, Toba, Japan, 1969, p. 66.
204 E. L. R. STOKSTAD, D. FORDHAM AND A. DE GRUNIGEN, *J. Biol. Chem.*, 167 (1947) 877.
205 E. S. BLOOM, J. M. VANDENBELT, S. B. BINKLEY, B. L. O'DELL AND J. J. PFIFFNER, *Science*, 100 (1944) 295.
206 O. H. LOWRY, O. A. BESSEY AND E. S. CRAWFORD, *J. Biol. Chem.*, 180 (1949) 389.
207 B. L. HILLCOAT, P. F. NIXON AND R. L. BLAKLEY, *Anal. Biochem.*, 21 (1967) 178.

208   J. M. WHITELEY, J. DRAIS, J. KIRCHNER AND F. M. HUENNEKENS, *Arch. Biochem. Biophys.*, 126 (1968) 956.
209   S. FUTTERMAN AND M. SILVERMAN, *J. Biol. Chem.*, 224 (1957) 31.
210   S. FUTTERMAN, *J. Biol. Chem.*, 228 (1957) 1031.
211   T. TAIRA, *Nature*, 189 (1961) 237.
212   J. A. BLAIR, *Biochem. J.*, 65 (1957) 209.
213   H. S. FORREST, E. GLASSMAN AND H. K. MITCHELL, *Science*, 124 (1956) 725.
214   S. NAWA, B. SAKAGUCHI AND T. TAIRA, *Rept. Natl. Inst. Genet. (Misima)*, 7 (1957) 32.
215   Y. HAYASHI, *Biochim. Biophys. Acta*, 58 (1962) 351.
216   H. REMBOLD AND W. GUTENSOHN, *Biochem. Biophys. Res. Commun.*, 31 (1968) 837.
217   F. BERGMANN AND H. KWIETNY, *Biochim. Biophys. Acta*, 28 (1958) 613.
218   E. G. KREBS AND E. R. NORRIS, *Arch. Biochem.*, 24 (1949) 49.
219   H. M. KALCKAR AND H. KLENOW, *J. Biol. Chem.*, 172 (1948) 349.
220   H. WIELAND AND R. LIEBIG, *Ann. Chem.*, 555 (1944) 146.
221   F. BERGMANN AND H. KWIETNY, *Biochim. Biophys. Acta*, 33 (1959) 29.
222   M. FRIEDKIN, *Ann. Rev. Biochem.*, 32 (1963) 185.
223   J. M. BUCHANAN, A. R. LARRABEE, S. ROSENTHAL AND R. E. CATHOU, in W. PFLEIDERER AND E. C. TAYLOR (Eds.), *Pteridine Chemistry*, MacMillan, London, 1964, p. 343.
224   F. M. HUENNEKENS AND K. G. SCRIMGEOUR, in W. PFLEIDERER AND E. C. TAYLOR (Eds.), *Pteridine Chemistry*, MacMillan, London, 1964, p. 355.
225   S. H. MUDD AND G. L. CANTONI, in M. FLORKIN AND E. H. STOTZ (Eds.), *Comprehensive Biochemistry*, Vol. 15, Elsevier, Amsterdam, 1964, p. 1.
226   S. KAUFMAN, in W. PFLEIDERER AND E. C. TAYLOR (Eds.), *Pteridine Chemistry*, MacMillan, London, 1964, p. 307.
227   S. KAUFMAN, *Biochim. Biophys. Acta*, 27 (1958) 428.
228   S. KAUFMAN, *J. Biol. Chem.*, 234 (1959) 2677.
229   S. KAUFMAN AND B. LEVENBERG, *J. Biol. Chem.*, 234 (1959) 2683.
230   S. NAKAMURA, A. ICHIYAMA AND O. HAYAISHI, *Federation Proc.*, 24 (1965) 604.
231   W. LOVENBERG, R. J. LEVINE AND A. SJOERDSMA, *Biochem. Pharmacol.*, 14 (1965) 889.
232   A. R. BRENNEMAN AND S. KAUFMAN, *Biochem. Biophys. Res. Commun.*, 17 (1964) 177.
233   T. NAGATSU, M. LEVITT AND S. UDENFRIEND, *J. Biol. Chem.*, 239 (1964) 2910.
234   P. M. NAIR AND L. C. VINING, *Phytochemistry*, 4 (1965) 161.
235   A. TIETZ, M. LINDBERG AND E. P. KENNEDY, *J. Biol. Chem.*, 239 (1964) 4081.
236   G. W. KIDDER AND V. C. DEWEY, *Biochem. Biophys. Res. Commun.*, 12 (1963) 280.
237   V. C. DEWEY AND G. W. KIDDER, *Arch. Biochem. Biophys.*, 115 (1966) 401.
238   N. A. NUGENT AND R. C. FULLER, *Science*, 158 (1967) 922.
239   R. C. FULLER, N. NUGENT, G. W. KIDDER AND V. C. DEWEY, *Federation Proc.*, 25 (1966) 205.
240   F. I. MACLEAN, H. S. FORREST AND D. S. HOARE, *Arch. Biochem. Biophys.*, 117 (1966) 54.
241   C. C. BLACK, A. SAN PIETRO, D. LIMBACK AND G. NORRIS, *Proc. Natl. Acad. Sci. (U.S.)*, 50 (1963) 37.
242   F. I. MACLEAN, Y. FUJITA, H. S. FORREST AND J. MYERS, *Science*, 149 (1965) 636.
243   F. I. MACLEAN, Y. FUJITA, H. S. FORREST AND J. MYERS, *Plant Physiol.*, 41 (1966) 774.
244   R. C. FULLER AND N. A. NUGENT, *Proc. Natl. Acad. Sci. (U.S.)*, 63 (1969) 1311.

Chapter I

Metabolism of Water-Soluble Vitamins

Section g

The Metabolism of the Cobalamins

L. MERVYN

Biochemistry Department, Glaxo Research Ltd., Greenford, Middlesex (Great Britain)

1. Introduction

The chemistry of vitamin B_{12}* and its coenzymes** has been well reviewed in *Comprehensive Biochemistry*, and is not discussed here.

For purposes of clarity in the present review the term vitamin B_{12} refers to any compound or group of compounds with a biological action resembling that of cyanocobalamin, but the chemical constitution of which is unknown or not known for certain. Thus the forms in urine and faeces are called vitamin B_{12} and not cyanocobalamin. Where the structure is known the appropriate chemical name has been used.

Although most of the studies on the vitamin, isolated in 1948, have been on cyanocobalamin (Fig. 1), it is now believed that the naturally occurring vitamin does not exist in this form. Vitamin B_{12}, in coenzyme forms, acts in a variety of biochemical transformations[1], most of which relate to micro-organisms. Two coenzyme forms have been isolated from animals and micro-organisms[1]. One, coenzyme B_{12}, has a 5'-deoxyadenosyl group covalently

---

* See A. F. Wagner and K. Folkers, Vitamin B₁₂, Vol. 11, Chapter VIII, p. 103.
** See H. Weissbach, A. Peterkofsky and H. A. Barker, The Cobamide Coenzymes, Vol. 16, Chapter IV, p. 189.

linked to the cobalt; in the other, methylcobalamin, the cyano group in cyanocobalamin is replaced by methyl (Fig. 1).

The present review deals with the absorption, distribution and excretion of vitamin B_{12} and its coenzymes in ill or apparently healthy animals and man.

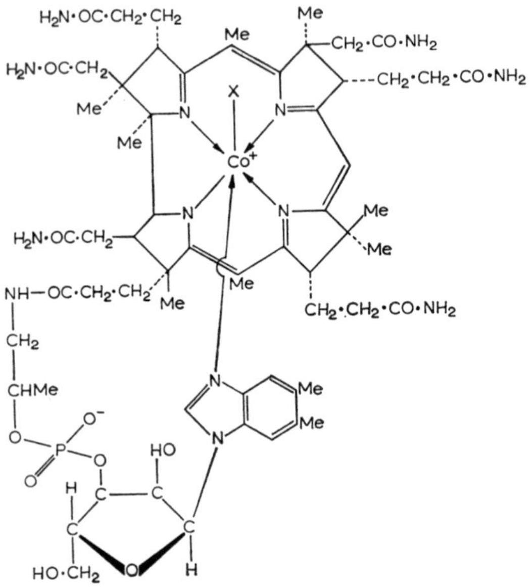

Fig. 1. Structure of the cobalamins. Cyanocobalamin, $X = CN$; hydroxocobalamin, $X = OH$; coenzyme B_{12}, $X = 5'$-deoxyadenosine; methylcobalamin, $X = CH_3$.

2. Sources and requirements

Vitamin B_{12} is biosynthesised exclusively by micro-organisms mainly as a coenzyme[2]. Its presence in higher plants can usually be traced to soil micro-organisms. In legumes coenzyme B_{12} is found in high concentrations in the root nodules where it is synthesised by bacteria. The vitamin is widely distributed in meats, fish and dairy products[2]. Carnivores and man obtain adequate amounts from their food. In many primates the serum B_{12} levels decrease considerably when the animals are captive[3]. This fall may reflect the change from a diet containing small insects and grubs to one that is purely vegetable.

In ruminants micro-organisms in the fore-stomach synthesise vitamin B_{12} for absorption from the gut. Usually deficiency in such animals results from lack of cobalt in the soil[4]. In some species, notably the rabbit, the site of synthesis is below that of absorption, but coprophagy prevents deficiency. The daily B_{12} requirements of various animal species have been reported[2]. On a body-weight basis man appears to need the least. The total B_{12} content of an adult human being is 2–5 mg, and this can decrease to 0.5 mg before signs of deficiency appear. The daily loss amounts to 0.1% of the body content[5], and a lapse of 6 months to 8 years is necessary before depletion produces megaloblastic changes. The normal daily intake of a healthy adult is 5–15 μg, and at least 5 μg are absorbed. The minimum daily intake required for health and normal haematopoiesis is probably 0.6–1.2 μg[5]. In growth and pregnancy this amount is probably insufficient.

In food vitamin B_{12}, probably as a coenzyme, is complexed with protein through peptide bonds[2], and these must be broken by cooking and proteolytic enzymes before absorption can occur. Recent results[5] suggest that vitamin B_{12} in food is no better absorbed than crystalline cyanocobalamin.

3. Uses of radioactive cobalamins in metabolic studies

In general metabolic studies on vitamin B_{12} have used cyanocobalamin labelled with radioactive cobalt. The minute amounts of the vitamin metabolised by tissues demands highly sensitive assays. The labelled vitamin is

TABLE I

PROPERTIES OF THE RADIOACTIVE ISOTOPES OF COBALT

Isotope	Half-life	Max. permissible human dose (μCi)	Relative dose delivered to liver	Main type of emission	Method of counting
Cobalt-57	270 days	200	1	γ	Well-scintillation counter
Cobalt-56	77 days	—	4	$\beta+,\gamma$	Geiger-Muller counter
Cobalt-58	71 days	30	2	$\beta+,\gamma$	Whole-body counter
Cobalt-60	5.26 years	10	29	$\beta-$	Well-scintillation counter

produced from radioactive cobalt in a fermentation medium. When incorporated in the molecule the cobalt does not exchange with other atoms. The properties and the methods of counting of the cobalt isotopes available are given in Table I.

Cobalt-56 is seldom used, whilst the hepato-toxicity of cobalt-60 limits its use to animal and *in vitro* experiments. The more widely used radioactive isotopes are cobalt-57 and -58. Both are weak emitters and are available at high specific activity (up to 300 μCi/μg). Such potency is essential for studies on plasma uptake in animals because the dilution factors are high. In addition cobalt-58 can be counted efficiently in all types of counters, but its short half-life limits its use to short-term studies. Cobalt-57 is not limited thus, but it must be counted in a well-scintillation counter.

Radioactive cyanocobalamin and hydroxocobalamin are available commercially. The degree of self-radiolysis of hydroxocobalamin is greater than that of cyanocobalamin, and material of high specific activity cannot be produced. Radioactive coenzyme B_{12} and methylcobalamin can be synthesised from labelled hydroxocobalamin[6].

Radioactive cobalamins find wide application in the diagnosis of malabsorption in man, and in absorption experiments on animals. For diagnosis an oral dose of 0.5–2.0 μCi of the labelled compound is administered and the degree of absorption is measured in various ways critically reviewed elsewhere[7]. Briefly they are:

(i) Urinary excretion test (Schilling)

This is the most widely used diagnostic test for malabsorption. After oral administration of 1 μCi of the labelled cobalamin, a "flushing" dose of 1000 μg of non-radioactive material is injected intramuscularly. Urine is collected for 24 h and the whole specimen is counted in a large well-scintillation counter. Healthy adults excrete 10% or more of the administered radioactive dose, but subjects with malabsorption excrete 5% or less.

(ii) Whole-body counting

This is generally acknowledged to be the most meaningful method. Shortly after administration of an oral dose of radioactive cobalamin (cobalt-58) a whole-body count is made. One week later the count is repeated and the difference gives a direct measure of absorption. The introduction of simpler, cheaper instruments[7] is popularising this technique.

(iii) Faecal excretion test

The difference between the radioactivity in the faeces and that administered is the amount retained. Normal adults absorb at least 50% of a 1 μCi dose, but pernicious anaemia patients absorb less than 30.

(iv) Hepatic uptake test

A dose of radioactive cyanocobalamin is given orally and the radioactivity accumulated in the liver is measured by surface-scintillation counting. A known amount of tracer vitamin is then injected and a second count made over the liver. The amount absorbed from the oral dose can then be calculated from the difference between the two counts.

(v) Plasma radioactivity test

Maximum radioactivity in the plasma occurs 8–12 h after an oral dose, and in a recent study[7] the peak levels in normal individuals were 0.9–10.4 $\mu\mu$g/ml, whereas pernicious anaemia patients with malabsorption had levels of 0–1.5 $\mu\mu$g/ml. A "flushing" intramuscular dose of 1000 μg of "cold" B_{12} removed the slight overlap between the two groups.

If malabsorption is indicated by any of the above tests they are repeated with the addition of intrinsic factor. If absorption is not then enhanced defective intestinal absorption is indicated.

4. Absorption of cobalamins

Active and passive mechanisms exist for the absorption of vitamin B_{12} from the gut. An active mechanism was first conceived 40 years ago by Castle, who suggested that an extrinsic factor in food combined with an intrinsic factor in the stomach to give a haemapoietic principle that was absorbed from the gut. Vitamin B_{12} is the extrinsic factor. Recent studies[8] have shown that the active mechanism dependent on intrinsic factor operates primarily when the amount of vitamin administered approximates that of a normal diet. The passive mechanism is probably one of simple diffusion becoming important with milligram quantities of the vitamin. Using whole-body counting procedures it has been shown that in man a maximum of 1.5 μg is absorbed by the intrinsic factor mechanism: 0.9% of the remainder is absorbed by simple diffusion[8].

The intrinsic factor has been obtained recently in a purified state from human gastric juice and hog pyloric mucosa. Reviews on intrinsic factor have been published by Glass[9] (1963) and Gräsbeck[10] (1968). Intrinsic factor is the first of many proteins to which B_{12} becomes attached during its metabolism.

By immunological and autoradiographic techniques, it has been shown that intrinsic factor is localised in the parietal cells of the fundus and body of the stomach in man, guinea pig, cat, rabbit and rhesus monkey. In the rat and pig it occurs in the chief pepsinogen-secreting cells of the pylorus and duodenum. The amount secreted is one hundred-fold that needed for absorption of the vitamin in the diet. Secretion of intrinsic factor is stimulated by food, histamine and gastrin. Guinea-pig intestine is particularly useful for studying the capacity of intrinsic factor to stimulate the *in vitro* uptake of radioactive cobalamins since it responds to the factor from other species[9].

Hog and human intrinsic factors are similar glycoproteins. Both have a molecular weight of around 60 000 but differ in carbohydrate content. In the presence of vitamin B_{12}, intrinsic factor probably dimerises to a complex containing 2 moles B_{12}. Studies with radioactive B_{12} have shown that 1 mg of the purified factor binds 25 μg of the vitamin. Intrinsic factor has a low affinity for cobalamin analogues with a base other than 5,6-dimethyl-benziminazole. Coenzyme B_{12} is bound despite the 5'-deoxyadenosyl group on the cobalt. These findings and immunological considerations suggest that the nucleotide of the cobalamin faces inward in the complex. The complex is smaller and more nearly spherical than intrinsic factor, suggesting that the latter accommodates the vitamin in a hole or pit[10]. When complexed together vitamin B_{12} and intrinsic factor afford mutual protection in the gut. When attached to intrinsic factor, B_{12} is not available to micro-organisms while the protein moiety is much more resistant to intestinal proteolysis than is the free factor[11]. The dog appears to be unique in respect of intrinsic factor: there is none in its stomach and intestinal absorption of vitamin B_{12} does not depend on intrinsic factor[12].

Experiments on patients with intestinal resection have indicated that the ileum is the main absorption site for the complex in the normal intestine[13]. Much confirmatory evidence is available for man and many other species[14]. In the rat the absorption site is more proximal. In the dog the ileum was still the best site of absorption even when transposed to a location nearer the stomach. Specific ileal receptors for the complex have been suggested on direct and indirect evidence. Rat gastric juice enhanced vitamin B_{12} absorption when given concomitantly but not if given much earlier. Preincubation of rat intestine with rat gastric juice confirmed that intrinsic factor alone can saturate the active sites making them unavailable to the complex. Intrinsic factor activity has been demonstrated in the supernate and sediment of a saline extract of intestine from guinea-pigs given labelled B_{12} bound by gastric juice[15].

Recent studies on the hamster have demonstrated that attachment of radioactive B_{12} to the brush borders and microvilli membranes of the ileum is through intrinsic factor[16]. The attachment was facilitated by divalent cations, particularly calcium, and resulted from adsorption rather than from an energy-dependent enzymatic reaction. A pH above 5.6 was essential for the binding. Species specificity of the intrinsic factor mechanism is probably resolved at this site. In the hamster intestine only intrinsic factor from the hamster, rat or rabbit promoted [57Co]cyanocobalamin uptake: human, hog and dog preparations were without effect. Crude preparations of a macromolecular factor that binds human intrinsic factor have been extracted from guinea-pig ileum[17].

The manner in which vitamin B_{12}, attached to the brush borders of the ileum, enters the mucosal cells is not known. The main problem is whether the vitamin disengages from the complex before absorption. Evidence suggesting that the complex itself is absorbed comes indirectly from studies on patients in whom malabsorption of cyanocobalamin is caused by failure to transport protein across cell membranes of villi and renal tubules[20]. In a study on patients given radioactive cyanocobalamin before resection of the ileum, Cooper[18] found that the excised mucosa contained intrinsic factor but its immunological properties were modified, probably by degradation. Others believe that intrinsic factor is not absorbed. Hines and Rosenberg[19] studied cyanocobalamin absorption by proximal and distal loops of guinea-pig intestine by incubating them first with [60Co]cyanocobalamin plus intrinsic factor, then with [57Co]cyanocobalamin alone. Absorption in the distal loop is known to depend on intrinsic factor but the uptakes of the two isotopes were similar. Adding an antibody that blocked complex formation abolished the uptake of the [57Co]cyanocobalamin in the distal loop, indicating that intrinsic factor was not absorbed in the distal mucosa but remained available for binding additional cyanocobalamin.

Although our knowledge of the absorption mechanism is incomplete, it is certain that a dose of radioactive cyanocobalamin is quickly removed by the intestinal mucosa, and after 2–3 h radioactivity appears in the portal blood and accumulates in the liver.

The relative efficiency of oral absorption of coenzyme B_{12} and cyanocobalamin varies with the species. In the rat, both forms of cobalamin were equally well absorbed when determined by faecal excretion of radioactivity[21]. The absorption of coenzyme B_{12} depended on intrinsic factor. When the compounds were administered together there was no indication of preferential absorption. In man studies on the hepatic uptake of coenzyme B_{12} labelled with cobalt-58 showed that the intestinal absorption of the compound depended on intrinsic factor, but the efficiency of absorption was less than that of cyanocobalamin[22]. Surface counting of the intestine indicated that radioactivity was retained longer in the gut after coenzyme administration, suggesting that this compound formed a stronger and easier attachment with the intestinal mucosa before absorption. These findings were confirmed and extended by Herbert and Sullivan[23] who measured the uptake of [^{60}Co]-coenzyme B_{12} in sick or pernicious anaemia patients by the urinary excretion method and by Heinrich and Gabbe[24] using whole-body counter techniques. Although coenzyme B_{12} was absorbed more slowly than cyanocobalamin from the gut its renal excretion was slower. *In vitro* studies on guinea-pig intestine homogenate also indicated that coenzyme B_{12} was less absorbed than cyanocobalamin at equimolar concentration. The degree of hydroxocobalamin absorption was intermediate. Sequential incubation of guinea-pig intestine with cyanocobalamin, hydroxocobalamin and coenzyme B_{12} in the presence or absence of intrinsic factor confirmed the preferential uptake of the intrinsic factor cyanocobalamin complex[25].

5. Malabsorption of cobalamins

Malabsorption may result from lack of secretion of intrinsic factor or failure to absorb the vitamin complex. The classic example of lack of intrinsic factor is Addisonian pernicious anaemia, in which disease gastric atrophy and loss of parietal cells decreases intrinsic factor secretion. Secretion of intrinsic factor may be suppressed by intrinsic factor antibody but its importance in the disease process is not certain. The antibody, (Type I or blocking antibody), which may be produced experimentally in the rabbit, is present in more than 50% of pernicious anaemia patients[26]. It combines with intrinsic factor preventing formation of the vitamin complex. The antibody is in the serum and gastric juice of infants born of mothers with pernicious anaemia[27]. Usually intrinsic factor appears in the gastric juice of infants in the first few days of life, but when antibody is present there is usually a delay of 1–3 months. Suppression of antibody formation by steroids may account for the improvement in vitamin B_{12} absorption in some pernicious anaemia patients on these drugs. Congenital pernicious anaemia is a rare condition: the only abnormality in an otherwise normal stomach is lack of intrinsic factor. Total gastrectomy eliminates the only source of intrinsic factor and results in complete loss of absorption of the vitamin. Partial gastrectomy may leave some functional parietal cells, but if these atrophy secretion of intrinsic factor ceases.

When malabsorption of vitamin B_{12} stems from defective gastric secretion of intrinsic factor, the latter may be administered with vitamin. Unfortunately patients often develop resistance to the intrinsic factor, which is not usually of human origin. When intrinsic factor does not achieve improvement, albeit temporary, malabsorption indicates defective absorption from the intestine.

A second type of antibody, (Type II or precipitating antibody), has been found in pernicious anaemia patients, but it is more rare than Type I. It binds the vitamin B_{12} intrinsic factor complex so preventing uptake by the villi of the ileum. Several methods are available for detecting these antibodies. One uses homogenate of guinea-pig intestinal mucosa, which takes up B_{12} intrinsic factor complex but not the free vitamin or the complex of vitamin B_{12} intrinsic factor and Type II antibody[29]. Differentiation of the two antibodies depends on the order of addition of the reactants. Type I antibody inhibits the uptake of [57Co]cyanocobalamin by the homogenate only when added first, whereas Type II antibody inhibits whatever the order of addition. An antibody to the distal microvillous membranes of the hamster has been induced in rabbits[30]. The partially purified antibody competes with the vitamin B_{12} intrinsic factor complex for the specific binding site of the hamster intestine.

Any generalised failure of intestinal function can prevent absorption of the vitamin B_{12} intrinsic factor complex: such failures occur in idiopathic steatorrhea, coeliac disease and tropical sprue. Lesions of the intestinal wall in regional enteritis and intestinal tuberculosis also prevent uptake as does the absence of the vitamin B_{12} absorbing mucosa of the ileum in ileitis or after resection. The malabsorption of vitamin B_{12} encountered in certain anatomical abnormalities (small intestine diverticulitis or blind-loop syndrome) is associated with bacterial growth in stagnant intestinal contents: because it is usually rectified by antibiotic treatment it is assumed that the vitamin is assimilated by the micro-organisms and is unavailable to the host. In people harbouring fish tapeworms vitamin B_{12} absorption is defective because the parasite absorbs the vitamin, produces a factor that splits the vitamin intrinsic factor complex, and causes gastric atrophy.

The use of labelled vitamin B_{12} in studies of the lesions described above has been reviewed[7].

6. Absorption of vitamin B_{12} by simple diffusion

The improvement in absorption of a single fixed dose of vitamin B_{12} by intrinsic factor is dose-dependent within certain limits. Normal subjects absorb 50–90% of an oral dose of 1 μg of the vitamin but only 25–30% of a 10-μg dose. With a 50-μg dose only 5% is absorbed. Thus the maximum amount absorbed from a single dose by the intrinsic factor mechanism is about 2 μg. Limitation is imposed not by the amount of intrinsic factor in the stomach but the capacity of the ileum to absorb vitamin B_{12}.

Many studies have revealed a passive diffusion mechanism for vitamin B_{12}. These were on pernicious anaemia patients given intrinsic factor by mouth: some had become resistant. Heinrich[8,31], using whole-body counting after oral administration of radioactive cyanocobalamin, concluded that a maximum of 0.9% was absorbed by diffusion mechanisms. This value applied for doses up to 1000 μg but thereafter it decreased: with a massive dose of 100 000 μg only 0.446% was absorbed. It was claimed that the amount retained (446 μg) sufficed a pernicious anaemia patient for 6 months. Berlin et al.[32] using radioactive vitamin measured cyanocobalamin absorption from high doses by the urinary excretion method. They concluded that 1.2% of an oral dose was absorbed by passive diffusion within a very wide dose range. The discrepancy in the results from the two schools probably reflects differences in methodology. The degree of direct uptake did not depend on the subject's ability to produce intrinsic factor: patients resistant to intrinsic factor or having other forms of disturbed B_{12} absorption also absorbed 1.2%. Nothing is known about the site of this type of absorption. What is certain is that a daily oral dose of 500–1000 μg of vitamin B_{12} is sufficient for treating patients with pernicious anaemia or other states of B_{12} deficiency. Faecal excretion and hepatic uptake studies on men dosed concomitantly with [57]Co-labelled cyanocobalamin and ^{60}Co-labelled hydroxocobalamin indicated that both analogues, although dose-dependent, were absorbed to a similar degree[33]. Similar findings were reported for the rat.

The whole concept of intrinsic factor has been challenged by Mooney and Heathcote[34]. They refute its existence, believe that vitamin B_{12} is absorbed as a complex with peptides, and suggest that the primary cause of pernicious anaemia is a lack of a gastric endopeptidase called cobalamase. This is said to initiate digestion of the protein attached to the vitamin in food, giving an absorbable vitamin B_{12}–peptide complex. A B_{12}–peptide complex, made by fermentation, has been tested on pernicious anaemia patients. 50 cases were given an average daily dose equivalent to 19 μg vitamin B_{12}, and complete remissions were claimed[34].

7. Vitamin B_{12} binders other than intrinsic factor

Many proteins in body fluids can bind vitamin B_{12}, but only intrinsic factor assists absorption from the gut. Three vitamin B_{12} binders, complexed with added radioactive cyanocobalamin, have been isolated from normal human gastric juice. They are termed binders S (slow), I (intermediate) and R (rapid), according to their electrophoretic mobilities. S has been characterised as intrinsic factor, and I, which has a lower molecular weight, also possesses intrinsic factor activity, and is believed to be a breakdown product of S[35]. Neither binder was detected in the gastric juice of pernicious anaemia patients. R was present, but had no intrinsic factor activity. Binding proteins similar to S and R have been isolated by gel filtration from the gastric mucosa of mice[36] and chimpanzees[37]. Immunological studies on the binders from man and chimpanzee showed that they behaved similarly as antigens.

R appears to be ubiquitous in body fluids. R-type glycoproteins have been isolated from saliva, bile, serum and cerebrospinal fluid: all were immunologically identical with that from gastric juice. With one exception all R-type binders have molecular weights of about 120 000: that from saliva is 60 000 (ref. 10). The differences among those of similar molecular weight derive from their sialic acid content. In serum the R-type binder, transcobalamin I, carries endogenous vitamin B_{12}. Transport of the vitamin in blood is described later. Although leucocytes and erythrocytes contain a single B_{12}-binding fraction similar to the body fluid binder R, a leucocyte binds some 2300 times more B_{12} than does a red cell. It has been suggested that the widespread binder R is a transport protein for vitamin B_{12} facilitating uptake by tissue cells[38].

8. Vitamin B_{12} transport in plasma

After absorption from the gut, vitamin B_{12} appears in the portal blood attached to specific proteins. There are at least two types, distinguished first by electrophoretic mobility. The one to which B_{12} is mainly bound is an α_1-globulin of the seromucoid fraction: the second migrates electrophoretically as a β-globulin. The two binders have been studied by Hall and Finkler[39-42], who named them transcobalamin I (TC I) and transcobalamin II (TC II) respectively.

To study vitamin B_{12} binding by the two proteins, normal subjects were injected intravenously with radioactive cyanocobalamin. DEAE-cellulose columns were used to fractionate the two binders, TC II being eluted before the main albumen fraction and TC I after, and the two fractions were counted. In experiments simulating physiological intake, labelled vitamin B_{12} was given orally, and blood samples were fractionated and the distribution between the two binders in the plasma was measured. Shortly after intravenous or oral administration more B_{12} was bound to TC II than to TC I, but after 24 h in the body the vitamin had disappeared from TC II. When TC II was isolated from normal plasma, complexed with cobalt-57 vitamin B_{12} and injected intravenously, a similar curve relating time with blood-complex concentration was obtained. In the experiments involving administration of labelled cyanocobalamin the vitamin became attached to TC I also but it disappeared from the plasma at a much slower rate.

In normal plasma TC I is almost saturated with B_{12} whereas TC II is unsaturated[42], thus the latter is available to bind exogenous B_{12} and carry it to the tissues. TC II acts primarily as a transport protein for the vitamin but TC I conserves the vitamin, releasing it sparingly when conditions dictate.

In disease the pattern changes. It has been known for some time that cyanocobalamin injected intravenously is cleared more slowly in pernicious anaemia and chronic myelogenous leukaemia (CML)[43]. In pernicious anaemia blood B_{12} levels are low and TC I is far from saturated: binding by TC II is deficient. Treatment with parenteral B_{12} saturates TC I and then the TC II binding capacity is restored[53]. In CML the capacity of TC I to bind cyanocobalamin is increased, whereas TC II is decreased in level, absent or non-functional. When CML patients were given radioactive cyanocobalamin orally or intravenously the vitamin was taken up by TC I only, a reversal of the normal pattern. Peak plasma levels were reached after 24 h (8 h in normal people) and the vitamin remained in the plasma much longer. In patients in remission the TC II binding capacity returned to normal[44]. Recently a binding protein, unknown in normal plasma, has been found in the plasma of patients with polycythaemia vera[45]. In molecular size and antibody potential the protein was similar to TC I, but it did not bind endogenous vitamin B_{12}.

Hall[76] has reviewed the plasma transport of vitamin B_{12} in man. The properties of TC I and TC II are summarised in Table II, their roles in disease have been reviewed by Herbert[46].

TABLE II

THE PROPERTIES OF TRANSCOBALAMINS[56]

Property	TC I	TC II
Mobility on paper at pH 8.6	α	β
Mobility on starch and geon at pH 4.5	anodal	cathodal
DEAE column	after albumen	before albumen
CMC column	eluted early	retained
Removal from plasma	slow	fast
Nature of protein	seromucoid	seromucoid
Transport function	carries endogenous vitamin B_{12}	carries exogenous vitamin B_{12} for a short period only
Molecular weight[105]	121 000	38 000
HeLa cell uptake	—	+
Reacts with:		
Anti-TC II	—	+
Anti-TC I	+	—
Anti-saliva R	+	—

9. Uptake of vitamin B_{12} by tissues

Protein binding is involved in the uptake of B_{12} by various organs and the catalytic effects of gastric juice and serum suggested that intrinsic factor might be involved[47,48], but present knowledge concerning serum TC I and TC II indicates that they are responsible for the catalytic effect. The effect of gastric juice derives probably from its content of R-type binders that resemble TC I[55], and it is likely that such binders contaminate the intrinsic factor preparations used in tissue uptake studies.

Uptakes by rat-liver homogenates and slices were greatly enhanced by serum, primarily by TC II[41]. Gastric juice was much less active than serum. TC II was found to be more effective than TC I in promoting the absorption of serum B_{12} by liver homogenates and the uptake by HeLa cells[49]. The

TC I-type binder in CML serum was even less efficient than that in normal serum.

Mature erythrocytes do not absorb significant amounts of vitamin B_{12}, but reticulocytes do. B_{12} uptake by these cells was studied by Retief et al.[50] using blood from patients with reticulocytosis. They postulated two distinct mechanisms, analogous to the two involved in intestinal absorption. In the first a glycoprotein carried the vitamin to specific receptors on the cell surface, and it required magnesium or calcium ions. It assumed great importance at physiological levels of B_{12}, but at high plasma vitamin B_{12} levels the vitamin was not bound and entered the reticulocyte by simple diffusion. TC II delivered more B_{12} to the reticulocytes than did TC I. Transfer of labelled cyanocobalamin to reticulocytes was much less in CML serum[51]. This did not result from excessive transfer of endogenous B_{12}, which is elevated in CML, but was associated with an α-globulin (TC I) binder unable to normally deliver the vitamin, hence the delayed clearance. Serum from pernicious anaemia patients, however, delivered slightly more exogenous cyanocobalamin to the red cells than did normal serum, probably because of less interference by endogenous B_{12}, which is low in pernicious anaemia. TC I and TC II in pernicious anaemia serum were as efficient in transferring B_{12} to reticulocytes as the binders from normal serum. TC II in CML serum was as efficient as that from pernicious anaemia or normal serum. Thus the α-globulin of CML was physiologically and chemically different from the α-globulins of normal pernicious anaemia serum[51].

There have been few studies on cobalamins other than cyanocobalamin. TC I and TC II bind hydroxocobalamin in vivo and in vitro[41,54]. In the rat, intravenously administered hydroxocobalamin and coenzyme B_{12} disappeared faster than cyanocobalamin from the blood and showed a stronger affinity for tissues[52]. Plasma more markedly enhanced the uptake of coenzyme B_{12} by erythrocytes than that of cyanocobalamin. The uptake of either cobalamin was facilitated more by B_{12}-deficient plasma than by normal plasma[57].

10. The distribution of vitamin B_{12} in tissues

Vitamin B_{12} is widely distributed in animal tissues where its concentrations vary considerably; in general microbiological assays have been used[58]. Usually the tissue is boiled with a cyanide buffer or proteolysed to liberate the vitamin as cyanocobalamin, which is then assayed.

In man, the principal storage organ for vitamin B$_{12}$ is the liver, which may contain several milligrams. Some recent values of vitamin B$_{12}$ concentration in human tissues are in Table III.

Liver concentration of B$_{12}$ appears to increase with the age of the subject. Lower liver values are encountered usually in diseases such as cirrhosis and hepatitis and in deficiency diseases like pernicious anaemia. In leukaemia patients levels are high[59].

TABLE III

VITAMIN B$_{12}$ CONCENTRATIONS IN HEALTHY HUMAN TISSUES[59]

Tissue	Number of samples	Average vitamin B$_{12}$ concentration $\pm$ S.D. ($\mu g/g$ wet tissue)
liver	6	0.653 $\pm$ 0.385
kidney	6	0.337 $\pm$ 0.338
adrenal	6	0.161 $\pm$ 0.051
heart	6	0.119 $\pm$ 0.041
pancreas	3	0.116 $\pm$ 0.100
spleen	5	0.055 $\pm$ 0.050
lung	3	0.050 $\pm$ 0.053
brain	2	0.027 $\pm$ 0.013

In the rat vitamin B$_{12}$ content of the kidney was higher than that of the liver[60]. The liver vitamin B$_{12}$ concentrations ($\mu g/g$ wet liver) of various animals were: cow 1.18, buffalo 1.30, sheep 1.33, goat 1.20, pig 0.59, rabbit 0.60, chicken 0.27, rat 0.052 and mouse 0.75 (ref. 61). The main storage organs in chicks are liver, kidney and spleen. When 3 μg radioactive B$_{12}$ was injected into eggs the dose was sufficient to promote growth in the hatched chicks for at least 12 weeks, when more than 30% of the radioactivity remained[2]. In fish the heart has a higher vitamin B$_{12}$ concentration than have liver and kidney: pollack heart contained 21 $\mu g/g$ dry weight[2].

Soon after the first isolation of a vitamin B$_{12}$ coenzyme from microorganisms (see Weissbach et al., Comprehensive Biochemistry, 16 (1965) 189), the main vitamin B$_{12}$ constituent of human, animal and avian liver was shown to be a vitamin B$_{12}$ coenzyme containing 5,6-dimethylbenziminazole as the nucleotide base (Fig. 1). The yield of coenzyme, determined spectroscopically, varied from 48 to 72% of the total B$_{12}$ present. The figures probably reflect the low extractibility of B_{12} derivatives when cyanide cannot be used: cyanide reacts with coenzyme B_{12} to produce cyanocobalamin. Additionally the high photo-sensitivity of coenzyme encourages degradation. In a recent study of the coenzyme content of animal tissues Yagiri[63] used a specific enzyme assay dependent on the ability of coenzyme B_{12} to act as an essential cofactor for dioldehydrase, which converts glycols to aldehydes. Within certain limits coenzyme B_{12} concentration was a function of the amount of aldehyde produced. From the results (Table IV) it is seen that coenzyme B_{12} was the main B_{12} constituent in the tissues studied.

TABLE IV

LIVER AND KIDNEY COENZYME CONCENTRATIONS IN VARIOUS ANIMALS[63]

Animal	Average liver concentration $\pm$ S.D. ($\mu g/g$ wet tissue)	Average kidney concentration $\pm$ S.D ($\mu g/g$ wet tissue)
normal rats	0.68 ± 0.35	1.88 ± 1.08
cyanocobalamin-saturated rats	0.62 ± 0.11	1.79 ± 1.28
cyanocobalamin-deficient rats	0.25	0.42
cyanocobalamin-deficient rats' offspring	0.15 ± 0.01	0.17 ± 0.03
pregnant rats (deficient)	0.19 ± 0.08	
offspring (not deficient)	0.88 ± 0.37	
human beings	1.23 ± 0.31	
guinea-pigs	0.69 ± 0.12	0.68 ± 0.28
rabbits	1.18 ± 0.25	1.65 ± 0.61
chick[62]	0.07	
lamb[62]	0.20	

Acute liver damage induced in rats by carbon tetrachloride released all the liver and kidney coenzyme B_{12} into the blood within 6–9 h, but after a further 12 h the coenzyme levels in these organs were normal[63].

Methylcobalamin has been isolated from calf liver[64] and detected in other tissues. In human liver it accounts for only 3% of the total B_{12}, whereas 70% is coenzyme B_{12}. In deficiency states such as pernicious anaemia the proportion of methylcobalamin in the depleted liver B_{12} increases[65]; the significance of this is unknown.

The distribution of radioactive cyanocobalamin within the liver cells of mice and rats has been studied[73]. The distribution of the vitamin within the cell varied with the time after injection suggesting a continual exchange of B_{12} among the cell components. Eventually more than 50% of the vitamin was in the mitochondrial fraction whereas the microsomal fraction contained less than 2%. The nuclear and supernatant fractions contained 16 and 15% respectively.

11. The distribution of vitamin B_{12} in serum

The results of the numerous investigations on serum vitamin B_{12} levels in health and disease have been reviewed[66]. Microbiological assays[58] have been used mostly but recently radioisotopic techniques have achieved some success[67]. Normal serum values are from 150 to 550 picograms/ml with an average of 300. Usually the serum level reflects the B_{12} status of the individual and is a convenient indicator of body resources. Serum levels below 100 pg/ml indicate B_{12} deficiency. Low levels are encountered in pernicious anaemia, megaloblastic anaemia, after gastrectomy, and in vegans. Very high levels are met in various liver diseases: in acute hepatitis they are 20–40 times normal. Liver damage by carbon tetrachloride in rats and dogs raised serum B_{12} by liberating the hepatic stores[68]. Such a liberation may account for the high serum B_{12} levels in liver disease. Very high serum levels have been recorded in leukaemias; they range from 6000 pg/ml in chronic granulocytic leukaemia to 1000 pg/ml in myelofibrosis[69]. The source of these elevated levels may be leucocytes dying in the blood stream[70].

The predominant form of serum B_{12} is methylcobalamin. It has not been isolated from serum but has been identified bioautographically with methylcobalamin isolated from liver and with the compound produced by partial chemical synthesis[71]. Coenzyme B_{12} has not been detected in normal plasma, despite the use of a specific enzyme assay. A study[72] of the distribution of plasma cobalamins in pernicious anaemia (low serum vitamin B_{12} levels) and in chronic myeloid leukaemia and acute hepatitis (extremely high levels) indicated that methylcobalamin predominates in the first two diseases but not in the third: overspill of coenzyme B_{12} from the damaged liver may account for this difference.

12. The turnover of exogenous cobalamins

(a) Tracer studies in rats

Early work on the metabolism of cyanocobalamin in rats has been reviewed[74]. Later studies compared the tissue distribution of cyanocobalamin, hydroxocobalamin and coenzyme B_{12} after intravenous[52,75], oral[83], and intramuscular and intraperitoneal administration[24]. Separate experiments on the three cobalamins injected intravenously into normal rats revealed a more rapid clearance of hydroxocobalamin and coenzyme from the blood and a higher uptake by liver and kidney. The differences between cyano-cobalamin and coenzyme increased with time suggesting conversion of cyanocobalamin to the coenzyme form. When cyanocobalamin and coenzyme were labelled with different cobalt isotopes, and given together, there was competition for uptake by the tissues. Plasma-clearance and tissue-uptake rates of normal and B_{12}-deficient rats given cyanocobalamin were similar. However, plasma clearance and uptake rates of hydroxocobalamin and coenzyme by the liver and kidneys were definitely higher in the deficient animals. In rats given large doses of cyanocobalamin before intravenous administration of the cyanocobalamin or coenzyme the disappearance of coenzyme was slow, presumably because the liver was saturated. This was confirmed by the reductions in hepatic uptake of both cobalamins. However, the renal uptake increased with time and was higher than normal, indicating a transfer of the cobalamins from the liver to the kidneys or, more likely, a greater renal uptake. In vitamin B_{12}-saturated animals uptake of coenzyme by the bone marrow was depressed also.

When [57Co]cyanocobalamin and [60Co]coenzyme B_{12} were injected together intravenously into rats with liver damage the blood levels of both cobalamins were higher than those obtained in normal rats. Liver uptake was delayed but renal uptake was normal. As the liver recovered uptake of both cobalamins was markedly higher than in normal rats, the coenzyme being preferentially absorbed[52].

Variation in route of administration (oral, intramuscular or intraperitoneal) did not affect the patterns of distribution of the cobalamins, but each had its own pattern[24]. About 50% more coenzyme than cyanocobalamin was absorbed from the gut. With all three cobalamins renal uptake was greater than hepatic uptake but the difference varied with the cobalamin. Liver uptakes for cyanocobalamin, hydroxocobalamin and coenzyme were 6, 17 and 20% respectively, and the ratios of kidney uptake/liver uptake were 6.6, 1.3 and 1.7. The quantity of hydroxocobalamin or coenzyme absorbed by the liver was 3–4 times that of cyanocobalamin. Levels of hydroxocobalamin and coenzyme were higher than that of cyanocobalamin in the pancreas but not in brain or skeletal muscle. The biological half-lives of cyanocobalamin and coenzyme were different for 6–8 weeks after uptake, but after a further

1–2 weeks they were similar, indicating complete conversion of cyano-cobalamin to coenzyme[24].

(b) Tracer studies in mice

Labelled cyanocobalamin injected intravenously into mice was found in the kidneys, liver, intestinal mucosa and endocrine glands[77]. As in the rat, the highest concentration appeared in the kidneys in the proximal convoluted tubules. In mice, the radioactive vitamin accumulates rapidly and abundantly in the placenta, from which it transfers to the foetuses in peak concentration after 24 h[78]. With a dose of 0.02 μg the concentration in the foetus was 130 times that in the mother. The kinetics of placental passage were very similar to those of intestinal absorption suggesting similar mechanisms[78].

(c) Tracer studies in dogs

Radioactive cyanocobalamin injected subcutaneously into dogs gives a pattern of tissue distribution similar to that of the endogenous vitamin[79]. Equilibration was complete after 14 days, a finding later obtained on rats[80]. The concentration in the dog pituitary was unexpectedly much higher than in the liver. The biological half-life of cyanocobalamin in the pregnant bitch, like that in the rat, was about 2 months[81]. 20% of the remaining cyanocobalamin was present in the litter, and the proportional distribution among the liver, kidneys and pancreas was similar in the mother and her puppies. Most of the radioactivity was in the liver but the kidneys had the highest concentration. Rosenblum et al.[82] administered [57]Co-labelled cyanocobalamin orally and ^{60}Co-labelled cyanocobalamin parenterally to dogs over a period of two weeks. 31 days later both isotopes were measured in 50 different tissues. The ratio [57]Co/^{60}Co was constant in all, indicating a common pool for cobalamins independent of the route of administration. The livers contained mainly hydroxocobalamin, probably derived from photosensitive coenzyme B_{12}.

13. Turnover studies in man

Small amounts of cyanocobalamin (up to 50 μg) injected intramuscularly were almost completely retained[84]. With higher doses appreciable amounts appeared in the urine during the first 8 h, the amount voided increasing with the dose. Urinary excretion of B_{12} is given by $E = D - 1.2 \, D^{0.89}$, where E is the amount excreted and D is the dose in micrograms[85]. When a dose of 1 μg of [^{60}Co]cyanocobalamin was injected intramuscularly the radioactivity disappeared from the injection site in 3–4 h and counts were maximal over the liver, spleen and kidney[86]. Doscherholmen and Hagen[89] analysed the organs of patients dosed orally with [^{60}Co]cyanocobalamin and intrinsic factor shortly before death. After a short stay in the intestine the vitamin rapidly appeared in the organs. Maximum radioactivity was reached faster in the spleen and kidneys than in the liver. The liver concentration increased slowly over several days whilst the spleen and kidney levels decreased, indicating a transfer to more permanent stores. In normal subjects intravenously administered radioactive cyanocobalamin disappeared rapidly from the blood but uptake by the liver lasted for many days[87]: the uptake was greater than could be accounted for by the blood level, and suggested the vitamin was being transferred from other organs. A similar exchange among tissues has been demonstrated for rats[88]. Doscherholmen and Hagen[89] have reviewed the evidence obtained in man that a sequential movement of vitamin B_{12} exists and involves five body compartments.

The biological half-life of cyanocobalamin in man is about a year[90]. The half-lives of [^{57}Co]hydroxocobalamin and [^{60}Co]cyanocobalamin in the liver are similar for the 30 weeks after administration. Thereafter the clearance of hydroxocobalamin slowed considerably compared with that of the cyano compound[91].

It is well established that hydroxocobalamin, compared with cyanocobalamin, is more slowly absorbed from an injection, is more slowly excreted in the urine, and gives higher and more prolonged blood B_{12} levels (see Boddy et al.[92] for a full list of references). Simultaneous intravenous administration of equal doses (15 or 1000 μg) of hydroxocobalamin and cyanocobalamin showed that hydroxocobalamin disappears from the blood faster than cyanocobalamin but less is excreted in the urine[52].

Only a few studies have been carried out on coenzyme B_{12}. When labelled cyanocobalamin, hydroxocobalamin and coenzyme B_{12} were compared in small oral doses the body retained 50, 86 and 77% respectively. With large intramuscular doses (100–1000 μg) slightly more coenzyme B_{12} than cyanocobalamin was retained, whereas retention of hydroxocobalamin was much more[24]. Boddy et al.[92] compared retentions of the three cobalamins by whole-body counting on patients without anaemia. [^{58}Co]Cyanocobalamin (0.5 μCi) was injected intramuscularly and one week later a similar dose of

[^{57}Co]hydroxocobalamin or [^{57}Co]coenzyme B$_{12}$ was likewise administered. After 3 days cyanocobalamin was retained less than either hydroxocobalamin or coenzyme B$_{12}$ which were retained to a similar degree. After
28 days, however, hydroxocobalamin was retained more than either of the
two cobalamins.

This behaviour of hydroxocobalamin may reflect its more basic character,
which allows it to attach more avidly to the proteins of serum, liver and
muscle[93]. Faster and more complete conversion to the storage form,
coenzyme B$_{12}$, may contribute also to its slower clearance[94].

In turnover studies involving labelled cyanocobalamin controversy has
arisen over the time required for the tracer to equilibrate with stores of the
endogenous vitamin. Equilibrium is assumed when the rate of loss of the
vitamin can be adequately expressed as a single exponential term. Equilibration periods from 10[95] to 250 days[96] have been calculated thus. The
problem has been discussed by Boddy and Adams[97]. After massive doses of
cyanocobalamin, hydroxocobalamin or coenzyme B$_{12}$ there were three
distinct periods of loss. The first involved considerable loss and lasted 3 days:
in the second the loss was slower but lasted 4–8 weeks, and it was followed
by a final period of steady low loss. In the last period the excretion rates of
all three cobalamins were identical[98].

14. The excretion of cyanocobalamin

After parenteral administration of [^{56}Co]cyanocobalamin to man large
amounts appear in the bile and faeces, but there is little in the urine[99]. In rat
experiments it was found that two-thirds of the vitamin in bile was reabsorbed from the intestine: a similar enterohepatic circulation probably
operates in man[100]. The nature of the vitamin excreted in the faeces is
unknown, and the faeces of many species contain a variety of analogues
produced by intestinal bacteria[2].

Most of the injected cyanocobalamin that appears in the urine, albeit a
small amount, is unchanged. When radioactive hydroxocobalamin was
injected into patients without pernicious anaemia less than half the urinary
cobalamin was unchanged, the remainder being cyanocobalamin and an
anionic cobalamin, probably sulphitocobalamin. These compounds were
found also when hydroxocobalamin was added to urine, hence they could be
artefacts[101]. Hydroxocobalamin and the coenzymes are sensitive to various
anions, making identification of the excretory form very difficult.

15. Interconversion of cobalamins

Body stores of vitamin B_{12} are present mainly as coenzyme B_{12}, so adminis-tered cyanocobalamin and hydroxocobalamin must be converted to this metabolically active form. In animal tissues hydroxocobalamin is more quickly converted to coenzyme B_{12} than is cyanocobalamin, suggesting that the limiting reaction is removal of the cyanide group[94]. Although reverse isotope dilution assays on tissues from dogs given radioactive cyanocobalamin indicated conversion to hydroxocobalamin over 5 weeks, no attempt was made to exclude light from these assays, and presumably the hydroxo-cobalamin was derived from coenzyme B_{12}. *In vivo* and *in vitro* experiments showed that guinea-pig liver can convert cyanocobalamin at the rate of 0.1–0.4 ng/g liver/day[102]. The kidneys and liver of men[94] or rats[103] given cyanocobalamin or hydroxocobalamin convert more than half the dose to the coenzyme in 48 h. If the liver was diseased or damaged with carbon tetrachloride, there was a distinct decrease in the rate of formation of coenzyme from cyanocobalamin but not from hydroxocobalamin, suggesting that these lesions affected mostly the removal of cyanide from the vitamin[94]. The suggestion that excessive amounts of cyanide in food or tobacco smoke may reconvert the active form of the vitamin to inactive cyanocobalamin has been reviewed recently by Smith[104]. Nothing is known about the origin of methylcobalamin in plasma, although small amounts accompany the coenzyme in most tissues.

For further reading I recommend *Vitamin B_{12} and Intrinsic Factor* (2nd European Symposium, Hamburg 1961, published by Ferdinand Enke, Stuttgart, 1962), Vitamin B_{12} Coenzymes (*Ann. N.Y. Acad. Sci.*, 112 (1964) 547–921), *Vitamin B_{12}* (E. Lester Smith, Methuen, London, 1965) and *Vitamins and Hormones* (various authors, 26 (1968) 319).

REFERENCES

1 H. P. C. HOGENKAMP, *Ann. Rev. Biochem.*, 37 (1968) 225.
2 E. LESTER SMITH, *Vitamin B_{12}*, Methuen, London, 1965, p. 1 *et seq.*
3 C. E. OXNARD AND W. T. SMITH, *Nature*, 210 (1966) 507.
4 E. LESTER SMITH, *Mineral Metabolism*, Vol. 2, Academic Press, London, 1962, p. 344.
5 R. M. HEYSSEL, R. C. BOZIAN, W. J. DARBY AND M. C. BELL, *Am. J. Clin. Nutr.*, 18 (1966) 176.
6 A. W. JOHNSON, L. MERVYN, N. SHAW AND E. LESTER SMITH, *J. Chem. Soc.*, (1963) 4146.
7 D. L. MOLLIN AND A. H. WATERS, *The Study of Vitamin B_{12} Absorption Using Labelled Cobalamins*, The Radiochemical Centre, Amersham, Bucks., England, 1968, p. 7.
8 H. C. HEINRICH AND E. WOLFSTELLER, *Med. Klin. (Munich)*, 61 (1966) 756.
9 G. B. J. GLASS, *Physiol. Rev.*, 43 (1963) 529.
10 R. GRÄSBECK, *Plenary Session Papers, XIIth Congress, Intern. Soc. of Haematol.*, New York, 1968, p. 124.
11 J. ABELS AND R. F. SCHILLING, *J. Lab. Clin. Med.*, 64 (1964) 375.
12 N. YAMAGUCHI, H. WEISBERG AND G. B. J. GLASS, *Gastroenterology*, 52 (1967) 1145.
13 C. C. BOOTH AND D. L. MOLLIN, *Lancet*, 2 (1959) 18.
14 V. RØNNON-JESSEN AND J. HANSEN, *Blood*, 25 (1965) 224.
15 A. BOASS AND T. H. WILSON, *Am. J. Physiol.*, 207 (1964) 27.
16 R. M. DONALDSON, I. L. MACKENZIE AND J. S. TRIER, *J. Clin. Invest.*, 46 (1967) 1215.
17 S. P. ROTHENBERG, *J. Clin. Invest.*, 47 (1968) 913.
18 B. A. COOPER, *Am. J. Physiol.*, 214 (1968) 832.
19 J. D. HINES AND A. ROSENBERG, *J. Clin. Invest.*, 46 (1967) 1070.
20 I. CHANARIN, *Gut*, 9 (1968) 373.
21 K. OKUDA AND V. TANTENGCO, *Proc. Soc. Exptl. Biol. Med.*, 110 (1962) 396.
22 D. H. LEE AND G. B. J. GLASS, *Proc. Soc. Exptl. Biol. Med.*, 107 (1961) 293.
23 V. HERBERT AND L. W. SULLIVAN, *Ann. N.Y. Acad. Sci.*, 112 (1964) 855.
24 H. C. HEINRICH AND E. E. GABBE, *Ann. N.Y. Acad. Sci.*, 112 (1964) 871.
25 G. B. J. GLASS AND Z. CASTRO-CUREL, *Ann. N.Y. Acad. Sci.*, 112 (1964) 904.
26 S. ARDEMAN AND I. CHANARIN, *Lancet*, 2 (1963) 1350.
27 S. BAR-SHANY AND V. HERBERT, *Blood*, 30 (1967) 777.
28 S. G. SCHADE, J. ABELS AND R. F. SCHILLING, *J. Clin. Invest.*, 46 (1967) 615.
29 L. A. E. ASHWORTH, J. M. ENGLAND, J. M. FISHER AND K. B. TAYLOR, *Lancet*, 2 (1967) 1160.
30 I. L. MACKENZIE, R. M. DONALDSON, W. L. KOPP AND J. S. TRIER, *J. Exptl. Med.*, 128 (1968) 375.
31 H. C. HEINRICH, *Plenary Session Papers, XIIth Congress, Intern. Soc. of Haematol.*, New York, 1968, p. 112.
32 H. BERLIN, R. BERLIN AND G. BRANTE, *Acta Med. Scand.*, 184 (1968) 247.
33 H. WEISBERG AND G. B. J. GLASS, *Proc. Soc. Exptl. Biol. Med.*, 122 (1966) 25.
34 F. S. MOONEY AND J. G. HEATHCOTE, *Brit. Med. J.*, i (1966) 1149.
35 R. GRÄSBECK, K. SIMONS AND I. SINKKONEN, *Biochim. Biophys. Acta*, 127 (1966) 47.
36 H. FLODH, B. BERGRAHM AND B. ODEN, *Life Sci.*, 7 (1968) 155.
37 G. D. FRENTZ, O. N. MILLER AND H. J. HANSEN, *Biochim. Biophys. Acta*, 147 (1967) 162.
38 K. SIMONS, *Soc. Sci. Fennica, Commentationes Biol.*, 27 (1964) 5.
39 C. A. HALL AND A. E. FINKLER, *J. Lab. Clin. Med.*, 60 (1962) 765.
40 C. A. HALL AND A. E. FINKLER, *Biochim. Biophys. Acta*, 78 (1963) 233.
41 C. A. HALL AND A. E. FINKLER, *J. Lab. Clin. Med.*, 65 (1965) 459.

42 C. A. HALL AND A. E. FINKLER, *Blood*, 27 (1966) 611.
43 E. A. BRODY, S. ESTREN AND L. R. WASSERMAN, *Blood*, 15 (1960) 646.
44 C. A. HALL AND A. E. FINKLER, *Nature*, 204 (1964) 1207.
45 C. A. HALL AND A. E. FINKLER, *J. Lab. Clin. Med.*, 73 (1969) 60.
46 V. HERBERT, *Blood*, 32 (1968) 305.
47 W. S. ROSENTHAL, H. WEISBERG AND G. B. J. GLASS, *Blood*, 30 (1967) 198.
48 W. S. ROSENTHAL, H. WEISBERG AND G. B. J. GLASS, *Blood*, 30 (1967) 210.
49 A. E. FINKLER, C. A. HALL AND J. V. LANDAU, *Federation Proc.*, 24 (1965) 679.
50 F. P. RETIEF, C. W. GOTTLIEB AND V. HERBERT, *J. Clin. Invest.*, 45 (1966) 1907.
51 F. P. RETIEF, C. W. GOTTLIEB AND V. HERBERT, *Blood*, 29 (1967) 837.
52 Y. YAGIRI, *J. Vitaminol.*, 13 (1967) 210.
53 C. LAWRENCE, *Blood*, 27 (1966) 389.
54 C. A. HALL AND A. E. FINKLER, *Proc. Soc. Exptl. Biol. Med.*, 123 (1966) 55.
55 R. GRÄSBECK, *Scand. J. Clin. Lab. Invest.*, 19 (1967) 7, Suppl. 95.
56 A. E. FINKLER AND C. A. HALL, *Arch. Biochem. Biophys.*, 120 (1967) 79.
57 V. HERBERT AND L. W. SULLIVAN, *Ann. N.Y. Acad. Sci.*, 112 (1964) 855.
58 W. H. C. SHAW AND C. J. BESSELL, *Analyst*, 85 (1960) 389.
59 J. M. HSU, B. KAWIN, P. MINOR AND J. A. MITCHELL, *Nature*, 210 (1966) 1264.
60 O. STEIN, Y. STEIN, J. ARONOVITCH, N. GROSSOWICZ AND M. RACHMILEWITZ, *Cancer Res.*, 18 (1958) 849.
61 K. G. SHENOY AND G. B. RAMASARMA, *Arch. Biochem. Biophys.*, 51 (1954) 371.
62 J. I. TOOHEY AND H. A. BARKER, *J. Biol. Chem.*, 236 (1961) 560.
63 Y. YAGIRI, *J. Vitaminol.*, 13 (1967) 197.
64 K. LINDSTRAND, *Acta Chem. Scand.*, 19 (1965) 1785.
65 K.-G. STÅHLBERG, *Scand. J. Haematol.*, 4 (1967) 312.
66 B. F. CHOW, R. L. DAVIS AND A. H. LAWTON, *Vitamin B_{12} and Intrinsic Factor*, Ferdinand Enke, Stuttgart, 1962, p. 330.
67 D. M. MATTHEWS, R. GUNASEGARAM AND J. C. LINNELL, *J. Clin. Pathol.*, 20 (1967) 683.
68 O. STEIN, Y. STEIN, J. ARONOVITCH, N. GROSSOWICZ AND M. RACHMILEWITZ, *J. Lab. Clin. Med.*, 54 (1959) 545.
69 K. OOMAE, *Bitamin*, 38 (1968) 299.
70 J. W. THOMAS AND B. B. ANDERSON, *Brit. J. Haematol.*, 2 (1956) 41.
71 K. LINDSTRAND, *Nature*, 204 (1964) 188.
72 K. G. STÅHLBERG, *Scand. J. Haematol.*, 1 (1964) 220.
73 G. E. NEWMAN, J. R. P. O'BRIEN, G. H. SPRAY AND L. J. WITTS, *Vitamin B_{12} and Intrinsic Factor*, Ferdinand Enke, Stuttgart, 1962, p. 424.
74 F. A. ROBINSON, *The Vitamin Co-Factors of Enzyme Systems*, Pergamon, London, 1966, p. 757.
75 H. UCHINO, S. UKYO, Y. YAGIRI, T. YOSHINO and G. WAKISAKA, *Ann. N.Y. Acad. Sci.*, 112 (1964) 844.
76 C. A. HALL, *Brit. J. Haematol.*, 16 (1969) 429.
77 H. FLODH, *Acta Radiol., Suppl.* 284 (1968) 1.
78 S. ULLBERG, H. KRISTOFFERSSON, H. FLODH AND Å. HANNGREN, *Arch. Intern. Pharmacodyn.*, 167 (1967) 431.
79 J. M. COOPERMAN, A. L. LUHBY, D. N. TELLER AND J. F. MARLEY, *J. Biol. Chem.*, 235 (1960) 191.
80 R. GRÄSBECK, R. IGNATIUS, J. JARNEFELT, H. LINDEN, A. MALI AND W. NYBERG, *Clin. Chim. Acta*, 6 (1961) 56.
81 A. L. LUHBY, J. M. COOPERMAN AND A. M. DONNENFELD, *Proc. Soc. Exptl. Biol. Med.*, 100 (1959) 214.

82 C. ROSENBLUM, P. G. REIZENSTEIN, E. P. CRONKITE AND H. T. MERIWETHER, *Proc. Soc. Exptl. Biol. Med.*, 112 (1963) 262.
83 C. ROSENBLUM, *Vitamin B₁₂ and Intrinsic Factor*, Ferdinand Enke, Stuttgart, 1962, p. 313.
84 C. LANG, R. A. HARTE, C. L. CONLEY AND B. F. CHOW, *J. Nutr.*, 46 (1952) 215.
85 D. C. CHESTERMAN, W. F. J. CUTHBERTSON AND H. F. PEGLER, *Biochem. J.*, 48 (1951) li.
86 L. M. MEYER, N. I. BERLIN, M. JIMINEZ-CASADO AND S. N. ARKUN, *Proc. Soc. Exptl. Biol. Med.*, 91 (1956) 129.
87 A. MILLER, H. F. CORBUS AND J. F. SULLIVAN, *J. Clin. Invest.*, 36 (1957) 18.
88 D. R. STRENGHT, W. F. ALEXANDER AND J. P. WACK, *Proc. Soc. Exptl. Biol. Med.*, 102 (1959) 15.
89 A. DOSCHERHOLMEN AND P. S. HAGEN, *Vitamin B₁₂ and Intrinsic Factor*, Ferdinand Enke, Stuttgart, 1962, p. 381.
90 L. L. SCHLOESSER, P. DESHPANDE AND R. F. SCHILLING, *Arch. Internal Med.*, 101 (1958) 306.
91 G. B. J. GLASS AND D. H. LEE, *Blood*, 27 (1966) 227.
92 K. BODDY, P. KING, L. MERVYN, A. MACLEOD AND J. F. ADAMS, *Lancet*, ii (1968) 710.
93 G. B. J. GLASS, H. R. SKEGGS, D. H. LEE, E. L. JONES AND W. W. HARDY, *Vitamin B₁₂ and Intrinsic Factor*, Ferdinand Enke, Stuttgart, 1962, p. 683.
94 Y. YAGIRI, *J. Vitaminol.*, 13 (1967) 228.
95 J. F. ADAMS AND K. BODDY, *J. Lab. Clin. Med.*, 72 (1968) 392.
96 P. REIZENSTEIN, G. EK AND C. M. E. MATTHEWS, *Phys. Med. Biol.*, 11 (1966) 295.
97 K. BODDY AND J. F. ADAMS, *Phys. Med. Biol.*, 13 (1968) 55.
98 K. BODDY AND J. F. ADAMS, *Am. J. Clin. Nutr.*, 21 (1968) 657.
99 R. GRÄSBECK, W. NYBERG AND P. REIZENSTEIN, *Proc. Soc. Exptl. Biol. Med.*, 97 (1958) 780.
100 J. B. STOKES, *Nature*, 191 (1961) 807.
101 E. H. KENNEDY AND J. F. ADAMS, *Clin. Sci.*, 29 (1965) 417.
102 P. REIZENSTEIN, *Blood*, 29 (1967) 494.
103 H. UCHINO, Y. YAGIRI, T. YOSHINO, M. KONDO AND G. WAKISAKA, *Nature*, 205 (1965) 176.
104 E. L. SMITH, *Plant Foods for Human Nutrition*, 1 (1968) 7.
105 B. HOM AND H. OLESEN, *Scand. J. Clin. Lab. Invest.*, 19 (1967) 269.

Trace Elements: Metabolism and Metabolic Function

B. L. O'DELL AND B. J. CAMPBELL

Departments of Agricultural Chemistry and Biochemistry, University of Missouri, Columbia, Mo. (U.S.A.)

1. Introduction

The term, trace element, is used here for historical reasons and because it is generally understood and accepted. When tissues were first analyzed, some elements could be detected but not measured precisely. These were reported as occurring in "traces" but there is no clear demarcation between trace elements and macro-elements. In general, those elements whose concentrations in tissues are expressed in parts per million or less are called trace or minor elements.

Almost all elements in an organism's environment will accumulate in the tissues to some extent and many could be described as trace elements. This discussion is concerned only with those minor elements which are essential for growth, health and completion of the life cycle in animals. In general such elements perform a biocatalytic function inasmuch as they occur in only catalytic quantities. Essential trace metals function as a component of, or in association with, an organic molecule, most commonly with an enzyme. For this reason the elements described in this chapter might best be described as enzyme cofactors although there are exceptions particularly among the non-metals.

There is a wide range of stability of the complexes formed between metals and proteins. From an operational point of view, it is convenient to divide the enzyme complexes into two groups, metalloenzymes and metal–enzyme complexes[1]. *Metalloenzymes* refer to catalytic proteins that contain a metal ion bound with sufficient firmness to remain intact during the usual isolation procedures. In general, the concentration of an essential ion increases with purification while that of non-essential ions decreases. When a metalloenzyme is pure, the metal should bear a stoichiometric relationship to the protein moiety. Although the metal of such a complex cannot be removed by simple dialysis, strong acids or chelating agents may remove it and cause loss of enzymatic activity. By *metal–enzyme complex* is meant a combination of protein and a loosely bound metal ion which can be readily removed by dialysis. Nevertheless, the metal is essential for maximum catalytic activity of the complex. If lost during the isolation procedure, it can be added, and in fact, must be added to restore complete activity. Obviously, the stability of protein–metal complexes will span a wide and continuous spectrum depending on the metal and the donor groups involved. Thus, these designations are arbitrary but useful in describing the catalytic complexes. Enolase, a typical metal–enzyme complex containing zinc, has a stability constant of $2 \cdot 10^5$ while Zn^{2+} carboxypeptidase, a metalloenzyme, has a constant of $3.2 \cdot 10^{10}$ (ref. 1).

For the most part, the trace elements which are essential components of enzyme systems are localized in one area of the periodic chart (see Fig. 1). They are concentrated in the first series of transition elements; molybdenum is found in the second series. Transition elements are defined as those elements with incompletely filled d orbitals. According to this definition, copper and zinc are not transition metals but the similarity of their ions

	I	II		III	IV	V	VI	VII	
	19	20	21–30	31	32	33	34	35	36
4	K	Ca	transition elements	Ga	Ge	As	$\boxed{Se}$	Br	Kr
	37	38	39–48	49	50	51	52	53	54
5	Rb	Sr	transition elements	In	Sn	Sb	Te	$\boxed{I}$	Xe

TRANSITION ELEMENTS

1st Series	21	22	23	25	25	26	27	28	29	30
	Sc^{1*}	Ti^2	V^3	$\boxed{Cr^5}$	$\boxed{Mn^5}$	$\boxed{Fe^6}$	$\boxed{Co^7}$	Ni^8	$\boxed{Cu^{10}}$	$\boxed{Zn^{10}}$
2nd Series	39	40	41	42	43	44	45	46	47	48
	Y^1	Zr^2	Nb^4	$\boxed{Mo^5}$	Tc^5	Ru^7	Rh^8	Pd^{10}	Ag^{10}	Cd^{10}

Fig. 1. Relationship of the essential trace elements to their neighbors in the periodic chart. Those proven essential are shown in blocks. *Number of electrons in the outer d orbitals indicated by superscripts.

to the transition-element ions makes their exclusion artificial. In Fig. 1, the number of electrons in the d orbitals of the element is indicated as a superscript. Upon ionization, the elements normally lose the first two electrons from the next higher s orbital, *e.g.*, Zn^{2+} is a d^{10} ion and Co^{3+} is a d^6 ion. The elements shown in blocks, Cr, Mn, Fe, Co, Cu, Zn, Mo, Se and I, are classed as essential trace elements but the possibility that others may be added to this list is not excluded. If there are unrecognized essential metals, one might predict that they have electronic structures similar to those already identified and, thus, belong to one of the transition series. Because of its importance in human health F is included in this chapter.

Enzymes possess catalytic activity primarily because they possess unique three-dimensional structures. The metal component of metalloenzymes may function by (*a*) helping maintain the structural integrity of the protein, (*b*) binding the substrate, coenzyme or both to the enzyme, or (*c*) becoming directly involved in the catalytic reaction and undergoing oxidation and reduction during electron-transfer reactions[1]. A common theory of enzyme catalysis visualizes the formation of an active enzyme–substrate complex. The attainment of this transition state would be facilitated by what Vallee and Williams[2] have termed a state of entasis; that is, the active center becomes an energetically poised domain. Some metal–protein complexes exist in irregular geometries and possess energies consonant with the entatic state. Removal of the metal ion results in partial, if not total, loss of enzymatic activity. In some cases the metal ion can be added to the apoenzyme with restoration of activity and in others irreversible changes occur in the protein structure.

The current theories of metal coordination as applied to enzyme action have been reviewed by Vallee and Coleman[1]. Metal–enzyme complexes are formed through coordinate bonds between the metal and ligands contained in the protein structure. The electron-donor atoms, O, S and N, in the protein show a certain degree of selectivity with regard to the metal. For example, zinc tends to coordinate with sulfur, copper with nitrogen and manganese with oxygen[3]. Virtually all of the reactive groups of amino acids in peptide linkage are postulated to serve as ligands: the carboxyl of glutamic and aspartic acids, the ε-amino of lysine, the imidazole of histidine, the phenoxy of tyrosine, the guanidinium of arginine, the sulfhydryl of cysteine and the atoms in the peptide linkage itself. Besides the ligands supplied by the protein, metals may be bound by prosthetic groups such as the porphyrin and corrin rings.

References p. 254

Coordination numbers of 2, 4 and 6 are most common, at least in model systems, and the most frequent geometry is a metal ion surrounded by six ligands forming an octahedral structure. Of the elements which have a coordination number of 4, Zn^{2+} prefers the regular tetrahedral while Cu^{2+} assumes the square planar structure. The geometry of some metal–protein complexes is irregular, and may involve 5 ligands[3]. Explanation of the physical properties of metal complexes is best provided by the molecular-orbital theory and references should be consulted for greater detail[1]. The nature of the metal ion and the binding site is commonly explored by absorption, optical rotatory dispersion, circular dichroism and electron-spin resonance. In recent years, X-ray diffraction and amino acid sequence studies have afforded more definitive descriptions of metal-binding sites.

Although much of the previous discussion relates to metal–enzyme interrelations, it is not implied that all trace elements function as enzyme components or cofactors. Iodine, long recognized as essential in trace quantities, functions entirely, so far as now known, as a component of the thyroid hormones. Furthermore, the importance of fluorine in bones and teeth may not be related to enzyme function. The mechanism by which selenium exerts its metabolic function is unknown at present as is the chemical nature of its active form.

Several reviews which discuss the interrelationship of metal ions and proteins have appeared[1,3,4,5]. The second edition of Underwood's mono-graph[6] gives the most complete coverage of nutritional aspects of all essential trace elements and a survey of biogeochemistry and distribution of the elements may be found in Bowen's monograph[7]. An attempt is made here to summarize knowledge concerning the metabolism and metabolic function of the trace elements essential for animals.

2. Iron

Iron is the fourth most abundant element in the earth's crust, and its presence in mammalian tissues has been recognized since early in the 18th century[8]. The iron content of a 70-kg man ranges[9] from 4 to 5 g and is distributed as follows: 70.5% in hemoglobin, 3.2% in myoglobin, 26% in storage compounds (ferritin and hemosiderin), 0.1% in transport form (transferrin), 0.1% in the cytochromes and 0.1% in catalase[10]. In biological materials iron occurs primarily in the form of complexes, so that free ionic iron occurs in only negligible amounts. Hence, the metabolism of iron consists of the biochemical transformation of iron complexes. Numerous reviews relating to iron metabolism and biochemical function have been published[6,8,10-15].

(a) Metabolism

(i) Absorption

The intestinal absorption of iron has been measured by a variety of techniques including balance experiments, measurement of serum iron following oral dosage, analysis of radioiron incorporation into hemoglobin, and determination of the radioactive iron content of the entire body. These techniques have been described in detail by Josephs[16] and by Bothwell and Finch[12]. It has been determined that, while absorption of iron can take place from the stomach and from any place in the intestinal tract, the metal is absorbed principally from the duodenum[17,18]. The ferrous form of iron is preferentially absorbed over the ferric form[19,20]. Ascorbic acid, which has the capacity of reducing iron to the ferrous state, increases absorption when it is administered with iron salts or iron containing foods[21]. Dietary factors that form insoluble iron complexes such as phosphates or phytates cause a reduction in iron absorption[22]. Besides Fe^{2+} other forms of iron can be absorbed; for example, hemoglobin iron is absorbed and its absorption is not decreased by phytate[12].

Absorption of food iron is influenced by a number of variables, including the nutritive state of the subject, the form of iron in the diet and the nature of other diet constituents. With these variables in mind it has been estimated that normal subjects retain 5–10% and iron-deficient subjects 10–20% of the iron in the food consumed. Thus if the average adult diet provides 12–15 mg of iron, the normal subject should absorb 0.6–1.5 mg iron and the iron-deficient subject 2–3 mg per day. Increase in iron absorption has been observed during the course of iron-deficiency anemia, the latter half of pregnancy, certain stages of hemochromatosis[23], and in general under conditions that produce increased erythropoiesis[24]. Iron absorption is reduced during diminished erythropoiesis[25], when iron stores are increased[26] and during certain infectious diseases[27].

Iron is absorbed directly into the intestinal mucosal cell, and there is essentially no excretion into the intestinal lumen[28]. From the mucosal cell iron passes into the blood stream directly without entering the lymphatic system[29]. Since animals exhibit only a very limited capacity to excrete iron[30], the role of iron absorption in controlling the iron content of the body assumes paramount importance. Studies by Dowdle et al.[31], who employed everted pouches of rat duodenum, indicate that iron uptake is an active transport process dependent upon oxidative metabolism and the generation of phosphate-bond energy. Other workers have questioned these results because of the long in vitro incubation periods which could produce tissue damage and because of the failure to rule out exchange of inert iron and radioiron[13,32]. Early theories introduced by Hahn and coworkers[33] and developed by Granick[18] suggest that control of iron absorption is maintained by the ferritin content of mucosal cells. This theory, known as the "mucosal block" theory, holds that when the mucosal cells become physiologically saturated with ferritin, further absorption is impeded until the iron is released from ferritin and transferred to plasma. Moore[11] has presented considerable data in conflict with this theory although he suggests that the intestinal mucosa does affect the activity of the absorptive mechanism. Studies involving the analysis of peripheral blood have suggested that iron absorption may be regulated by the level of unbound plasma transferrin[34]. This suggestion is supported by the observations that iron absorption is increased in iron-deficiency anemia and in late pregnancy when transferrin levels are high and serum iron and transferrin saturation values are low[13]. Conrad et al.[35] have proposed that absorption is controlled by the concentration of iron in the mucosa which in turn depends on the amount of "messenger iron" entering these cells from the plasma. Fletcher and Huehns[36] suggest that this "messenger iron" represents the transferrin molecules that have two bound iron atoms. Furthermore, the degree to which transferrin molecules are saturated with iron depends upon the rate at which the protein is synthesized and catabolized, both rates being related to the erythropoietic rate[37,38]. It must be concluded that present experimental data are insufficient to establish the mechanism for regulation of iron absorption.

(ii) *Transport*

Transport of iron occurs by way of the plasma. Iron from degraded red cells supplies most of the iron entering the plasma, and most of the iron leaving the plasma is taken up by the bone marrow for hemoglobin synthesis. As stated above only a small proportion of dietary iron is absorbed into the blood, and small quantities of iron are excreted. Exchange between tissue enzymes and tissue stores contributes to plasma iron, but only to a limited extent. The iron content[10] of the plasma in normal adult males ranges from 90 to 100 and for females from 70 to 150 μg/100 ml. The total amount of iron in the circulating plasma of normal subjects[10] at any one time is 3–4 mg and the plasma concentration undergoes a diurnal variation with the level decreasing by as much as 57% from morning to evening[39]. Variations in plasma iron level also result from a number of disease states[12].

Although there is a small amount of hemoglobin circulating in the plasma (0.16–0.57 mg containing 0.5–2 μg of iron per 100 ml)[40], by far the greatest amount of plasma iron is complexed to the specific iron-binding, β_1-globulin, transferrin. This plasma protein was discovered independently by Holmberg and Laurell[41] and by Schade and Caroline[42]. The stability constant[43] of the iron–transferrin complex is between 10^{26} and 10^{30}. Thus binding of iron by transferrin at physiological conditions essentially eliminates significant binding by chelating agents such as phosphate or citrate. Normal plasma contains from 0.24 to 0.28 g of transferrin per 100 ml. In terms of iron binding capacity, this corresponds[11] to about 200–450 μg per 100 ml. In the normal situation only about 30% of the transferrin is saturated with iron; the remaining 70% represents a latent or unbound reserve[44].

The degree of saturation of transferrin affects the deposition of iron in liver stores and the supply of iron to red cell precursors. At saturation levels above 60%, much of the iron is deposited in the liver. Incubation of iron bound to transferrin with liver slices in the presence of ATP and ascorbic acid causes the transfer of iron to the storage form, ferritin[45]. It is suggested that during the course of this reaction 2 moles of ATP, 1 mole of ascorbic acid, and 1 mole of transferrin form a complex. Ferric iron is reduced to ferrous iron at the expense of ascorbic acid and the reduced iron is released from transferrin and incorporated into ferritin. The process is dependent on oxidative metabolic reactions that generate ATP[45].

At levels of transferrin saturation between 30 and 60%, transfer of iron to reticulocytes occurs primarily in the bone marrow[12]. Jandl et al.[46] showed that iron is transferred directly from transferrin to receptor sites on the membranes of reticulocytes, and that competition occurs between the reticulocyte membrane and the transport protein. Regulation of iron delivery at this stage may be controlled by the nature of the iron binding to transferrin and by conformational changes which occur upon binding protein to the receptors[36,46,47]. Iron does not become free during transfer, and it cannot be eluted by apotransferrin or other chelators[46]. Furthermore, iron is transferred selectively from transferrin to red cell precursors which are still actively synthesizing hemoglobin rather than to mature cells[48,49].

This suggests that receptor sites on the membranes of red cell precursors exhibit a specificity for transferrin and that this specificity is lost upon maturation of the red cell. Thus regulation of iron absorption and iron delivery may be influenced by the structure of the transport protein, transferrin[36,49].

(iii) Storage

In the normal man storage iron constitutes approximately 25% of the total. Estimates indicate that between 600 and 1500 mg of iron are stored intracellularly in liver, bone marrow, spleen, and other tissues[11]. These iron stores, primarily ferritin and hemosiderin, provide a reserve of iron for metabolically functional compounds such as hemoglobin, myoglobin and the iron-containing enzymes. Under normal conditions iron is stored primarily in the form of ferritin[50]; however, as the concentration of iron in tissues rises, the concentration of hemosiderin increases to a greater degree than that of ferritin[51].

Ferritin is an iron–protein complex containing approximately 20% iron and consisting of a core of ferric hydroxide, ferric phosphate micelles surrounded by a protein shell[52]. The protein portion, apoferritin, is a homogeneous protein[53] of molecular weight 460 000–480 000, and the core has the approximate composition[54], $[(FeOOH)_8FeO_2PO_3H_2]$.

Hemosiderin is an amorphous substance of variable and ill-defined composition. The percentage of iron reported varies[8] from 8 to 45%. It can be observed as golden yellow granules in tissue sections and smears, and it has been suggested that the term hemosiderin should be restricted to those granules that are water-insoluble[55]. These granules contain several different organic constituents including protein, such as apoferritin[56,57]. That the *de novo* synthesis of apoferritin is induced by iron salts has been demonstrated in intact animals[58,59], in tissue slices[60], and in tissue cultures[61]. As the iron level increases intracellularly, dispersed ferritin becomes concentrated in vacuoles, is denatured, and loses water solubility[62]. Degradation of apoferritin then takes place presumably by intracellular proteases, and finally aggregation of iron micelles leads to the formation of mature hemosiderin. Hemosiderin-like material can also be produced from ferritin by denaturation followed by trypsin digestion[63].

(iv) Excretion

Since the work of McCance and Widdowson[64] it has been known that iron is not excreted in significant amounts in the urine and that fecal iron represents primarily unabsorbed iron. That animals hold iron tenaciously is shown by the fact that in hemolytic anemia and during treatment of polycythemia with phenylhydrazine, when large amounts of iron are liberated from lysed red cells, less than 0.5% of this iron appears in the urine and feces[65]. Small amounts of iron are lost from the gastrointestinal tract because of the rapid turnover of the mucosa which results in the loss of 50–80 g of exfoliated cells per day[66]. In normal individuals there also occurs a small loss of red cells through the gastrointestinal tract[67]; biliary losses are small but measurable. Dermal loss occurs, and may contribute to iron-deficiency anemia in hot humid regions of the world[68]. Finch[69] has estimated the daily iron losses in an adult 70-kg male as follows: blood 0.35 mg, mucosa 0.10 mg, bile 0.20 mg, urine 0.08 mg, skin 0.20 mg. The importance of the conservation and reutilization of iron in the body is realized if one considers that from a normal intake of 12–15 mg iron per day, the normal adult absorbs from 0.6 to 1.5 mg and during the same period of time catabolism of hemoglobin releases 20–25 mg of iron. If this catabolized iron were not conserved, the iron loss could not be restored by any reasonable diet[11].

(v) Turnover

Plasma iron turnover has been estimated from the rate at which an intravenous dose of ^{59}Fe-transferrin leaves the plasma. The normal half-time[11] is 90–100 min. From this estimate between 20 and 42 mg of plasma iron turns over per day[70,71], although at any one time there are only 3–4 mg of iron in the whole plasma volume[11]. Almost all of this iron is taken up by the erythroid bone marrow for incorporation into hemoglobin. Other tissues that assimilate measurable quantities are the liver[12] and the placenta[72]. The incorporation of radioiron into myoglobin, cytochrome c, and catalase was reported by Theorell et al.[73], following its intraperitoneal injection into guinea pigs. In the case of myoglobin measurable specific activities were not obtained until one month after injection. The specific activity of liver catalase reached a maximum after 4–5 days, and that of cytochrome c in skeletal and cardiac muscle increased steadily over 59 days.

Although mechanisms for deposition and mobilization of iron stores have not as yet been established, it is known that the iron in both ferritin and hemosiderin can be mobilized when needed for hemoglobin synthesis[11].

It is likely that ferritin is synthesized by filling apoferritin shells with iron rather than by aggregation of apoferritin around a preformed iron-core template[74]. Most ferritin iron is in the ferric state, but iron apparently enters and leaves the molecule as ferrous iron[75]. Under anaerobic conditions the enzyme, xanthine oxidase, participates in the release of iron from ferritin[76,77].

(vi) Physiological function

Iron deficiency has been ranked second only to protein malnutrition in the numbers of people affected[14]. Iron deficiency in human adults results in listlessness and fatigue, palpitation on exertion, and sometimes in a sore tongue, angular stomatitis, dysphagia, and koilonychia[78]. Depressed growth and reduced resistance to infection are also observed in children. Among domestic animals iron-deficiency anemia occurs frequently in suckling pigs and the pathology and hematology of this disease has been described by Seamer[79].

The principal manifestation of iron deficiency is hypochromic microcytic anemia, a type of anemia in which the erythrocytes have a reduced concentration of hemoglobin[10]. The disease is accompanied by a normoblastic hyperplastic bone marrow that contains little or no hemosiderin. Serum iron is subnormal and total iron-binding capacity levels are above normal. The course of the deficiency follows a sequence of depletion of iron stores, rise in transferrin levels, fall in serum iron, decrease in hemoglobin, and appearance of hypochromia[10]. There have also been demonstrated decreases in the levels of myoglobin, catalase, and cytochrome c in various tissues as a result of experimental iron deficiency[10].

(b) Chemistry

From a physiological point of view, the two ions, Fe^{2+} and Fe^{3+}, are the most important oxidation states of iron. The oxidation potential of $Fe^{2+} \rightarrow Fe^{3+} + e^-$ is such that molecular oxygen can convert ferrous to ferric ion in both acidic and basic aqueous solutions. The ferrous ion forms salts with all stable anions, and most of its complexes are octahedral. Since quite strong ligand fields are necessary to cause spin-pairing in Fe^{2+} complexes, practically all Fe^{2+} complexes are high spin (paramagnetic). Two common spin-paired or diamagnetic complexes of divalent ion are $Fe(CN)_6^{4-}$ and $[Fe(o\text{-phen})_3]^{2+}$. Heme, the porphyrin complex with Fe^{2+}, is the most abundant and physiologically important complex. Four heme complexes are present in each hemoglobin molecule, and they are bonded to the globin portion by coordination of imidazole nitrogen of histidine to the metal ion. In the heme complex in Fig. 2, the sixth coordination position is occupied by either a water or oxygen molecule. The magnetic moment of each ferrous ion in the hydrated complex is 5.4 Bohr magnetons showing the presence of four unpaired electrons. Combination of the complex with oxygen leads to the replacement of the coordinated water molecule, and the complex simultaneously becomes diamagnetic. This process of oxygenation takes place without a valence change of the ferrous ion. When the iron in heme is oxidized to Fe^{3+}, oxygen is no longer coordinated[80].

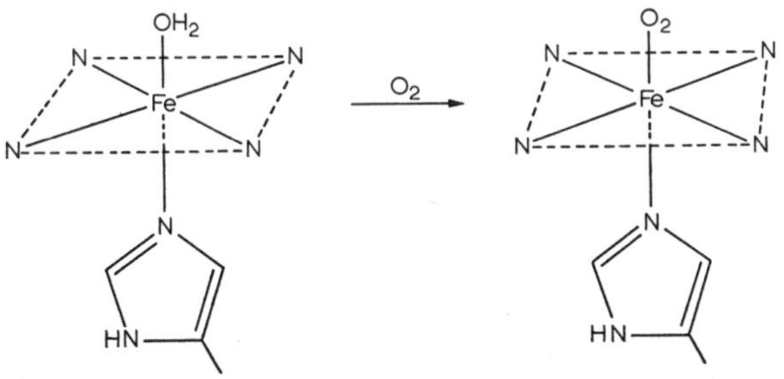

Fig. 2. Oxygenation of iron (Fe^{2+}) in the heme of hemoglobin. The octahedral complex changes from high-spin to low-spin and becomes diamagnetic.

(c) Biochemical function

The functional properties of biological iron compounds are integrated into the metabolism of virtually all organisms, both plant and animal. Exploration of the biochemistry of these compounds has provided an enormous literature including several recent reviews[81–85]. Iron compounds that facilitate electron transport include the cytochromes which are generally associated with oxidative phosphorylation in mitochondria[86–88], NAD-linked dehydrogenases[89,90], flavoproteins[86,91], and the ferredoxins[91,92] which are found in plants and microorganisms but not in animals. Peroxidases and catalases are iron-containing oxidative enzymes found in both animal and plant tissues[93,94]. The biochemistry of these iron compounds, summarized in Table I, has been treated elsewhere in *Comprehensive Biochemistry*[95,96].

TABLE I

BIOLOGICALLY IMPORTANT IRON COMPOUNDS

	Typical source	Function	Molecular weight	Fe content (g. atoms/mole)	Refs.
Heme compounds					
Hemoglobin	Mammalian red blood cells	oxygen carrier	67 000	4	95
Myoglobin	Skeletal muscle	oxygen carrier	16 500	1	95
Heme enzymes					
Cytochrome oxidase (EC 1.9.3.1)	Heart muscle	electron transport	290 000	3	96, 141
Cytochrome c	Heart muscle	electron transport	12 400	1	96
Catalase (EC 1.11.1.6)	Horse erythrocyte	peroxide breakdown	250 000	4	96
Peroxidase (EC 1.11.1.7)	Horseradish	peroxide breakdown	40 000	1	96
Non heme compounds					
Succinic dehydrogenase (EC 1.3.99.1)	Heart	electron transport	200 000	4	96
Reduced NAD dehydrogenase (EC 1.6.99.3)	Heart	electron transport	550 000	16–18	96
Xanthine oxidase (EC 1.2.3.2)	Milk	electron transport	290 000	8	97
Ferritin	Liver	iron storage	450 000 (apoferritin)	3000	96
Hemosiderin	Liver	iron storage			95
Transferrin	Blood plasma	iron transport	90 000	2	95
Conalbumin	Egg white	iron transport	87 000	2	95
Haemerythrin	Golfingia (Phascolosoma) gouldii	oxygen carrier	107 000	16	95a
Ferredoxin	Clostridium pasteurianum	electron transfer	12 000	10	91

3. Copper

In general copper occurs in relatively high concentration in soil (20 p.p.m.), but there are areas of the world in which the soil copper is so low or unavailable that deficiency diseases occur among the indigenous plants and animals[7]. In spite of its early observed presence in plants and animals it was not until 1924, that copper was established as a dietary essential element[98]. The literature related to the metabolism of copper in man[99-102] and animals[6] has been reviewed.

(a) Metabolism

(i) Absorption

An adult man consuming a normal diet will have a daily copper intake of 2–5 mg, of which approximately 30% is absorbed. A large proportion of the absorbed copper is excreted by way of the bile so that a negligible amount appears in the urine. Cartwright and Wintrobe[100] have depicted these pathways schematically as shown in Fig. 3.

Radioactivity appears promptly in blood after oral ingestion of radio-copper, suggesting that absorption occurs in the upper digestive tract. The mechanism involved is not known but there is evidence in the mouse for

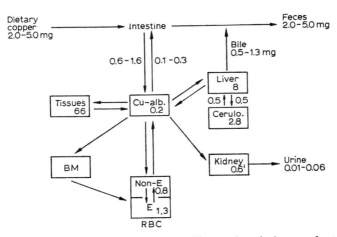

Fig. 3. Metabolic pathways of copper in man. The numbers in boxes refer to mg of copper in the pool and those next to arrows refer to daily turnover. Cu-alb., direct-reacting fraction; cerulo., ceruloplasmin; non-E., nonerythrocuprein; BM, bone marrow. Reproduced from Cartwright and Wintrobe[100].

both active and passive transport. In the chick, ^{64}Cu given orally is bound to duodenal protein having molecular weight[103] of approximately 10000. It is postulated that this protein plays a role in absorption of copper. Support for this hypothesis was provided by the fact that zinc and cadmium, inhibitors of copper absorption, displace copper from the duodenal protein. Endogenous biliary copper is poorly absorbed in rats, probably because it is bound to a specific protein which impairs absorption[104].

Absorption of copper is dependent to a large extent upon its chemical form and thus upon other dietary constituents. The copper of copper sulfide and copper porphyrin is poorly absorbed as is that in raw meat. High levels of zinc depress the absorption of copper from ligated loops of rat duodenum and the evidence indicates that it is a direct effect on the intestine[105]. Complexes of some amino acids with copper allow a higher rate of absorption[106] than that of $CuSO_4$. Such complexes may be absorbed intact.

(ii) Transport

Absorbed copper is transported principally by the plasma where it is first loosely bound to albumin. This copper reacts directly with sodium diethyl-dithiocarbamate and is termed direct reacting copper. After a ^{64}Cu dose, the radioactivity in the direct reacting fraction soon declines and activity appears in the copper protein, ceruloplasmin. This has been observed in both man[100] and the rat[107]. Ceruloplasmin copper, which does not react with diethyldithiocarbamate, constitutes approximately 95% of the element found in plasma. It does not exchange with ionic copper and thus ceruloplasmin does not fulfill a role for copper transport analogous to that of transferrin for iron transport. There is an extremely small fraction of plasma copper which is bound to amino acids. In spite of the small amount of this fraction, it may play a significant role in the transport of copper across membranes. Complexes with histidine, threonine and glutamine and with both histidine and threonine are present in normal serum[108].

In the rat, maximal concentrations of ^{64}Cu are quickly reached in liver, kidney and gastrointestinal tract, organs which then slowly release radioactivity. In other organs, accumulation of copper begins after ceruloplasmin makes its appearance in the plasma[107].

The highest concentrations of copper are found in liver (6.6 μg/g) and brain (5.4 μg/g) while the whole-body average for most vertebrates is 1.5–2.5 μg per g of fat-free tissue. The total body copper of man has been estimated at 80 mg, and plasma copper ranges[3] from 80 to 150 μg per

100 ml. Plasma levels in avian species are much lower, the mean for turkeys[109] being 5.0. No ceruloplasmin is found in turkey and peacock blood, and little or none in normal chicken blood[110]. The level in mammals varies[109] as follows: pig > man > rat > sheep > cow > dog. The copper concentration in most organs is quite uniform among species but that in liver is highly variable[6]. In rat liver, the copper is distributed among the subcellular fractions as follows: soluble fraction, 64%; mitochondria, 8%, microsomes, 5% and nuclei and debris, 20%. A similar distribution has been observed in human liver[102].

(iii) *Excretion*

The chief excretory pathway for copper is *via* the feces. Less than 1% of the intake is excreted in the urine and a negligible quantity in sweat[102]. In general, milk is low in copper; the non-fat solids of cow's milk after processing averages less than one p.p.m.

(iv) *Interrelationships*

Copper exhibits strong physiological interactions with other elements. An interrelationship between zinc and copper was first revealed by the observation that copper reverses zinc toxicity in rats. More recent studies have demonstrated that zinc in excess of the requirement accentuates copper deficiency[111]. There is a copper component of cadmium toxicity in addition to the better known zinc component[111]. Silver ions but not Hg^{2+} ions are antagonistic to copper in the nutrition of the growing chicken[112]. The interaction that has attracted widest attention involves copper, molybdenum and sulfate[6]. Dietary sulfate intensifies the harmful effect of molybdenum in ruminants but alleviates molybdenum toxicity in rats. One explanation for this dichotomy relates to the copper status of the animal[113]. Molybdenum fed to copper-deficient rats produces toxic symptoms which are intensified by sulfate, but prevented by copper. When copper stores are adequate or in surplus, sulfate completely prevents molybdenosis, whereas copper has little effect. *In vitro* $CuSO_4$ and Na_2MoO_4 react to form an insoluble complex which is believed to render the copper biologically inert[114].

(v) *Physiological function*

Pathology associated with copper deficiency includes anemia in all species studied, although its nature and severity varies widely. Other pathology includes neonatal ataxia or swayback in lambs, achromotrichia or lack of pigmentation, and connective tissue defects, including bone disorders and cardiovascular failure[6]. The manner in which copper influences erythrogenesis is not known, but it is doubtful that it directly involves hemoglobin synthesis. Absorption of orally administered iron from the gastrointestinal tract is impaired in copper-deficient swine[115]. Intramuscular iron increases iron storage in the reticuloendothelial and parenchymal cells of the liver without relieving the hypoferremia of copper deficiency. These observations have led to the hypothesis that copper is required for the release of iron from tissues, including the intestinal mucosa which accumulates iron in copper deficiency.

Neonatal ataxia in lambs, although preventable by dietary copper, is difficult to identify with simple copper deficiency. Pathologically, it is characterized by failure of myelination of the spinal cord and brain stem. An analogous syndrome has been produced in neonatal guinea pigs in which there is a decrease in brain phospholipid suggesting delayed myelination[116]. A defect in phospholipid synthesis has also been reported in the copper-deficient rat[117].

Spontaneous bone fractures occur occasionally among farm animals that consume copper-deficient forage. Dogs, pigs and chicks fed deficient diets frequently suffer fractures and severe bone deformities. These bone disorders appear to reflect a defect in the organic matrix rather than failure of calcification. Connective tissue integrity is impaired in the pig[118] and the chick[119,120] both of which experience dissecting aneurysms and spontaneous rupture of major vessels when deprived of dietary copper. The primary defect in the arteries is a failure to form normally crosslinked elastin but collagen crosslinking is also adversely affected[121]. The desmosines, compounds derived from peptidyl lysine and involved in covalently linking polypeptide chains in elastin[122], are not formed in the absence of copper[121,123]. Copper is an essential component of an enzyme which catalyzes the oxidation of the ε carbon of lysine. Fig. 4 depicts a possible mechanism by which desmosine in peptide linkage might arise from oxidative deamination of three moles of lysine followed by aldol condensation and reaction with a fourth lysine residue[122,123].

Analogous to the lathyritic state, a higher proportion of the collagen in the aorta and tendon of copper-deficient chicks is salt-soluble, indicating decreased intermolecular crosslinking[124]. Intramolecular bonding is also impaired[125]. An aldehyde derived from peptidyl lysine, α-aminoadipic-δ-semialdehyde, is the intermediate involved in the crosslinking of both

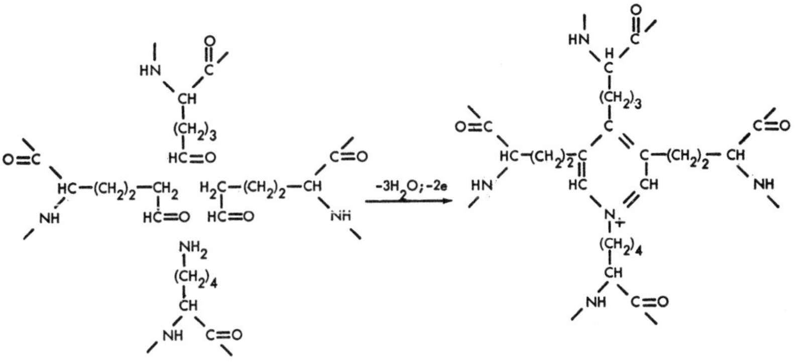

Fig. 4. Formation of desmosine from 3 molecules of α-aminoadipic-δ-semialdehyde and one of lysine as they exist in peptide linkage. The reaction involves dehydration followed by oxidation.

elastin and collagen and the aldehyde concentration of isolated collagen is decreased by copper deficiency[125,126]. A recent review of the mechanisms involved in the crosslinking of these proteins should be consulted for details[127]. The first oxidative step is catalyzed by a copper-dependent amine oxidase and there is an inverse relationship between the benzylamine oxidase activity in connective tissues of chicks and the proportion of the collagen which is salt-soluble[124]. Addition of cupric ion to an homogenate prepared from a copper-deficient aorta restores amine oxidase activity to near the normal level[128]. At least in avian species, the amine oxidase found in connective tissues is critical to life and is one of the first copper-dependent enzymes to be critically affected by nutritional deficiency.

Although copper deficiency can be readily produced in a variety of experimental animals and occurs naturally among domestic animals under certain environmental conditions, there is no clear evidence that an uncomplicated copper deficiency occurs in man[129]. A more important problem for man is the toxicity of copper and there must be a mechanism to regulate its retention. Although this mechanism is not known, there is an inherited disease in man, Wilson's disease, in which the mechanism does not function and copper accumulates in tissues, particularly brain, liver and kidney while plasma levels are lower than normal[101,102,130].

(b) Chemistry

Copper exists in two principal ionic states, Cu^+ and Cu^{2+}. The first, is a

d^{10} ion, *i.e.*, all of the 3 d orbitals are filled whereas Cu^{2+} is a d^9 ion. Cuprous ion has a radius of 0.93 Å, a size sufficient to easily allow four ligand atoms to be associated with it. The usual coordination number is four but it may coordinate with 2 or 6 ligands. Complexes of Cu^+ normally have a tetrahedral geometry while those of Cu^{2+} are square planar or tetragonal. The latter geometry, which involves four short and two long metal ligand bonds, is due to the Jahn–Teller effect commonly associated with d^9 ions. Because of the d^{10} structure of Cu^+, its compounds are diamagnetic and generally colorless. The d^9 electronic configuration of Cu^{2+} confers upon it the property of paramagnetism and in general its complexes are colored blue or green. Cu^{2+} forms more stable complexes with most ligands than the other transition metals of the first series[80].

Only with advent of electron-spin resonance spectroscopy was it possible to demonstrate that copper participates directly in the electron transfer catalyzed by cuproproteins[131]. Cu^{2+} reacts reversibly with mercaptides to form Cu^+ and disulfide.

$$Cu^{2+} + RS^- \rightleftharpoons Cu^+ + 1/2\ (RS)_2$$

If there is an excess of RS^-, CuSR is formed[132].

(c) Biochemical function

Cupric ion itself serves as oxidative catalyst, *e.g.*, in the oxidation of ascorbic acid by molecular oxygen to form dehydroascorbic acid. However, the catalytic activity of the cuproprotein, ascorbate oxidase, is 1000 times as great as that of an equivalent amount of cupric ion. Cupric ion also catalyzes the hydrolysis of the disulfide bond of model compounds[133]. Essentially all copper-containing metalloenzymes are concerned with catalysis of oxidation–reduction type reactions in which O_2 is the electron acceptor. Whether or not copper atoms undergo alternating valence change in all cases while catalyzing such reactions is not clear. Certainly not all of the copper in metalloenzymes undergoes such change. Some of the copper atoms may be involved in maintenance of an essential conformation of the protein, some with binding of the substrate and some may have no identifiable role. These functions will be discussed in connection with the metalloenzymes, listed in Table II, which are physiologically important.

(i) Cytochrome oxidase (EC 1.9.3.1)

Although cytochrome oxidase was first recognized as a hemoprotein, it

TABLE II

COPPER METALLOENZYMES

Enzyme	EC number	Source	Molecular weight	Cu content		Refs.
				%	g. atoms/mole	
Cytochrome oxidase	1.9.3.1	Beef heart mitochondria	290 000	Cu: Heme = 1	3	139, 140, 141, 171
Ceruloplasmin		Blood plasma	160 000	0.32	8	145, 172
Ascorbate oxidase	1.10.3.3	Squash	140 000	0.34	8	173
Tyrosinase	1.10.3.1	Mushroom	32 000	0.20	1	148
Laccase	1.10.3.2	Lac tree (Rhus)	120 000	0.25	4	174
Galactose oxidase	1.1.3.9	Mold	75 000	0.85	1	175
Uricase	1.7.3.3	Liver and kidney	120 000	0.05	1	150
Dopamine-β-hydroxylase	1.14.2.1	Adrenals	290 000	0.05	2	151
Diamine oxidase	1.4.3.6	Pea seedlings	96 000	0.09	1	176
Diamine oxidase	1.4.3.6	Kidney	185 000	0.07	2	155, 156
Plasma amine oxidase	1.4.3.4	Blood plasma	195–255 000	0.09–0.1	4	161, 162

is now clear that it contains copper as well as iron. The first suggestion of copper's involvement came from nutritional studies[98,117,134] which showed that the cytochrome oxidase activity of heart muscle in copper-deficient rats, pigs, cattle and sheep is greatly reduced. In the classic paper on cytochrome a_3 Keilin and Hartree[135] observed that their heart preparation contained copper and postulated that cytochrome oxidase is composed of cytochromes a and a_3. There is now no reasonable doubt[136-138] that cytochrome oxidase contains copper and in a 1:1 ratio with heme a but the concept of two components, a and a_3, has been challenged[139]. Because it is bound to mitochondrial membranes and is lipoprotein in nature, cytochrome oxidase cannot be solubilized without the use of a detergent. For this reason it has been difficult to obtain a homogeneous enzyme in the native state, and estimates of its molecular weight vary widely. Using the detergent, Emasol 1130, Okunuki et al.[140] obtained a species, molecular weight of 530000, which is believed to be a tetramer. Morrison[141] found a 580000 and a 290000 molecular weight fraction. According to his model, cytochrome oxidase consists of two cytochrome a units each with a molecular weight of 120000 and one cytochrome a_3 unit with 60000 molecular weight. Each unit contains one heme a and one copper atom.

The nature and function of the copper in cytochrome oxidase is not entirely clear. Only 30–40 % of the copper is accounted for as cupric copper in the electron-spin resonance spectrum[138]. Both cytochrome a_3 and the copper component of cytochrome oxidase lose their unique characteristics at elevated pH. Beinert and Palmer[142] have raised the question whether or not the unusual properties of a_3 may not be due to an interaction of cytochrome a and copper. Under anaerobic conditions and in the presence of cytochrome c, copper can be removed from cytochrome oxidase with loss of enzymatic activity[143]. The apoenzyme contains about 45 % of the original copper and possesses 15 % of the activity. Copper can be restored by $[Cu(CH_3-CN)_4]ClO_4$ and activity is regained. Thus, at least part of the copper in the enzyme undergoes valence change during catalytic activity and part is essential for enzyme activity.

(ii) Ceruloplasmin

Although the physiological role of ceruloplasmin is not known with certainty, this cuproprotein does possess oxidase activity and is here classed as a metalloenzyme. Its more critical function may relate to copper balance and transport[144]. Generally, patients with Wilson's disease possess low levels of plasma ceruloplasmin but this is not an absolute relationship. Mammalian ceruloplasmin[145] has a molecular weight of approximately 160000 and contains 8 copper atoms, 4 as Cu^+ and 4 as Cu^{2+}. The copper atoms are readily exchangeable under reducing conditions and apoceruloplasmin will combine with copper to reconstitute ceruloplasmin. The protein contains carbohydrate, equivalent to 12200 molecular weight, including 9 moles of sialic acid[144]. Ceruloplasmin isolated from the serum of growing chickens infected with *Salmonella gallinarum* has a molecular weight of 158000 and contains 0.2% copper, 5 g.atoms per mole[146].

The oxidase activity of ceruloplasmin is usually measured by oxygen consumption or absorbance change at 530 mμ when *p*-phenylenediamine serves as substrate. Transferrin and apotransferrin are potent inhibitors of the reaction, which may be related to the fact that Fe^{2+} stimulates ceruloplasmin activity. It is generally believed that iron is absorbed from the intestine in the Fe^{2+} state and in plasma this is rapidly oxidized to Fe^{3+} before it is incorporated into transferrin. A study of the oxidation of Fe^{2+} by serum[147] showed a significant catalytic activity in serum which correlates with ceruloplasmin levels. A postulated role for ceruloplasmin is the promotion of iron utilization by stimulation of iron oxidation and transferrin saturation.

(iii) Tyrosinase (EC 1.10.3.1)

This enzyme, which occurs in both plant and animal tissues, is also known as polyphenoloxidase, phenolase, cresolase and catecholase. It appears to catalyze two types of reactions, hydroxylation of monophenols and dehydrogenation of catechols[148].

Cresol has been commonly used as a substrate for the first reaction and catechol for the second, hence, the names cresolase and catecholase. Preparations may have high, low or intermediate ratio of catecholase to cresolase activities and this has been explained on the basis of two active sites[148]. Copper is readily displaced by hydrogen ion with concomitant loss of enzymatic activity, the two types of activity showing different susceptibility. The essential copper in mushroom polyphenoloxidase appears to be entirely in the Cu^+ state[149]. Copper dissociates from the enzyme during storage and during catalysis. Copper can be readily restored by adding Cu^{2+} but it is reduced to Cu^+. The ease of dissociation of copper from this enzyme raises a question whether or not it should be classed as a metalloenzyme.

(iv) Uricase (EC 1.7.3.3)

Although uricase has been known since the early years of biochemistry, it was only recently recognized as a cuproprotein[150]. Copper is bound tenaciously to the protein and is not dissociated by trichloroacetic acid or strong chelating agents. The enzyme is inhibited by cyanide and other complexing agents which are also reductants. These inhibitions are overcome by urate at low concentrations.

(v) Dopamine-β-hydroxylase (EC 1.14.2.1)

This enzyme is a monooxygenase concerned with side-chain hydroxylation of phenylethylamine and its derivatives. Dopamine-β-hydroxylase derived its name from its role in biosynthesis of norepinephrine. It catalyzes the hydroxylation of 3,4-dihydroxyphenylethylamine (dopamine) according to the following equation:

$$Dopamine + O_2 + ascorbate \rightarrow norepinephrine + H_2O + dehydroascorbate$$

The reaction is catalyzed by fumarate and other dicarboxylic acids and catechol may substitute for ascorbate. The enzyme contains 2 g.atoms of Cu^{2+} per mole and it appears that these are reduced to Cu^+ by ascorbate before the reaction is catalyzed[151].

(vi) Amine oxidases (EC 1.4.3.4, 1.4.3.6)

Enzymes of this class exhibit a wide range of substrate specificity and contain at least one of three prosthetic groups, copper, pyridoxal phosphate or flavin adenine dinucleotide. All catalyze the oxidative deamination of amines according to the following stoichiometry:

$$R\text{-}CH_2\text{-}NH_2 + O_2 + H_2O \rightarrow R\text{-}CHO + NH_3 + H_2O_2$$

Those enzymes that have the greatest affinity for short-chain diamines, such as putrescine and cadaverine, have been generally classed as diamine oxidases and those that preferentially catalyze the oxidative deamination of aliphatic or arylalkyl monoamines as monoamine oxidase. This classification is not entirely satisfactory because of the considerable overlap of activity among the two classes. Classification as to the nature of the organic prosthetic group, flavin or pyridoxal, is more definitive, but because of the historical relationship of enzyme name and its substrate, monoamine and diamine oxidase will be retained here. Reviews of the literature before 1963 relating specifically to diamine oxidase[152] and monoamine oxidase[153] as well as more recent advances in amine oxidase research[154] have appeared.

The amine oxidase found in pea seedlings is typical of the diamine oxidases. It contains one g.atom of copper per mole and probably contains pyridoxal phosphate as a prosthetic group[152]. Treatment with diethyl-dithiocarbamate removes copper and renders the enzyme inactive. Activity is restored by Cu^{2+}. Pig kidney is the most common animal source of diamine oxidase where it is found in the supernatant fraction. It has been obtained as a homogeneous, crystalline, pink-colored protein[155,156]. Pig-kidney diamine oxidase does not contain flavin but spectrophotometric studies suggest the presence of pyridoxal phosphate[155]. The metabolic function of kidney diamine oxidase has not been defined more specifically than as a regulator of the cellular concentration of diamines and polyamines.

Monoamine oxidases are typically intracellular enzymes bound to mitochrondrial membranes. They are not inhibited by carbonyl reagents and probably do not contain pyridoxal phosphate[153,157]. When isolated in a highly purified state, mitochondrial monoamine oxidases have a bright yellow color and contain a flavin prosthetic group[157-159], probably flavin adenine dinucleotide. There is some evidence[160] that mitochondrial monoamine oxidase contains copper, but most investigators[157,158] find an insignificant quantity in purified preparations. It seems likely that these enzymes are flavoproteins, not dependent upon copper.

There is a group of extracellular amine oxidases which cannot be easily classified as either mono- or diamine oxidases. They have been isolated from blood plasma and are probably best described as plasma amine oxidases[154]. They differ somewhat in substrate specificity but have common prosthetic groups. The amine oxidases found in plasma of ruminants catalyze the oxidative deamination of spermine and spermidine, whereas plasma amine oxidase of non-ruminants generally have no affinity for the polyamines[153]. Those found in human, rabbit and pig plasma, catalyze the oxidation of benzylamine, tyramine, histamine, and the aliphatic mono-amines[154]. Activity in avian blood is negligible[128]. Crystalline plasma amine oxidases from beef[161] and pig plasma[162] are pink proteins with absorption maxima at about 480 mμ. They contain 4 g.atoms of copper and at least 2 moles of pyridoxal phosphate per mole of enzyme. Copper deficiency in pigs[163] and sheep[164] results in decreased plasma amine oxidase activity.

The metabolic function of the plasma amine oxidases is not known, but one may speculate that they are found in plasma because they have an extracellular function and simply leak into the tissue fluids. Connective tissue, including skin, cartilage, aorta and tendon, from copper-deficient animals contains much less than the normal level of benzylamine oxidase activity[124,126,128]. It is postulated that this enzyme is concerned with the oxidative deamination of the ε-amino group of peptidyl lysine, forming the aldehydic function which is involved in the crosslinks of both elastin and collagen[124,125]. As discussed above there is failure of maturation of these proteins in copper[121] and pyridoxine deficient animals[165]. β-Amino-propionitrile, a lathyrogen which also decreases elastin and collagen cross-linking, is a competitive inhibitor of the aortic amine oxidase[128].

(vii) Less-well defined copper metalloproteins

There is no doubt that other proteins in nature qualify as copper metallo-enzymes. Some of these are outside the scope of this review, and for some, the evidence is equivocal. Mention should be made of δ-aminolevulinate dehydratase, mercaptopyruvate transsulfurase, and plastocyanin. The δ-aminolevulinate dehydratase of *Ustilago sphaerogena* has been purified 40-fold and there is a correlation between copper content and enzyme activity[166]. There are unconfirmed claims that this enzyme from other sources contains copper[167] and plastocyanin, first isolated from a green

TABLE III

NON-CATALYTIC CUPROPROTEINS

Protein	Source	Molecular weight	Cu content		Refs.
			(%)	(g. atoms/ mole)	
Hemo-cuprein	Bovine erythrocytes	35000	0.34	2	177
Erythro-cuprein	Human erythrocytes	33000	0.32–0.36	2	178, 179
Hepato-cuprein	Mammalian liver	?	0.34	—	177, 180, 181
Cerebro-cuprein	Human and bovine brain	30000	0.29	2	181, 182
Mitochondro-cuprein	Human and bovine neonatal liver	—	3–5	—	181
Hemocyanin	Snail	900000	0.23	32	183

alga[168] and found generally in chloroplasts is commonly classed as a copper protein.

A group of cuproproteins for which no catalytic function is known is listed in Table III. Their chemical nature has been reviewed previously[95]. The soluble proteins erythrocuprein, hepatocuprein and cerebrocuprein, contain approximately 0.3% of copper and have molecular weights of about 30000. In spite of their highly similar physical properties they are probably not identical. These proteins constitute well over one-half of the copper found in the respective tissues from which they are isolated. This fact suggests that their chief function is one of copper storage[169]. Neonatal hepatic mitochondrocuprein, which is insoluble and contains more than 3% of copper, probably serves as a storage form of copper for the newborn infant. Hemocyanin is a unique cuproprotein in that it performs a specific physiological function, namely, oxygen transport. It is found in the hemolymph of mollusks and arthropods[170]. One mole of O_2 is bound for each two g.atoms of copper. Copper is easily removed by dialysis against acid or cyanide.

4. Zinc

Zinc, the 25th most abundant element, was first shown to have biological significance in 1869, when it was found to be required for the growth of *Aspergillus niger*. Since then specific requirements for zinc have been shown in several species of higher plants and animals[6]. Although Bertrand and coworkers[184] had attempted to demonstrate that zinc is essential for mice, success was first achieved using this species at about the same time as it was shown to be required by the rat[185]. Evidence that zinc is required by avian species[186] and man[187] appeared much later. A metabolic role for zinc was first provided by the demonstration that carbonic anhydrase (EC 4.2.1.1) contains 0.33% zinc and that the element is essential for the activity of the enzyme[188]. Since that observation zinc has been found to be an essential component of several metalloenzymes of key metabolic significance. Recent reviews of zinc metabolism have provided excellent coverage of the literature[6,189–192].

(a) Metabolism
(i) *Absorption*
The normal human intake of zinc ranges from 10 to 15 mg per day[193],

and its absorption occurs chiefly in the distal portion of the small intestine[194]. While most work suggests that zinc is poorly absorbed from natural diets[6], it has been shown that when low levels of zinc salts are added to purified diets as much as 80% of the element is absorbed by rats[195]. By use of [65]Zn and a whole-body counting technique, zinc absorption in growing chickens fed practical type diets has been found to range from 20 to 30%[196]. In ruminants a dietary deficiency of zinc increases its absorption and decreases endogenous fecal losses[197]. The results of a study involving 13 young college women indicate that of an average intake of 13.2 mg of zinc per day, 6.6 mg is retained, 5.6 excreted in the stool, and 1.0 mg excreted in the urine[198]. In another investigation[199] involving 15 human subjects, the average apparent absorption of [65]Zn was 50.8%.

Relatively little is known about the mechanism or sites of zinc absorption. Studies with everted gut sacs from rats show that zinc is taken up by the mucosa but not transported across the intestinal wall[194,200]. Mucosal uptake was not affected appreciably by anaerobic conditions or cyanide and the mechanism should not be considered active transport. Ligated portions of the intestinal tract absorb [65]Zn at different rates[201]. The duodenum has the highest rate of absorption, followed by ileum and jejunum in that order. This pattern is similar to that of iron absorption.

The absorption of zinc from the gastrointestinal tract is markedly affected by other dietary components. Naturally occurring metal complexing compounds, such as phytate decrease absorption while certain amino acids and peptides, as well as synthetic chelating agents, such as ethylenediaminetetraacetate, increase absorption under some conditions[202]. The zinc in plant proteins such as soybean protein is not absorbed as efficiently as that in animal proteins. It was first demonstrated in chickens that dietary phytate, a compound which is largely restricted to plants, decreases zinc absorption[203]. This observation has been confirmed in several species[200,204,205]. High levels of phosphate also decrease zinc absorption as does excess calcium when added to diets based largely on plant protein. Calcium is believed to exert its effect only when phytate is present, giving rise to a highly insoluble complex of calcium, zinc and phytate[200]. When added to diets that contain phytate, ethylenediaminetetraacetate and amino acids such as cysteine increase zinc absorption[202,206].

(ii) Transport

In human blood 75–85% of the zinc is contained in erythrocytes, 12–22%

in the plasma, and 3% in the leucocytes. Nevertheless, a single leucocyte contains 25 times as much zinc as a single erythrocyte[6]. Zinc in the plasma is bound to protein in two forms: (a) firmly bound zinc which appears to be in combination with globulin makes up 34% and (b) loosely bound zinc complexed with albumin makes up 66% of the total[207]. Following oral administration of [65]Zn to the dog, 20–30% is bound to α_1- and α_2-globulins, 20–30% to serum albumins, and the remainder to the β_1- and γ-globulin fractions. All serum fractions contain both firmly and loosely bound combinations, but zinc seems to be most strongly bound to the globulins[208]. In the human several serum proteins, albumin, α-, β-, and γ-globulins, bind zinc with the α-globulin binding most tenaciously[191]. After intravenous injection of [65Zn]glycine into rabbits, 90% of the labeled zinc is cleared from the bloodstream in 3 h. During the 3 to 24 h period all of the zinc in the plasma is complexed and approximately one-half is bound to α-globulin[209]. The quantitative aspects of zinc binding to albumin have been reported by Gurd and Wilcox[4], and the combination of zinc with the α_2-glycoprotein of human plasma has been described by Burgie and Schmid[210]. In vitro studies indicate that zinc can combine with transferrin[211], but there is not convincing evidence that this protein is responsible for transport of zinc[6]. A specific zinc carrier in plasma has not been identified.

As regards the cellular components of blood, it is likely that almost the entire zinc content of erythrocytes occurs in carbonic anhydrase[212]. A zinc-containing protein is also present in human leucocytes[213], but no physiological role has been found for this metalloprotein. Although leucocytes concentrate zinc more effectively than erythrocytes, they do not exchange [65]Zn back to plasma, whereas there is an appreciable erythrocyte-plasma exchange[209]. Thus the red blood cell serves as a source of mobilizable zinc in the animal body[208,214].

(iii) Storage

The zinc content of a normal 70-kg man ranges from 1.4 to 2.3 g[215]. Most human organs contain between 20 and 30 μg of zinc/g wet tissue; however, the metal is found in higher concentrations in bones, the choroid of the eye, the prostate gland and its secretions, and the skin and its appendages[216]. It is possible to raise the level of zinc slightly in most tissues by increasing the dietary intake, but storage in organs other than the bones is limited. The concentration of zinc in bone has been used as a measure of zinc absorption and nutritional status in young growing animals[204]. The high zinc content of the choroid of the eye has been shown to be due to a high concentration of the metal in the *tapetum lucidum cellulosum* which is thought to be responsible for the iridescence of the eyes of carnivores[217].

On a dry weight basis the dorsolateral rat prostate contains[218] 874 ± 63 μg Zn/g; the rabbit prostate 1296 ± 31 μg/g, and the normal human prostate, 859 ± 96 μg/g. The zinc concentration in human semen ranges[219] from 100 to 200 μg/g. Selective accumulation of ^{65}Zn occurs in the dorsolateral prostate of the rat[220]. The transitional zone of human epidermis concentrates zinc, and the data suggest that the metal is combined with protein-bound histidine[221]. The zinc concentration in the hair of both man and swine has been employed as an index of zinc status[222].

Autoradiographic and cytological techniques demonstrate that ^{65}Zn is not localized in any particular region of the liver and that it occurs in higher concentrations in the cortex than in the medulla of kidney and adrenal gland[223]. When injected into dogs, mice and rats ^{65}Zn accumulates first in the pancreas, liver and spleen, and subsequently a large portion is transferred to bone[224]. Three investigations[225–227] of intracellular zinc in rat liver indicate the following distribution, expressed on the basis of μg Zn per mg nitrogen: nuclei, 0.77; mitochondria, 0.42; microsomes, 0.65; supernatant 2.0; and reconstituted whole liver, 1.05.

(iv) Excretion

Zinc whether ingested or injected is excreted primarily by way of the feces. Thus, fecal zinc consists of the unabsorbed portion in food as well as the endogenous zinc from the pancreatic juice, bile and other digestive secretions[228]. In rats approximately 70% of a single intravenous dose of ^{65}Zn is excreted by way of the intestine[229]. A similar observation in mice led Cotzias[230] to suggest that zinc metabolism is controlled by at least two homeostatic mechanisms, which act at intestinal sites, one of absorption and one of excretion. Miller[197] has reviewed the literature on absorption and excretion of zinc by ruminants and concluded that the intestine is the dominant route of excretion in these animals. In growing chickens, surgically prepared so as to allow separation of urine and feces, less than 35% of the zinc is excreted in urine regardless of diet, but a higher proportion appears in feces when phytate is fed[231].

Some zinc is lost in sweat. In a study of human subjects in Egypt the zinc content of whole sweat was 115 ± 30 μg/100 ml and that of cell-free sweat was 93 ± 26 $\mu g/100$ ml. It is evident from these results that in a hot climate 2–3 mg of zinc per day may be lost in sweat.

(v) Turnover

When [65]Zn is administered intravenously to man most of it disappears from the blood within 24 h, though a minor fraction can be detected in the blood for many weeks[232]. In the rat only 5% of a dose remains in the plasma[233] after 30 min. An appreciable portion is stored in bone. 3 days after an intravenous dose of [65]Zn, 11–15% of the isotope is present in the skeleton and it remains as long as 37 weeks[229]. In steers 10% of a [65]Zn dose could be found in the serum 4 h after injection but only 3% after 10 h, when equilibrium was reached[234]. Equilibrium of zinc between soft tissues and plasma zinc is established within a few hours following injection[229,234]. Accumulation and turnover of retained zinc are most rapid in the pancreas, liver, and kidney of dogs[228]; in the pancreas, liver, kidney, pituitary, and adrenals in steers[234]; and in the liver, spleen, and kidney in humans[235]. Although the dorsolateral prostate of rats concentrates a single dose of [65]Zn 15–25 times as much as other tissues, the turnover rate of zinc in this tissue is slower than in other tissues[236]. Rates of turnover of zinc in muscle, red cells and testis are intermediate between those of bone and hair which are extremely slow, and those of the glandular organs mentioned above[229].

(vi) Interrelationships

Under some dietary conditions high levels of calcium aggravate zinc deficiency[237] but as discussed in the section on absorption this seems to be related to the level of phosphate, and more particularly, to the level of phytate in the diet[202]. Only when the diet contains natural foodstuffs or added phytate is there clear evidence that calcium is antagonistic to zinc absorption and utilization. In vitro, calcium, zinc and phytate combine to form a complex that is less soluble than either zinc or calcium phytate[200,238]. When such a complex is formed mucosal uptake of [65]Zn is decreased and it is postulated that, by this mechanism, excess calcium decreases the absorption of zinc from the intestine[200].

As described in the section on copper, excess zinc accentuates copper deficiency. In view of this physiological interaction one might expect an effect of excess copper on zinc metabolism. Using ligated intestinal loops in rats, Van Campen[239] showed that simultaneous addition of Cu^{2+} depressed the absorption of [65]Zn. This observation confirms the existence of a copper–zinc interaction at the site of the intestinal mucosa. Other investigations of copper–zinc interrelations are controversial. Copper at high levels (125–250 p.p.m.) has been reported to alleviate the symptoms of zinc deficiency in pigs[240], but others have observed either no effect[241] or an increased incidence of parakeratosis when 0.1 % copper sulfate was fed to pigs[242].

The Cd^{2+} ion, which has a similar configuration, interacts physiologically with zinc. This is dramatically demonstrated by the fact that Zn^{2+} administered simultaneously with Cd^{2+} counteracts the toxic effect of subcutaneously administered cadmium. Cadmium alone causes rapid destruction of rat and mouse testis[243,244]. In avian species dietary cadmium accentuates zinc deficiency signs which are largely overcome by zinc[245,111]. When ^{65}Zn is given parenterally, cadmium depresses both intestinal excretion of zinc[246] and its uptake by testis and dorsolateral prostate[247].

(vii) Physiological function

The pathology of zinc deficiency has been described for several species, including rats[248], mice[249], pigs[250], calves[251,252], lambs[252,253], goats[254], dogs[255], monkeys[256], chickens[257], turkeys[258], Coturnix quail[259] and man[191]. Although the deficiency signs manifested are somewhat species-dependent, there are several common features that can be defined as typical of zinc deficiency. These, in order of frequency of occurrence, are (1) lesions of the skin and its appendages, (2) abnormalities of the skeleton, particularly related to the long bones and epiphyseal cartilage and (3) defects in the reproductive organs, particularly testicular development in the male. In all species growth rate is depressed, but this is directly related to loss of appetite and decreased food intake. Unless controls, whose food intake is carefully monitored are provided, it is difficult to delineate those signs that are specific for zinc from those due to inanition.

Pathology of the skin, and the epithelium in general, is the most common sign of zinc deficiency. This is observed most dramatically as parakeratosis of the skin in the pig, but parakeratosis of the esophagus occurs in all species examined. In some species a dermatitis more correctly described as hyperkeratosis is common. Associated with the skin lesion there is loss of hair and in the case of birds, poor feathering.

Skeletal defects are especially common among the avian species subjected to zinc deficiency. In the developing embryo the formation of long bones may fail entirely[260] and normal chicks fed a zinc-deficient diet from hatching develop short thick bones with exceptionally large cartilage caps which give a typical enlarged hock appearance. Such chicks also have a stilted or arthritic gait which is relieved in part by histamine and antiarthritic drugs[206]. Dwarfism is typical of the zinc-deficiency syndrome observed in man[261].

Failure of testicular development is commonly associated with zinc deficiency at least in mammals. In the male rat there is atrophy of germinal epithelium and retarded development of the epididymes and prostate[219]. Hypogonadism has been observed in calves and is part of the syndrome observed in the zinc-deficient human male. Pathology of reproductive system in the female is not as obvious as in the male but female rats have abnormal estrous cycles and eventually fail to reproduce[262]. Furthermore there is a high incidence of congenital abnormalities among their offspring. Zinc stores are not sufficient to support an adequate pregnancy if none is supplied during gestation and at parturition the female exhibits an abnormal, depressed behavior and shows a lack of the normal maternal instincts[263].

(b) Chemistry

The elemental form of zinc has two s electrons outside a filled 3d shell and upon oxidation loses both of these electrons to give the divalent ion, the biologically important form. There is no evidence for any other oxidation state and the ion does not undergo oxidation–reduction nor does it transfer electrons during enzymatic action. Zinc ions in aqueous solution at elevated pH normally form a hydroxide precipitate unless a complexing species such as an amino acid, peptide, protein, or organic chelator, e.g., o-phenanthroline, is present.

There are no ligand field stabilization effects in the zinc ion because of the completed d shell. Therefore, the stereochemistry of zinc complexes is determined solely by size, electrostatic forces, and covalent bonding forces. Zinc occasionally has a coordination number of six forming octahedral complexes, but the more common coordination number is four as it exists in the tetrahedral complexes, $[Zn(CN)_4]^{2-}$ and $[Zn(NH_3)_4]^{2+}$. A few cases of square-planar zinc complexes, for example, bis(glycinyl)zinc are known.

(c) Biochemical function

The role of zinc in promoting and controlling metabolic events at the molecular level is becoming increasingly evident, but it is not yet possible to directly relate biochemical function to the pathology of deficiency. The participation of zinc as catalyst in a great variety of biochemical systems has been observed. In some of these systems it serves as part of metallo-enzymes and in others as a component in conjunction with metal-activated enzymes and nucleic acids. A list of currently recognized zinc metalloenzymes is provided in Table IV. This table was adapted from a manuscript in preparation[264] made available to the authors by Dr. Vallee who has made major contributions to this area[1]. Those enzymes of particular importance in higher animals are discussed below.

Carboxypeptidase A (EC 3.4.2.1) is an enzyme which digests polypeptide chains starting from the carboxyl terminal end and preferentially attacking peptide bonds that involve aromatic residues. The enzyme also catalyzes the hydrolysis of esters. Variants of this enzyme designated carboxy-peptidase A, α, β, γ, and δ contain different N-terminal sequences[265]. Zinc can be removed from carboxypeptidase A with subsequent loss in activity in direct proportion to the amount of zinc removed[266]. Total activity can be restored to the apoenzyme by dialyzing against zinc containing buffers. It is possible to replace zinc with other metals, and the activity and specificity are dependent upon the particular metal at the active site[267]. One g.atom of Co^{2+}, Ni^{2+}, or Fe^{2+} per mole of apoenzyme will restore both peptidase and esterase activities[267], but when Zn^{2+} is replaced by Cd^{2+}, Hg^{2+}, or Pb^{2+}, the resulting enzymes catalyze the hydrolysis of esters but not of peptides[268]. Acetylation of carboxypeptidase A with either acetic anhydride or acetylimidazole increases esterase activity 7- to 8-fold, while peptidase activity is abolished completely[269]. It has been shown that apocarboxy-peptidase can bind dipeptide substrates without the presence of zinc, and indeed prior binding of apoenzyme with peptide substrate blocks restoration of activity by the addition of zinc ions[270]. In contrast esters and N-acetyl-amino acids require zinc for binding and do not form complexes with the apoenzyme[271].

Chemical sequence work[272,273] and high resolution X-ray diffraction analyses[274-279] have provided considerable insight into the role of zinc as it participates in the mechanisms of action of carboxypeptidase. Recent reviews of this work include those presented by Dickerson and Geis[280] and by Lipscomb[280a]. The carboxypeptidase molecule is roughly spherical, about $50 \times 42 \times 38$ Å, with 307 constituent amino acid residues. The zinc atom resides in a broad depression in the surface of the molecule. This depression is the active site, and the substrate specificity is produced by a

TABLE IV

ZINC METALLOENZYMES

	EC number	Cofactors	Mol. wt.	g.atom Zn/mole	Source	Refs.
Peptidases and esterases						
Carboxypeptidase A	3.4.2.1	None	34 300	1	Bovine pancreas	208, 268
Carboxypeptidase B	3.4.2.2	None	34 300	1	Porcine pancreas	281, 283
Carbonic anhydrase	4.2.1.1	None	30 000	1	Bovine erythrocytes	188, 296
Alkaline phosphatase	3.1.3.1	None	89 000	4	E. coli	330, 331
Neutral protease		None	44 700	1–2	B. subtilis	332, 333
Renal dipeptidase		None	47 200	1	Porcine kidney	306, 334
Dehydrogenases						
Alcohol dehydrogenase	1.1.1.1	NAD	150 000	4	Yeast	335, 336
Glutamate dehydrogenase	1.4.1.2	NAD	1 000 000	2–6	Beef liver	320, 337
Malate dehydrogenase	1.1.1.37	NAD	40 000	1	Beef heart	324
D-Lactate cytochrome reductase	1.1.2.4	NAD	50 000	4–6	Yeast	338, 339

nearby hydrophobic pocket into which the aromatic side-chain of the substrate is inserted. The zinc atom seems to be tetrahedrally coordinated to nitrogens of His 69 and His 196 and to an oxygen of Glu 72. The fourth ligand to zinc is a coordinated water molecule which is replaced by the carbonyl oxygen of the substrate's susceptible peptide bond. Three other side-chains, Arg 145, Tyr 248, and Glu 270, are arranged around the edge of the active site and appear to interact directly with the substrate swinging to a new configuration in the process. Tyr 248 changes its position 14 Å to place its phenolic group near the peptide nitrogen of the bond to be split. Arg 145 moves 2 Å to provide interaction with the substrate's terminal carboxyl group. Zinc coordinates with the oxygen of the carbonyl group of the substrate, and Glu 270 moves nearer the carbonyl carbon atom. It is proposed that Glu 270 acts as a nucleophile which forms a transient anhydride with the carbonyl group of the substrate[277]. At the same time the zinc atom both orients the substrate and polarizes the carbonyl group, aiding in the formation of the anhydride. The Tyr 248 proton attacks the nitrogen of the peptide bond. Subsequent reaction with water cleaves the anhydride and restores the tyrosine proton producing complete hydrolysis of the peptide bond and restoring the enzyme side groups to their original state. This postulated mechanism is in accord with amino acid sequence and X-ray diffraction data and provides a precise role for zinc in the mechanism of action of a metalloenzyme.

Carboxypeptidase B (EC 3.4.2.2) differs from carboxypeptidase A in that it preferentially catalyzes the hydrolysis of *C*-terminal basic amino acids. It has been isolated from porcine and bovine pancreas and shown to be a zinc metalloenzyme[281,282]. Removal of zinc results in a loss in activity that can be restored by zinc or by other metals[283,282]. It has been indicated that a thiol group acts to bind zinc in the active center as a metal-mercaptide[284].

Carbonic anhydrase (EC 4.2.1.1) is widely distributed in animals and has been isolated in pure form from bovine and human erythrocytes. The enzyme catalyzes the hydration of CO_2 and the dehydration of carbonic acid. It also catalyzes the hydrolysis of esters, for example, *p*-nitrophenyl acetate, and the hydration of aldehydes. Zinc is essential for activity and upon reaction of the enzyme with complexing reagents activity is lost. Aromatic sulfonamides strongly inhibit the activity of carbonic anhydrase[285], and zinc is essential for the firm binding of the sulfonamide inhibitors[286]. Several monovalent anions also inhibit the enzyme[287,288], and it has been suggested that these ions, like sulfonamides, bind to the metal or to an adjacent site. Kernohan[289] has suggested that a proton dissociation (pK=6.9) from the active center of the enzyme accounts for the reversible catalysis. The hydration reaction is presumably catalyzed by the unprotonated form, and dehydration by the protonated form. Subsequent work has shown that binding of a single proton to the active site abolishes CO_2 hydration[290,291]. After measuring the binding of azide, sulfide, and cyanide to the enzyme, Coleman suggested that the active form of the enzyme for hydration is E–Zn–OH[292]. Infrared spectroscopic measurements indicate that bicarbonate ion is coordinated to zinc through its negatively charged oxygen atom during the dehydration reaction[293] and it has been suggested that a hydroxyl group coordinated to zinc attacks bound CO_2 during hydration and converts[291] it to HCO_3^-. Inactivation of carbonic anhydrase by reaction with bromoacetate and iodoacetamide implicates a histidine residue in the catalytic mechanism[294], and identification of histidine as a component of the active center of the molecule has been confirmed[295].

Zinc can be replaced by cobalt in carbonic anhydrase and the resulting enzyme is active[296]. Binding of acetazolamide and cyanide to the cobalt enzyme can be followed by measurement of changes in the absorption spectrum and in optical rotatory dispersion[291,297]. X-Ray diffraction data currently available for carbonic anhydrase indicate that the active site exists as a deep depression with a catalytically essential zinc atom at the bottom[298].

Alkaline phosphatase (EC 3.1.3.1) catalyzes the hydrolysis of a variety of phosphate esters including *p*-nitrophenylphosphate, β-glycerophosphate, 5′-AMP and ATP. It can also act as phosphotransferase. The enzyme isolated from *E. coli* has a molecular weight of 89 000 and is composed of two identical polypeptide chains[299]. The dimer can be dissociated into enzymatically inactive monomers[300,301]. The complete enzyme contains 4 g.atoms of zinc, two of which are required for activity and two for maintenance of quaternary structure[302].

During the course of hydrolysis the enzyme forms a phosphoseryl intermediate suggesting that a serine residue is involved at the active center[303]. Two ionizable groups (pK 7.1 and pK 8.6) are important for the binding of the substrate to the active center, and it has been suggested that one of these dissociations may be the ionization of a zinc-coordinated water molecule[304]. Photooxidations of the apoenzyme in the presence of Rose Bengal renders it incapable of reactivation by zinc. There is a loss of 3 histidine residues that are not photooxidized when the native zinc enzyme is subjected to the same treatment[305]. Alkaline phosphatases from swine kidney and human leucocytes have not been purified to the extent necessary for investigations of mechanism of catalysis.

Renal dipeptidase isolated from particulate fractions of swine kidney cortex, has been characterized as a zinc metalloenzyme, which contains[306] one zinc atom per molecular weight of 47 200. More recent results obtained by sedimentation-equilibrium and by gel-filtration methods indicate that the enzyme exists in a form with molecular weight 87 000 to 93 000. Renal dipeptidase acts upon a variety of dipeptides but not when the terminal amino group is blocked by groups such as the carbobenzoxy group. The enzyme requires that the NH_2-terminal amino acid be in the L-configuration, but the C-terminal amino acid may be in the D- as well as the L-configuration. The enzyme has no esterase activity, nor does it act against leucinamide, tripeptides or proteins. Inhibition by monovalent anions as well as by inorganic phosphate and nucleotides has been observed. It has been suggested that a proton dissociation of pK 8.5 which takes place from the enzyme–substrate complex results from the loss of a proton from a water molecule coordinated to zinc at the active center of renal dipeptidase[307].

Alcohol dehydrogenase (EC 1.1.1.1) is responsible for the catalytic oxidation of a number of alcohols and the reduction of aldehydes using NAD as a cofactor. It has been obtained in crystalline form from yeast and has a molecular weight of 151 000 and 4 independent catalytic sites. An analogous crystalline enzyme of approximately one-half this molecular weight has been isolated from mammalian liver. Recent analyses of horse-liver alcohol dehydrogenase indicate a zinc content of from 3.1 to 4.3 g.atoms of Zn per mole enzyme, based on a molecular weight of 80 000[308]. The zinc atoms of the alcohol dehydrogenases cannot be removed without irreversible loss of enzyme activity. Removal of zinc from yeast alcohol dehydrogenase causes a dissociation of the apoenzyme into subunits[309]; protein denaturation and aggregation occurs when zinc is removed from horse-liver alcohol dehydrogenase[310]. It has been reported that the horse-liver enzyme contains two classes of zinc atoms. Two atoms are involved at or near the two coenzyme binding sites, and two are required to maintain the polymeric structure[311].

Comparison of optical rotatory dispersion effects produced by NAD binding, to those produced by the formation of an enzyme–Zn–o-phenanthroline complex indicates that the zinc atoms in alcohol dehydrogenase function in coenzyme binding[190]. Kinetic and fluorometric studies suggest the formation of a ternary complex of the enzyme with coenzyme and substrate in close juxtaposition to the zinc site[312,313]. Alcohol dehydrogenase from horse liver has two cysteine groups per mole which are essential for the activity of the enzyme[314,315]. Carboxymethylation of liver alcohol dehydrogenase with iodoacetate has led to the isolation of active center peptides having the sequence[316] Met–Val–Ala–Ser–Ile–Cys–Arg and a longer peptide containing this sequence[317]. Anions inhibit liver alcohol dehydrogenase, and the pK of the group that binds anions has been estimated[318] to be 8.9–9.1. It has been suggested that this ionization is that of a zinc-bound water molecule[312].

Glutamate dehydrogenase (EC 1.4.1.2) is an almost universally distributed enzyme which catalyzes the reversible oxidative deamination of glutamic acid.

$$\text{Glutamate} + \text{NAD}^+(\text{NADP}^+) \rightleftharpoons$$
$$\alpha\text{-ketoglutarate} + \text{NH}_3 + \text{NADH(NADPH)} + \text{H}^+$$

The most recent investigations show that the molecular weight of the smallest enzymatically active subunit of beef-liver glutamic dehydrogenase is 280 000 and the associated molecule (molecular weight about 2.2 million) is composed of 8 subunits[319]. The presence of 2 to 6 g.atoms of zinc in the enzyme has been reported to be essential for its activity[320], but a functional role for zinc has been questioned[321]. Although glutamic dehydrogenase is inhibited by a variety of metal complexing agents[322], this inhibition may be caused by interaction of these agents with non-polar side-chains of the enzyme rather than with zinc atoms[323].

Malate dehydrogenase (EC 1.1.1.37) catalyzes the oxidation of malic acid in the presence of NAD to yield oxaloacetic acid and the reduced coenzyme. This constitutes a component reaction of the citric acid cycle and the enzyme is of further critical importance in that it provides a shunt for hydrogen transport into mitochondria. Malic dehydrogenase of beef heart contains one g.atom of zinc per mole, and the enzyme is inhibited by *o*-phenanthroline which removes the essential zinc atom[324]. Upon loss of zinc the enzyme dissociates into monomers.

Metal–enzyme complexes. A number of enzymes have been described which are activated by zinc as well as by other divalent ions. A list of these enzymes previously tabulated by Vallee[189] is presented in Table V. Since these enzymes do not meet the criteria established for their characterization

TABLE V

METAL–ENZYME COMPLEXES ACTIVATED BY ZINC[a]

Enzyme	EC number	Activating metals
Glycylglycine dipeptidase	3.4.3.1	Zn^{2+}
Arginase	3.5.3.1	Zn^{2+}, Mn^{2+}, Fe^{2+}, Co^{2+}, Ni^{2+}, Cd^{2+}
Alanyl- and leucylglycine dipeptidase		Zn^{2+}, Pb^{2+}, Cu^{2+}, Mn^{2+}, Sn^{2+}, Cd^{2+}, Hg^{2+}
Tripeptidase		Zn^{2+}, Co^{2+}
Glycyl-L-leucine dipeptidase	3.4.3.2	Zn^{2+}, Mn^{2+}
Carnosinase	3.4.3.3	Zn^{2+}, Mn^{2+}
Aminopeptidase	3.4.1.2, 3.4.1.3	Zn^{2+}, Co^{2+}, Mn^{2+}
Histidine deaminase	4.3.1.3	Zn^{2+}, Hg^{2+}, Cd^{2+}
Lecithinase	3.1.1.4,5; 3.1.4.3,4	Zn^{2+}, Cu^{2+}, Mg^{2+}, Co^{2+}, Mn^{2+}
Enolase	4.2.1.11	Zn^{2+}, Mg^{2+}, Mn^{2+}
Yeast and *Clostridium* aldolase	4.1.2.7,13	Zn^{2+}, Fe^{2+}, Co^{2+}
Oxaloacetate decarboxylase	4.1.1.3	Zn^{2+}, Mn^{2+}, Co^{2+}, Cd^{2+}, Pd^{2+}, Ni^{2+}, Fe^{2+}, Mg^{2+}, Ba^{2+}, Cu^{2+},
Dihydro-orotase	3.5.2.3	Zn^{2+}
L-Mannosidase		Zn^{2+}

[a] Other metals which increase the observed activity are also listed.

as metalloenzymes, and they lack metal specificity, it is difficult to explain the biological function of zinc in these molecules.

Nucleic acids. Zinc complexes are formed with nucleotides, but the degree of association is much less than that observed for amino acids[325]. The fact that zinc is found in ribonucleic acids from a variety of sources, has led to the suggestion that it plays a role in maintaining the configuration of the RNA molecule and thus may have a function in protein synthesis and the transmission of genetic information[326]. More direct evidence for the involvement of zinc in nucleic acid metabolism is provided by the results of Fujioka and Lieberman[327] who reported that zinc is specifically required for the synthesis of DNA by regenerating rat liver after partial hepatectomy when the animals are perfused with ethylenediaminetetraacetate. Dietary zinc deficiency in young rats also decreases the incorporation of thymidine into the nuclear DNA of their liver cells[328]. However, no impairment of DNA synthesis is observed in the testes of zinc-deficient rats[329]. As regards specificity of zinc, experiments with intact animals are difficult to interpret because of the effect of food intake upon growth and hence upon the production of new cells and DNA. Any effect of zinc that stimulates cell proliferation would increase DNA biosynthesis without necessarily implicating zinc directly in DNA metabolism.

5. Cobalt

Cobalt is unique among the essential trace elements because there is no evidence that the cobalt ion either in the free form or as a simple protein complex is needed by any organism. Cobalt in the form of a specific complex, vitamin B_{12} or one of the cobamides, is essential for animals and many bacteria[340]. Although plants contain cobalt, there is no evidence that it occurs there as a cobamide or that it is of direct physiological significance to plants[341]. Vitamin B_{12} is unique among vitamins in that neither it nor its precursor is produced by plants. Rather it is a microbial product and animals are directly or indirectly dependent upon microbes for their supply of this vital compound. Ruminants provide their own fermentative process for the production of the B vitamins but cobalt must be present for the biosynthesis of vitamin B_{12}. Microorganisms in the rumen are wasteful of cobalt, even when it is supplied as the vitamin, in the sense that not all the element is incorporated into a form useful to animals. The role of rumen microflora in the economy of ruminant animals makes ionic cobalt of particular significance to this group of animals. Although one can not dismiss the possibility that some organisms require cobalt other than that in a corrinoid complex, there is no such evidence at present and this discussion will hinge primarily around the metabolism and metabolic function of cobalt as it exists in the cobamides.

(a) Metabolism

When administered orally in physiological doses, ionic cobalt is excreted largely in the feces. A small portion of the fecal cobalt represents endogenous cobalt excreted *via* the bile and to a lesser extent that excreted directly through the intestinal wall. Parenterally administered cobalt is excreted primarily in the urine[6]. Whereas supplementation of practical type diets with ethylenediaminetetraacetate increases the absorption of zinc, this chelating agent has an adverse effect on the absorption of ^{60}Co by the growing chicken[342].

In man and other non-ruminants small oral doses of vitamin B_{12} are absorbed by a mechanism involving the intrinsic factor. Large doses are absorbed, in part, by diffusion, independent of the intrinsic factor[343]. The chemical nature of the intrinsic factor is outside the scope of this discussion, but it appears to be a β-globulin with a molecular weight[344] of at least 4000. Vitamin B_{12} is poorly absorbed, presumably because of its large molecular weight, and it is difficult to visualize how it could be absorbed through the intestinal wall in combination with a still larger molecule. The intrinsic factor must function by interacting with both the vitamin and the gut wall. The amount of vitamin B_{12} that can be absorbed by this mechanism in man is limited. Doses larger than 2 μg, even when given with intrinsic factor, are not efficiently absorbed. Dietary sorbitol increases the absorption of vitamin B_{12} in experimental animals and normal human subjects but not in pernicious anemia patients[343]. However, pernicious anemia can be effectively treated with large oral doses of cyanocobalamin (300–1000 μg/day) without supplementary intrinsic factor[345]. Physiological doses of coenzyme B_{12} are as effectively absorbed as cyanocobalamin[340] and man absorbs vitamin B_{12} as it occurs naturally in food as efficiently as cyanocobalamin. Adults consuming a diet that supplies 5–15 μg of vitamin B_{12} absorb 5 μg/day[346]. The total body pool of vitamin B_{12} has been established at approximately[347] 3 mg, and there appears to be an obligatory daily loss equivalent to 0.1 % of the body pool. This value correlates reasonably well with the estimated absorption of a normal adult with good stores of the vitamin. In the dog the lower small intestine is the primary locus of vitamin B_{12} absorption[348] and the same appears to be the case in man[349].

Little is known about the transport of ionic cobalt in the blood plasma or of the tissue storage forms other than the cobamides. There are many proteins that will bind vitamin B_{12}, but well defined specificity is lacking[340]. It has been postulated that the intrinsic factor is absorbed along with the vitamin but the evidence is indirect[350]. The storage and excretion of cobalt, when it is injected into young rabbits as $^{60}Co^{3+}$, [^{60}Co]cyanocobalamin, [^{60}Co]hydroxocobalamin, and [^{60}Co]5,6-dimethylbenzimidazoylcobamide coenzyme, have been determined[351]. Kidney tissue accumulated more radioactivity than liver whereas in man and ruminant the liver is the chief storage organ for cobalt. Cobalt in the form of cyanocobalamin was most efficiently stored in kidney, followed, in order, by hydroxocobalamin, the coenzyme and finally by Co^{3+}. Nearly 50 % of the Co^{3+} dose was excreted in the urine within 3 days compared to 3.6 % of the coenzyme. In general coenzyme B_{12} is the chief storage form of cobalt in animals while hydroxo-cobalamin constitutes a small component. Essentially no cyanocorrinoids are found in tissues even when cyanocobalamin is administered[340,352].

In normal human subjects small injected doses of cyanocobalamin are completely retained, but with doses above 50 μg appreciable proportions appear in the urine[340]. In man a considerable proportion of orally ingested cobalt may be excreted in the urine. In one balance study involving young women, 73–97% of the dietary cobalt was absorbed and 67% appeared in the urine[6]. Inorganic cobalt injected into rabbits is rapidly excreted in the urine[351]. It may be concluded that excess cobalt and the cobamides are eliminated from animals largely *via* the kidneys.

No metal ion antagonism involving cobalt has been reported and no ions other than cobalt have been found in nature complexed with the corrin ring. There is an unexplained interaction of cobalt and a constituent of the perennial grass, *Phalaris tuberosa*. In some areas cattle and sheep that consume this grass develop Phalaris staggers, a disease characterized by muscular tremors, tachycardia and incoordination. Chronically affected animals show demyelination in the spinal cord and the lower brain. The similarity of this pathology to that of pernicious anemia suggests vitamin B_{12} deficiency, but administration of the vitamin is ineffective against Phalaris staggers. There seems to be no doubt about the efficacy of cobalt supplementation[6].

(i) Physiological function

Only in ruminant animals is it appropriate to speak of cobalt deficiency and here only because the rumen flora synthesize vitamin B_{12} efficiently when cobalt is present but, for obvious reasons, not in its absence. Cobalt deficiency in ruminants is in essence a vitamin B_{12} deficiency and all signs can be relieved by parenteral administration of the vitamin. In monogastric animals there is usually not sufficient synthesis of vitamin B_{12} by intestinal microorganism to meet the vitamin requirement. Although slight response to cobalt supplementation has been observed on occasion, in general, only vitamin B_{12} is effective.

Severe cobalt deficiency in ruminants is characterized by listlessness, anemia and emaciation, the latter resulting from loss of appetite. At necropsy such animals exhibit fatty livers, hemosiderosis of the spleen and hypoplasia of the bone marrow. The vitamin B_{12} contents of blood and liver drop precipitously[6]. Signs of vitamin B_{12} deficiency in monogastric animals largely mimic those reported for cobalt deficiency in ruminants. Although non-specific, failure of growth is the most common sign of deficiency, anemia is a cardinal sign of vitamin B_{12} deficiency in man, but it is usually mild or non-existent in laboratory animals such as rats and chicks. In man severe vitamin B_{12} deficiency due to lack of the intrinsic factor results not only in macrocytic anemia but the untreated pernicious anemia patient eventually exhibits severe central nervous system disorders which are associated with demyelination in the spinal cord[340]. Vitamin B_{12} deficiency impairs reproduction in all species studied and commonly results in congenital anomalies among the offspring[353].

(b) Chemistry

In aqueous solution cobalt exists in two oxidation states, Co^{2+} and Co^{3+}, the former being highly favored unless a complexing agent is present. Cobaltous ion has a d^7 electron configuration and forms complexes with both octahedral and tetrahedral geometry. Complexes of Co^{3+}, the d^6 ion, have been extensively studied because their exchange rates are favorable for investigation. All Co^{3+} complexes are octahedral and most involve nitrogen as the electron donor. Only a few complexes of Co^+ have been synthesized and all involve π-bonding ligands[80]. The existence of Co^+ complexes became evident mainly from work on vitamin B_{12} and its reduction products. The chemistry of vitamin B_{12} is simulated by that of bis-dimethylglyoximatocobalt compounds (cobaloximes)[354]. Both the cobamides and cobaloximes can be reduced to Co^+ complexes.

The skeletal structure of the cobamides may be visualized by representing the corrin ring of cobinamide with a planar ring and the ligands by X and Y.

The central cobalt atom in cyanocobalamin (X = CN, Y = dimethylbenzimidazole) is trivalent and makes up approximately 4.3% of the complex. In coenzyme B_{12}, α-(5,6-dimethylbenzimidazolyl)-Co-5'-deoxyadenosylcobamide, the 5-methylene carbon of the deoxyadenosyl moiety is linked directly to cobalt, constituting a carbon ligand[355]. Methyl and other alkyl groups can also serve as carbon ligands in this position. The presence of a stable carbon–cobalt bond is a notable feature of coenzyme B_{12} and largely accounts for its unusual catalytic function. The bond is cleaved not only by cyanide and acid but also, by visible light[352,356]. Cleavage by light is a homolytic reaction which gives rise to an organic free radical and vitamin B_{12r} (Co^{2+}). Spectroscopic and nuclear magnetic resonance studies suggest that Co^{3+} compounds, containing methyl or the carbanion of deoxyribose, are essentially 5-coordinate complexes[357]. According to Williams and coworkers[357] the Co–CH_3 linkage can be viewed as an anion bound to Co^{3+} or a cation bound to Co^+.

Hydroxocobalamin can be reduced stepwise as follows:

$$\begin{array}{ccccc} \overset{\displaystyle OH}{\overset{\displaystyle |}{Co^{3+}}} & \xrightarrow{e} & Co^{2+} & \xrightarrow{e} & Co^+ \\ B_{12a} & & B_{12r} & & B_{12s} \end{array}$$

Vitamin B_{12s} is an oxygen-sensitive gray-green species which decomposes slowly in water with the evolution of hydrogen. This reduced form reacts readily with alkyl halides to form alkyl derivatives with the carbon–cobalt bond[354]. Coenzyme B_{12} is synthesized by animal tissues from cyano- or hydroxocobalamin and adenosine triphosphate with the release of tripolyphosphate[352].

(c) Biochemical function

The major metabolic defects that result from vitamin B_{12} (cobalt) deficiency are failure of propionate metabolism, failure of methyl-group transfer and neogenesis of the labile methyl group, and failure of deoxyribose reduction and thus of deoxyribonucleic acid synthesis. Some organisms have alternate pathways that obviate the need for the coenzyme B_{12} in the latter two processes but, at least in higher animals, there is an absolute requirement for vitamin B_{12} in propionate metabolism. In fact, failure to metabolize propionate is the primary abnormality in cobalt-deficient ruminants. These animals depend largely on the bacterial fermentation products, acetate and propionate, for their source of energy; the conversion of propionate to succinate requires coenzyme B_{12}. For the most part, the gross pathology of vitamin B_{12} deficiency can not be identified with specific enzymatic reactions. However, the transmethylase reaction involved in methyl-group neogenesis is related to fatty livers and the methionine-sparing effect. There are other coenzyme B_{12} dependent reactions which have been observed only in microorganisms. A discussion of the coenzyme B_{12} dependent enzymatic systems most likely to be of significance in higher animals follows[352,355,358].

Methylmalonyl-CoA mutase (EC 5.4.99.2) plays a key role in the metabolism of propionate:

$$\text{Propionyl-CoA} + CO_2 \xrightarrow{\text{Carboxylase}} \text{methylmalonyl-CoA}$$

$$\text{Methylmalonyl-CoA} \xrightarrow{\text{Mutase}} \text{succinyl-CoA}$$

Coenzyme B_{12} is the only cobamide that serves as a cofactor for the mammalian mutase and its dissociation from the holoenzyme requires rigorous conditions such as acid ammonium sulfate treatment. In the mutase reaction a hydrogen atom moves from the methyl-C to C-2 of malonyl-CoA while the thio-ester carbonyl moves in the opposite direction. The hydrogen atom transferred does not exchange with the solvent, suggesting that it moves as a hydride ion bound to the coenzyme. An analogous reaction catalyzed by a bacterial enzyme, glutamate mutase, requires a cobamide coenzyme and involves interconversion of methylaspartate and glutamate[352].

N^5-Methyltetrahydrofolate-homocysteine cobalamin methyltransferase is a coenzyme B_{12}-dependent enzyme involved in the methylation of homocysteine to form methionine. This reaction is also dependent upon adenosylmethionine (AMe) which converts the original cobalamin enzyme to the active methylcobalamin enzyme[359]. As indicated in the following scheme, the methylcobalamin–enzyme complex catalyzes the reaction[359].

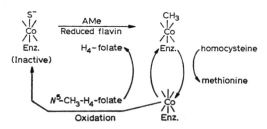

A *ribonucleotide reductase* which requires coenzyme B_{12} occurs in some bacteria and may be important in animals[360]. This enzyme catalyzes the reduction of C-2 of ribonucleoside triphosphates to give the corresponding deoxyribonucleotides. The reaction requires a reductant such as dihydrolipoate, but the more likely reductant in bacteria is thioredoxin which is maintained in the reduced state by NADPH and thioredoxin reductase. In this reaction C-5 of the adenosyl moiety of coenzyme B_{12} functions as the hydrogen-transferring agent, but the hydrogen atom comes to equilibrium with the solvent water[361]. Other coenzyme B_{12} dependent enzymes found in bacteria include dioldehydrase, glyceroldehydrase and ethanolamine deaminase, all of which catalyze the formation of aldehydes by a similar mechanism. The reactions catalyzed by these enzymes differ from the ribonucleotide reductase reaction in that hydrogen is transferred intermolecularly in the latter reaction and intramolecularly in the dioldehydrase reaction[361].

A reasonable hypothesis for the general mechanism of action of coenzyme B_{12} in the cobamide-dependent reactions is that the cobalt–carbon bond is so polarized that it readily dissociates

$$Co^{2+}-CH_2-Rib \rightleftharpoons Rib-CH^+ + {}^-Co^+$$

The latter participates in nucleophilic displacement at a vulnerable carbon atom to produce a new carbonium ion and a hydride ion which is trapped by the coenzyme carbonium ion[362].

The carbonium ion rearranges and the hydride ion returns to a new position.

6. Manganese

Manganese is widely distributed in the earth's crust and is found in all living plants and animals. The uniform appearance of manganese in biological systems suggested its significance in living organisms, but it was not until 1931 that the element was shown to be an essential dietary constituent[6]. This discovery stimulated research to elucidate its catalytic role in biochemical systems but specific functions have been found only recently.

(a) Metabolism

The earlier literature relating to absorption, excretion and turnover of manganese has been reviewed[6,363]. Although the precise loci in the gastrointestinal tract concerned with absorption are not known, it appears that manganese is efficiently absorbed but also efficiently excreted. Whereas the absorption rate is proportional to the concentration presented for absorption, animals possess an effective homeostatic mechanism for maintenance of constant tissue concentrations. Homeostatic control is effected by excretion rather than by selective absorption, and excretion occurs almost exclusively by way of the intestine[364]. Rats injected with $^{54}Mn^{2+}$ excrete the isotope from the duodenum, jejunum and ileum at several different exponential rates which vary with time after treatment. Excretion from the proximal segments is augmented by administration of stable Mn^{2+}. Rectal obstruction essentially abolishes total body loss of the isotope while bile-duct ligation only diminishes the loss[365]. Endogenous manganese enters the intestine primarily by way of bile, and under most circumstances biliary excretion constitutes the chief regulatory mechanism for maintenance of manganese homeostasis, but with overloading, auxiliary gastrointestinal routes are mobilized[365]. Manganese may be reabsorbed and become involved in enterohepatic circulation. Because of the effective excretory mechanisms which control manganese balance most animals tolerate high dietary intakes of the ion.

When the dietary supply of manganese becomes limiting, the excretory mechanisms fail to maintain balance and deficiency symptoms appear. Under conditions of limited intake other dietary factors affect the true absorption by interacting with manganese in the intestinal lumen. Abnormally high levels of calcium and phosphorus in the diet increase the manganese requirement[6] and this is believed to be due to absorption of Mn^{2+} onto the surface of insoluble calcium phosphate[366].

Manganese in animal tissues is in a highly dynamic state of equilibrium and the turnover rate is directly related to the level of manganese in the diet or to that administered parenterally. Of importance is the fact that this equilibrium is not appreciably affected by any of a large number of other metal ions that have been tested[363]. After parenteral administration of ^{54}Mn to mice, the concentration of isotope in various tissues is directly related to the dietary level of manganese[367]. The concentration of manganese in most animals is highest in bone (3.5 p.p.m.) followed closely by pituitary, mammary gland and liver. Radioactive isotopes administered to rats tend to concentrate in liver, pancreas and similar tissues, rich in mitochondria. Turnover of manganese appears to be more rapid in liver mitochondria than in the nuclei[363].

The failure of metallic ions in general to affect the turnover of manganese suggests a lack of interaction with other elements. This concept is substantiated by the paucity of literature relative to specific interactions involving manganese, but an antagonism between iron and manganese has been suggested. Excess manganese (1000–2000 p.p.m.) retards hemoglobin formation in lambs and baby pigs and results in anemia[6]. Furthermore, liver-cell fractions rich in iron tend to be poor in manganese.

(i) Physiological function

The biological significance of manganese was first established by the demonstration that it is essential for normal ovarian and testicular function in mice and rats. Manganese-deficient male rats and rabbits lose libido and fecundity and show seminal tubular degeneration and lack of spermatogenesis. Reproduction in poultry is also impaired by manganese deficiency. In spite of these early observations, a specific role for manganese in the reproductive process has not been elucidated[6].

Another deficiency sign is that of skeletal abnormalities. These defects are manifested primarily as shortening and bowing of the forelegs in rats, mice and rabbits and as lameness and enlarged hocks in pigs. A similar shortening of the long bones, observed in chick embryos is termed chondrodystrophy. Growing chicks and turkey poults fed a manganese-deficient diet develop perosis or slipped tendon, an abnormality characterized by enlarged hocks, short and twisted bones, and slipping of the gastrocnemius tendon from its condyles. Calcification is not appreciably affected in manganese deficiency but development of the organic matrix is impaired. The concentrations of hexuronic acid and hexosamine in the epiphyseal cartilage of deficient chicks are severely reduced. Most of the reduction in hexosamine can be attributed to galactosamine which is decreased from approximately 3.7 to 1.3 % of the dry lipid-free cartilage. Derangement of mucopolysaccharide metabolism also occurs in bone, articular cartilage and intestine[368]. Similar observations have been made in newborn manganese-deficient guinea pigs[369]. Chondroitin 6-sulfate and chondroitin 4-sulfate were equally and significantly reduced in epiphyseal cartilage.

Closely allied biochemically with the failure of long bone development is a typical congenital ataxia observed in the offspring of manganese-deficient dams in both mammalian and avian species[370]. Ataxic rats and mice can not right themselves while falling in air or while swimming in water. In this connection it is noteworthy that a genetic mutant in mice, pallid, exhibits a similar defect in coordination. Supplementation of the dam's diet with a high level (1000 p.p.m.) of manganese prevents the abnormality in the first generation but not in subsequent generations fed a normal stock diet. The offspring of manganese-deficient guinea pigs are permanently damaged and continue to exhibit head retraction, incoordination and tremors throughout life. The defect is due to failure of otolith development in the maculae of the inner ear[371]. Otoliths are composed of many otoconia, calcium carbonate crystals, imbedded in an organic matrix. Acid mucopolysaccharides are normally present in high concentration in the otolithic membrane but in manganese deficiency the membrane does not give the typical metachromatic staining for mucopolysaccharides.

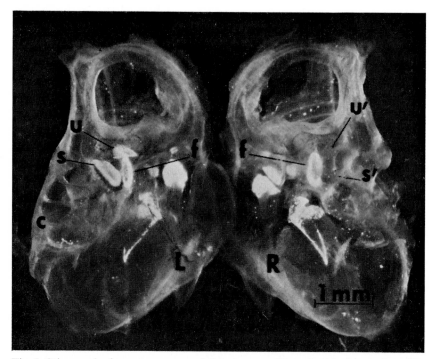

Fig. 5. Otic capsules from a manganese-deficient mouse. The animal had normal otoliths in the left utriculus and sacculus but they were absent from the right ear. U and S refers to normal utricular and saccular otoliths, respectively, while U′ and S′ refers to the absence of otoliths in these areas. Reproduced from Erway et al.[372].

Fig. 5 shows a photograph of otic capsules from a manganese-deficient mouse[372]. It is concluded that in the ear, as in long bones, there is a failure of mucopolysaccharide formation and consequently of calcification. The same is true of the genetic defect, pallid, in which there is clearly a genetic-nutrition interaction[370].

(b) Chemistry

The characteristic oxidation state of manganese in solution is Mn^{2+}, a d^5 ion. The highest oxidation state, Mn^{7+}, corresponds to the loss of all 3d and 4s electrons. Under the usual conditions manganese complexes are composed of Mn^{2+} coordinated with six ligands to give an octahedral geometry. Although Mn^{2+} forms diverse complexes, their formation constants are less than those of succeeding elements in the first transition series. This results from the fact that, as a d^5 ion, its complexes have no ligand-field stabilization energy[80]. For this reason one might predict the existence of fewer metalloenzymes that contain manganese than of other transition metals. In most synthetic complexes, manganese has 5 unpaired electrons and is consequently paramagnetic in nature. Because of the resulting large magnetic moment, Mn^{2+} has a large effect on the longitudinal proton relaxation rate of water protons, i.e., on the rate of attainment of equilibrium of nuclear-spin states induced by pulsed nuclear magnetic resonance. The increased rate is effected by promotion of proton–electron dipolar magnetic interaction[373].

(c) Biochemical function

Manganous ion has long been recognized as an activator of enzymes which require the presence of a divalent ion in the assay medium, but generally the activation is non-specific, Mn^{2+}, Mg^{2+}, Ca^{2+} and Zn^{2+} being more or less interchangeable. Furthermore, these ions are readily dissociated from their metal–enzyme complexes. These non-specific manganese activated enzymes include a variety of hydrolases, kinases, decarboxylases and trans-ferases[1]. Arginase (EC 3.5.3.1) is the prototype of the metal-activated enzymes which utilize Mn^{2+}.

The only clear evidence for a manganese metalloenzyme is that of pyruvate carboxylase (EC 6.4.1.1)[374]. This enzyme, isolated from chicken liver mitochondria, catalyzes an adenosine triphosphate-dependent CO_2 fixation according to the following stoichiometry.

$$\text{Pyruvate} + \text{ATP} + \text{HCO}_3^- \underset{}{\overset{\text{acetyl-CoA, } Mg^{2+}}{\rightleftharpoons}} \text{oxaloacetate} + \text{ADP} + \text{P}_i$$

Manganous ion can replace Mg^{2+} as the activating ion but, in addition, the

References p. 254

enzyme contains 4 g.atoms of firmly bound manganese in stoichiometric proportions to 4 moles of bound biotin. That manganese is an integral part of the enzyme was first detected by the observation that pyruvate carboxylase contains a paramagnetic metal ion which enhances the longitudinal relaxation rate of water protons by a factor of more than 4 times that of the unbound metal ion. Radioactive ^{54}Mn has been incorporated into the enzyme *in vivo* and the specific activity of the purified enzyme, related to enzymatic activity, was constant during the later stages of purification.

Experimental evidence supports the hypothesis that manganous ion in pyruvate carboxylase functions in the second or transcarboxylation portion of a two-step carboxylation reaction:

$$\text{Enzyme–biotin–CO}_2 + \text{pyruvate} \rightleftharpoons \text{enzyme–biotin} + \text{oxaloacetate}$$

The first step, which involves carboxylation of biotin bound to the enzyme surface, requires[375] the addition of a divalent metal ion, either Mg^{2+} or Mn^{2+}. It has been proposed that the bound manganese of pyruvate carboxylase facilitates the carboxylation of pyruvate by increasing both the nucleophilic character of the methyl group of pyruvate and the susceptibility of $1'$-N-carboxybiotin to nucleophilic attack. This mechanism and the general occurrence of biotin in carboxylases suggest that manganese, along with biotin, might be an integral part of other carboxylases. However, analyses of acetyl-CoA carboxylase (EC 6.4.1.2) from chicken liver and propionyl-CoA carboxylase (EC 6.4.1.3) from pig heart have indicated the absence of manganese, zinc and cobalt[376].

The role of manganese in the function of two other enzymes concerned with CO_2 fixation, phosphoenolpyruvate carboxykinase (EC 2.7.1.40) from pig-liver mitochondria and phosphoenolpyruvate carboxylase from peanut cotyledons, has been studied[377]. Although there is no evidence that manganese is an integral part of these enzymes, the results show that Mn^{2+} binds tightly to the free phosphoenolpyruvate carboxykinase to form a binary complex. A ternary complex composed of enzyme, Mn^{2+} and substrate may be formed with both phosphoenolpyruvate and inosine diphosphate. The latter compound binds to the enzyme more tenaciously in the absence of Mn^{2+}, but phosphoenolpyruvate is tightly bound only in the presence of Mn^{2+}. The enzyme enhances the effect of Mn^{2+} on the proton relaxation rate and the enhancement is reduced by phosphoenolpyruvate. Similar results were obtained with phosphoenolpyruvate carboxylase.

There are other enzymes that are no doubt dependent upon manganese for optimal function under physiological conditions. Historically, manganese has been associated with arginase, although as a group arginases cannot be classed as metalloenzymes. Arginase activity in tissues is decreased by manganese deficiency[6], and the enzyme is activated[1] by Mn^{2+}, Co^{2+} and Ni^{2+}. The optimum pH and specific activity of rat-liver arginase depend on the activating ion used[378]. Alteration in pH, removal of Mn^{2+} or replacement of Mn^{2+} with Co^{2+} does not produce a measurable change in molecular size as measured by sedimentation equilibrium. The results of this[378] and earlier studies show that arginase activity is highly dependent upon manganese.

Specific cations, Mn^{2+} or Mg^{2+}, are required for the catalytic activity and stabilization of the quaternary structure of the glutamine synthetase (EC 6.3.1.2) isolated from *Escherichia coli*. This enzyme is an aggregate of 12 subunits arranged in two hexagonal layers. Taut and relaxed forms can be detected by electron microscopy, and addition of Mn^{2+} to the fully dissociated enzyme promotes reconstitution of the double hexagon structure[379]. Studies with ^{54}Mn have revealed 3 sets of binding sites, 12 of high, 12 of moderate and 48 of low affinity. The affinity for Mg^{2+} was 400-fold less than that[380] for Mn^{2+}.

The manganese ion (Mn^{2+}) can also be substituted for Mg^{2+} in ribonucleic acid polymerase systems[381,382], although the role of the requisite cation in these systems is not yet understood.

With present knowledge it is not possible to relate specific manganese-dependent enzymes directly to gross pathology of manganese deficiency. Such relationships have begun to emerge in mucopolysaccharide metabolism as it relates to derangements of glycosaminoglycans in cartilage, bone and related tissues. Since animals deficient in manganese do not synthesize chondroitin sulfate at the normal rate, several enzyme-catalyzed reactions involved in the synthesis of chondroitin sulfate and its protein complex have been studied[383]. Two enzyme systems found in the 105 000 g sedimentable particulates of epiphyseal cartilage homogenates appear to be sites for manganese function. One is a polymerase which catalyzes formation of polysaccharide from UDP–N-acetylgalactosamine and UDP–glucuronic acid; the other is galactotransferase which incorporates galactose into the galactose–galactose–xylose trisaccharide which links the mucopolysaccharide to protein. Although Mn^{2+} stimulates these enzyme systems *in vitro*, Co^{2+} also activates them.

References p. 254

7. Chromium

Chromium is the most recent addition to the family of essential trace elements. While it meets the usual criteria required to establish essentiality, chromium differs from other essential elements, such as iron and copper, in the extent of its essentiality. Chromium appears to be analogous to the fat-soluble vitamins in the sense that the latter are essential only for higher animals whereas iron, copper and most of the B-vitamins are required by all species. An extension of this generalization is that chromium and the fat-soluble vitamins function for the most part in relation to control mechanisms in multicellular organisms rather than as enzyme cofactors.

The first evidence that chromium is an essential nutrient for animals was presented in 1959 when Cr^{3+} was identified with the glucose-tolerance factor[384]. This and related functions of chromium have been extensively reviewed[385,386,6]. Biological interest in chromium has been stimulated in recent years not only because of its physiological significance but also because of the use of chromic oxide as non-digestible marker for absorption studies and the use of chromate as an erythrocyte label in hematology.

The concentration of chromium in soil ranges from 5 to 3000 p.p.m. with an average of 100 p.p.m.[7]. Land plants contain approximately 0.2 p.p.m. while animals on the average contain less than 0.1 p.p.m. Lung tissue frequently contains the highest concentration, probably because chromium is taken up from the air breathed. Exceptionally high concentrations are found in brain tissue, particularly in the caudate nucleus[386].

(a) Metabolism

In the early studies of chromium metabolism, because of analytical difficulties, high levels of the element were used but, in general, the results agree with more recent investigations in which physiological doses have been employed[6,386]. Chromium is poorly absorbed from the gastrointestinal tract. Between 0.5 and 3 % of an oral dose of trivalent chromium is absorbed, while the absorption of chromate is slightly higher. The site and mechanism of absorption are unknown, but clearly the chemical form of chromium has a marked effect upon absorption. Unless Cr^{2+} is complexed with suitable ligands, the alkaline pH of the intestine will cause precipitation of insoluble compounds as the result of olation. When Cr^{3+} is given to the rat in a soluble form, the blood level reaches a maximum within 1 h after an oral dose[387].

After chromium reaches the blood stream it is specifically bound and transported by transferrin[387]. This β-globulin has two distinct metal-binding sites and forms complexes with Cr^{3+}, Mn^{3+} and Co^{3+} as well as with Fe^{3+} and Cu^{2+} (ref. 388). It binds Cr^{3+} in a pattern similar to iron and, in fact, Cr^{3+} competes with iron for binding sites. The binding of Cr^{3+} to plasma proteins is in contrast to that of Cr^{6+} which is absorbed preferentially by erythrocytes where it is bound within the cells, although probably not as Cr^{6+} (ref. 389). In rats given $^{51}Cr^{3+}$ intravenously the dose level, the previous diet, or the sex has little or no effect on the rate of clearance from the blood, the tissue distribution or the excretion rate[390]. 4 days after dosing, the concentration of radioactivity in heart, pancreas, lung and brain had dropped to 70–90% while that in spleen, testis and epididymis remained constant or increased up to 200% of the initial concentration. Mature testes show the most dramatic uptake of chromium and allow slow release of it, but the metabolic significance of this fact is unclear.

Chromium is present in embryos and newborn animals but $^{51}Cr^{3+}$ administered to pregnant rats is not found in the newborn pups. However, $^{51}Cr^{3+}$ incorporated into brewer's yeast and administered to the mother in the form of an extract is efficiently transferred to the fetuses[391]. These observations suggest that a naturally occurring complex of chromium is transported across the placenta whereas the ion and simple complexes of it are not. This observation has an important connotation, namely, that a specific chromium complex is required at the functional level and that its physiological or nutritional essentiality is intermediate between that of a stable metal complex such as vitamin B_{12} and a simple metallic ion. In this sense it may not be possible to assess the nutritional value of a chromium source by a simple chemical analysis. Furthermore, labeling an animal with $^{51}Cr^{3+}$ does not provide a physiologically valid model[386].

Chromium is excreted in both urine and feces, but the kidney is the major pathway accounting for at least 80% of injected doses. The dialyzable portion of chromium in blood is filtered at the glomerulus but up to 63% is reabsorbed in the tubules[386]. The site of chromium excretion into the intestine is unknown.

(i) Physiological function

Rats fed a diet based on 30% Torula yeast or 10% soybean protein and containing 0.07–0.08 p.p.m. of chromium develop progressive impairment of glucose tolerance[386]. At times commercial laboratory chows also produce

the syntrome in both squirrel monkeys[392] and rats. The rate of blood glucose
decline following a glucose load in the rat is normally 4–5% per minute;
without chromium the rate decreases to 2–3%. Rats housed in dust-protected
plastic cages and fed a 10% soybean protein diet low in chromium grow
slowly and weigh approximately 118 g in 5–9 weeks. Supplementation of
the drinking water with 2 p.p.m. of chromium as $CrCl_3 \cdot 6H_2O$ improves the
growth rate about 10%, a highly significant increment[393]. Animals subjected
to exercise or hemorrhage show greater differences in weight gain. To date
the only gross pathology observed in rats fed diets low in chromium is
corneal vascularization and opacity[394].

The effect of chromium on glucose tolerance is directly related to and
dependent upon insulin. Fasting blood levels of glucose in young rats are
not influenced by chromium, but the element accentuates and prolongs the
effect of injected insulin as shown by lowered blood glucose levels and
increased tissue uptake of a non-metabolizable amino acid[395]. Chromium-
supplemented rats also maintain higher glycogen stores in liver and muscle
than controls. Fasting hyperglycemia and glycosuria have been observed
in old rats and 3rd and 4th generation rats maintained in an environment
carefully controlled to exclude chromium[396].

(b) Chemistry

Chromium exists in solution in two principal oxidation states, Cr^{3+} and
Cr^{6+}. Hexavalent chromium is almost always linked to oxygen and is a
strong oxidizing agent. The trivalent state, a d^3 ion, forms hexacoordinate
or octahedral complexes. These complexes are relatively inert kinetically
and persist in solution for appreciable periods of time under thermodynamically unstable conditions[80]. Generally, Cr^{3+} complexes are stable
at pH 4 or less but at high pH they undergo hydrolysis. The resulting
hydroxo complexes form polynucleate bridge complexes by a process known
as olation. Olation, enhanced by alkali, may proceed with time to form
large insoluble macromolecules that precipitate from solution. At the pH
of blood simple Cr^{3+} compounds undergo olation and may become biologically inert. Trivalent chromium coordinates preferentially with oxygen
and nitrogen and the slowness of the exchange rate is exceeded only by that
of cobalt complexes[386].

Biological activity is restricted to complexes of Cr^{3+} of limited stability;
highly stable complexes such as acetylacetonate are inactive. Active com-

plexes include chromium perchlorate, chrome alum, trioxalato chromate and chromium salicylate. Although such complexes possess glucose-tolerance factor activity, none allows chromium to come to immediate equilibrium with the body pool. A chromium-containing factor extracted from brewers' yeast is immediately active biologically[386].

(c) Biochemical function

With present knowledge, it appears that the chief function of chromium relates to insulin and the membrane transport of cell metabolites. Insulin requires chromium at the site of action to exert its maximal effect. On the other hand, without insulin chromium and its complexes are inert. The mechanism by which insulin and chromium interact to promote glucose uptake by tissues is obscure. In the presence of insulin, chromium also increases the rate of galactose entry into epididymal fat tissue *in vitro* and enhances the utilization of glucose for fat synthesis[386]. Pretreatment of donor animals with chromium increases glucose uptake of isolated lens tissue in response to insulin[397].

There is no evidence that chromium serves as an essential component of a metalloenzyme or even as a specific enzyme activator. It is of interest that phosphoglucomutase (EC 2.7.5.1), which is involved in glucose metabolism, requires activation by magnesium and a second metal ion. Chromium is the most active of several effective second ions and is the only ion that maintains activity in the absence of magnesium[398].

Chromium interacts strongly with nucleic acids and this relationship has been investigated in depth by Vallee and coworkers. They found from 260 to 1080 p.p.m. chromium in a ribonucleoprotein from beef liver[326]. More highly purified RNA from this source contained 50 to 137 p.p.m. and that from other sources varied from 18 to 400 p.p.m. Whether or not this high concentration of chromium associated with ribonucleoprotein has any biochemical significance is not known.

8. Molybdenum

The uniform occurrence of molybdenum in low concentrations in plant and animal tissues was first pointed out by Ter Meulen in 1932. However, the essentiality of molybdenum as a trace element was difficult to establish because most organisms have a relatively low requirement for the element and most soil and water contain quantities that make naturally occurring deficiencies rare. The first evidence for its biological significance was provided by Bortels who demonstrated that molybdenum is required for nitrogen fixation by *Azotobacter*. It was subsequently shown to be required by all nitrogen-fixing organisms and by the higher plants[6]. Attention was called to the physiological significance of molybdenum in animals when Ferguson *et al.*[399] reported that excess intake of molybdenum by cattle results in a syndrome called "teart". This disease, which is characterized by severe diarrhea and loss of condition, has been observed in several areas where excessive molybdenum is found in the herbage, including parts of The Netherlands, England, New Zealand, Sweden, Canada, and the United States. That molybdenum plays a role in normal animal metabolism was suggested by the fact that trace amounts of molybdenum markedly increased the tissue levels of xanthine oxidase when the diet is limiting in the element[400,401]. Reviews of the literature concerning molybdenum have been provided[6,402-405].

(a) Metabolism
(i) Absorption
Hexavalent Mo is readily absorbed from the intestinal tract of rats, rabbits and guinea pigs, although the precise locus of absorption has not yet been determined. Water-soluble forms, such as sodium and ammonium molybdate, and even insoluble compounds, such as MoO_3 and $CaMoO_4$, are well absorbed, but Mo in the form of MoS_2 is poorly absorbed[406].

(ii) Transport
The level of Mo in blood is directly related to the dietary intake as has been observed in sheep and cattle[407]. Dick[408] reported that the values for whole blood in sheep increased from 2 μg to 495 μg/100 ml when the Mo intake was changed from 0.4 to 96 mg/day. The level of Mo in blood is not only proportional to Mo intake, but is also inversely dependent on the inorganic sulfate intake[408]. Little is known concerning the chemical form in which Mo occurs in the blood, although Scaife[409] has shown that both red cell and plasma Mo is readily dialyzable and is wholly present as an anion, probably molybdate.

(iii) Storage
The concentration of Mo in animal tissues is generally low, and this uniformly low level has been found in several different species[6]. The element

is not normally concentrated in any particular organ, although the content in liver and kidney is generally higher than that in other tissues[410]. Dietary intake and the presence of inorganic sulfate or copper in the diet have been shown to influence tissue concentrations of molybdenum. The molybdenum concentrations in liver, kidney, bones, and skin of several species are increased by increasing the dietary intake[406,411]. Tissue levels of Mo in sheep are sensitive to sulfate and are markedly decreased by an increased intake of inorganic sulfate[6]. In cattle, addition of Cu to the diet decreases the liver level of Mo, although the dietary intake of Mo remains the same[407].

Excretion of Mo occurs primarily in the urine. In experiments using [99]Mo, Comar[412] found that 34% of a dose fed to a steer appeared in the feces and 45% in the urine over a 2-week period. When the [99]Mo was administered intravenously, 11% of the dose appeared in the feces and 37% in the urine over a 6-day period. Studies in man have shown that 24–29% of an injected dose of [99]Mo is excreted in the urine over 10 days while 1.0–6.8% is excreted in the stool[413]. The form excreted in sheep's urine is an uncomplexed anion, probably molybdate[409]. Urinary excretion of the element depends upon the inorganic sulfate content of the diet. When a high sulfate-diet is substituted for a low sulfate-diet, there is a highly significant increase in urinary Mo with little change in fecal Mo. The increased excretion of Mo caused by inorganic sulfate is not a result of increased urine output since several diuretics have been shown to increase urine volume without increased Mo excretion[409]. Other anions such as tungstate, selenate, silicate, permanganate, phosphate, malonate, and citrate do not produce the effect on Mo excretion that is observed with sulfate[408,409].

(iv) Turnover

There is a rapid uptake of orally ingested [99]Mo by all tissues with a slightly preferential accumulation in the liver, kidney, and bone[414,415]. Measurements of Mo turnover in human plasma and whole blood show that Mo disappears from the vascular component very rapidly and that after 1 h the concentration of [99]Mo in whole blood and plasma ranges from 2.5 to 5% of the injected dose[413]. The levels in both compartments continue to decrease and less than 0.5% remains after 24 h.

(v) Interrelationships

A physiological antagonism between tungsten and molybdenum has been demonstrated and this might have been predicted from the similarity of these elements. Molybdenum deficiency is difficult to induce and has been achieved for the most part only by feeding tungstates[416,417]. Another interaction, which involves copper, molydenum and sulfate, has attracted more scientific attention[6]. This three-way interrelationship was discussed briefly in the section on copper. Although there is a degree of variation among species, sulfate intensifies the toxicity of molybdenum in ruminants but alleviates its harmful effects in rats. These differences may be related in part to the copper status of the animals[113], but there may also be a true species difference. In the rabbit high dietary levels of molybdenum plus sulfate decrease liver uptake of copper, ceruloplasmin synthesis and biliary excretion of copper[418]. Under these conditions copper accumulates in the brain of rabbits[419] whereas, in general, tissue copper tends to be depleted in ruminants[6]. In sheep fed a combination of sulfate and molybdenum there is slower clearance of copper from blood than in controls or in animals fed sulfate only[420]. This results from a decrease in both copper uptake by the liver and in ceruloplasmin synthesis. The results are best explained by the hypothesis that molybdenum and sulfate interfere with ceruloplasmin synthesis and that ceruloplasmin is essential for tissue utilization and biliary excretion of copper. Thus excess molybdenum induces a copper deficiency by preventing its utilization. The toxicity of molybdenum is overcome in part by copper but sulfate accelerates the removal of molybdenum from tissues[421].

(vi) Physiological function

Although a relationship between dietary molybdenum and tissue levels of xanthine oxidase has been clearly demonstrated in the rat[400,401], the essentiality of the element was not established because neither the growth nor the purine metabolism of molybdenum-depleted rats is affected. The growth rate of lambs fed diets containing a low level of Mo is increased by supplements of molybdate. This has been interpreted as an indirect effect mediated by rumen microorganisms[422]. Molybdenum deficiency in chickens has been produced by adding tungstate, a specific inhibitor of Mo, to a Mo-deficient diet[416,417]. In these experiments there was an increased excretion of xanthine and hypoxanthine, a decreased excretion of uric acid, growth retardation and finally death. These effects were overcome by feeding adequate Mo. Molybdenum deficiency may be more readily achieved in the chicken because this species metabolizes all nitrogen substrates to uric acid, a pathway which involves the obligatory participation of xanthine dehydrogenase and therefore Mo.

(b) Chemistry

Molybdenum exhibits oxidation states from $2+$ to $6+$ and is found in a wide variety of stereochemistries in its various compounds. The ability of Mo to coordinate with a large number of ligands provides the versatility required for interaction with biological macromolecules. The most important oxidation states from the biochemical point of view are the $5+$ and $6+$ states. The thermodynamic reduction potential for the Mo^{5+}–Mo^{6+} couple is of critical importance in its biochemical reactions because it lies in the right range for interaction with flavins[402]. A valuable analytical probe for Mo is the electron paramagnetic resonance signal of the Mo^{5+} state which is exhibited at $g = 1.97$ in biological systems.

(c) Biochemical function

The biological significance of molybdenum at the molecular level became apparent when it was discovered that the flavoproteins, xanthine oxidase, and nitrate reductase contain molybdenum and that the element is necessary for their catalytic activity. A recent review of these enzymes has been provided by Spence[402], and some of the enzyme properties tabulated in Table VI are taken from this review.

Xanthine oxidase (EC 1.2.3.2) catalyzes the oxidation of xanthine and other purines to uric acid, and it also exhibits aldehyde oxidase activity.

TABLE VI

MOLYBDENUM-CONTAINING ENZYMES

Enzymes	EC number	Other cofactors	Substrate	Product
Xanthine oxidase	1.2.3.1	FAD, Fe	Xanthine, purines	Uric acid
Aldehyde oxidase	1.2.3.1	FAD, Fe	RCHO	RCOOH
Assimilatory nitrate reductase	1.9.6.1	FAD, cyt. b	NO_3^-	NO_2^-
Respiratory nitrate reductase	1.9.6.1	FAD, cyt. c	NO_3^-	NO_2^-
Nitrogenase		?	N_2	NH_3

Three cofactors, flavin, molybdenum, and iron, participate in the enzymatic action. Using sensitive EPR equipment Beinert and his coworkers[423,424] have been able to trace the electron transfer process from substrate to oxygen. Four different EPR signals were observed and two of them were assigned to Mo^{5+}. The asymmetric nature of the signals was interpreted as an indication that the Mo was bound by at least three-non-equivalent pairs of ligands, assuming octahedral coordination. These workers regarded the two molybdenum species as existing in the same oxidation state but differing in their coordination spheres. A change in EPR signal was accompanied by a rearrangement of ligands in the molybdenum complex.

Investigations of the chicken-liver enzyme, xanthine dehydrogenase, have clarified earlier observations of species differences in the electron-transfer sequence[425]. The purified chicken-liver enzyme has a molecular weight similar to that of milk xanthine oxidase (300 000) and an identical cofactor content with a flavin to molybdenum to iron ratio of 1:1:4. However, milk xanthine oxidase utilizes molecular oxygen as an acceptor, and will not interact with NAD or clostridial ferredoxin; the avian enzyme uses NAD as its physiological electron acceptor and has poor reactivity toward oxygen and ferredoxin. For this reason the avian enzyme is referred to as a dehydrogenase. The preferred electron acceptor for the enzyme from *M. lactilyticus* is ferredoxin, with oxygen being a poor substitute and NAD being completely ineffective[426]. Rajagopalan and Handler[425] suggest that the species differences lie in the geometric relationships of the electron-carrier components within the enzyme complex. Binding of arsenite, cyanide or methanol to xanthine dehydrogenase, xanthine oxidase and liver aldehyde oxidase results in absorption spectra changes. The difference spectra indicate that the ligands bind to molybdenum in all three enzymes[427].

Aldehyde oxidase (EC 1.2.3.1). The catalytic oxidation of aldehydes to acids has been observed in biological preparations from several sources. Studies of the electron-transfer sequence in rabbit-liver aldehyde oxidase have utilized reflectance spectra, EPR measurements and inhibitor studies[428]. Following complete reduction of the enzyme by excess substrate and partial autoxidation, the EPR signal at $g = 1.94$ (iron) disappears first, followed by the signal at $g = 2.00$ (flavin), and finally by the loss of the signal at $g = 1.97$ (Mo). These data suggest the following sequence.

$$Substrate \rightarrow Mo \rightarrow FAD \rightarrow Fe \rightarrow 0$$

Based on the inhibition studies it appears that a metal sulfide linkage is involved at the binding site of the enzyme[428], and this suggestion is supported by molybdenum-complex measurements in model systems[429].

Nitrate reductase (EC 1.9.6.1). The reduction of nitrate in plants and microorganisms is catalyzed by two types of reductases. The first type, assimilatory reductase, catalyzes reduction of nitrate to nitrite as the first step in the utilization of nitrate for anabolic purposes. Garrett and Nason[430] have described the purification and properties of an assimilatory nitrate reductase from *Neurospora crassa* which requires FAD, Mo, and cytochrome b_{557} as cofactors. On the basis of data obtained from this preparation they proposed the following electron-transfer scheme:

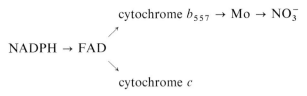

The molecular weight of the nitrate reductase complex was estimated to be approximately 228 000, and the authors suggest that a molecular weight of this size could easily accommodate a model of several intimately linked proteins or polypeptides, each responsible for a single step in the over-all electron-transfer sequence.

The second type of nitrate reductase, respiratory reductase, utilizes nitrate as a terminal-electron acceptor instead of oxygen and does not make use of the reduction products. The enzyme requires FAD, cytochrome *c*, and molybdenum as cofactors. Results from EPR analyses led Nicholas and Walker[431] to propose the following sequence for the respiratory nitrate reductase from *Pseudomonas aeruginosa*.

$$NADH \rightarrow FAD \rightarrow \text{cytochrome } c \rightarrow Mo \rightarrow NO_3^-$$

Nitrogenases. Nitrogenase refers to the enzymatic activity involved in the biological fixation of nitrogen. A specific well-characterized enzyme system responsible for this process has not as yet been obtained, but it seems likely that molybdenum participates in the reduction. Molybdenum enriched fractions obtained from *Azotobacter* exhibit high activity in fixing nitrogen, and the characteristic EPR signal ($g = 1.97$) has been observed in a preparation of *Azotobacter vinelandii* particles which was actively fixing nitrogen[432]. The molecular basis for Mo participation in this extremely important biological reaction should become clearer as the properties of the enzyme systems involved in the catalytic reduction are characterized.

9. Selenium

Selenium first attracted the attention of biologists because of its toxicity and the fact that two diseases of livestock known as "alkali disease" and "blind staggers" are the result of chronic and acute selenium poisoning, respectively. Only since 1957 has its significance as a dietary essential been recognized. In that year Schwarz and Foltz[433] showed that minute quantities of selenium prevent liver necrosis in rats fed diets based on Torula yeast. Selenium is widely distributed in the earth's crust, but its concentration in top soil and highly diverse availability to plants result in both seleniferous areas that produce toxic plants and low-selenium areas where animals suffer deficiency diseases. Parts of the northern plains area of the United States produce seleniferous vegetation[434] whereas in the Pacific Northwest the soil may contain less than 0.2 p.p.m. of selenium, and animals fed the vegetation grown on this soil respond to selenium supplementation[6,435]. The dominant form of the element in plants is selenomethionine and the selenium content is normally 0.2 p.p.m. or less[435]. Several reviews relating to the metabolism and function of selenium have appeared recently[6,434–437].

(a) Metabolism

When administered either as a soluble salt or as it exists in seleniferous grain, selenium is readily absorbed from the gastrointestinal tract of both ruminant and monogastric animals. To study absorption under near physiological conditions ^{75}Se-labeled compounds are commonly used. In both sheep and swine maximum absorption occurs in the lower small intestine with a slight secretion occurring in the duodenal portion. Neither absorption nor secretion occurs in the stomach, cecum or colon. Net absorption in sheep was found to be 35% and in swine 85%[438]. In a comparison of [75Se]selenomethionine and [75Se]selenite absorption and excretion in lambs, it was found that the excretion of radioactivity in the feces was approximately the same, 35–40% of that consumed during 10 days[439]. Those fed selenite excreted slightly more ^{75}Se in the urine and retained less in the blood and tissues. Using the everted sac technique, McConnell and coworkers[440] have measured the movement of selenite, selenocystine and selenomethionine across intestinal membranes. Only selenomethionine was transported against a concentration gradient. In this system membrane transport was energy-dependent and the most active sacs were prepared from the ileum, activity increasing with distance from the duodenum. Selenomethionine and methionine appear to be transported by the same mechanism and to compete for the same absorption site.

Selenium moves readily across the placenta, but in the sheep, fetal selenium does not reach equilibrium with the maternal burden[438]. Selenium also passes the mammary and oviduct barriers and appears in milk and eggs[6]. [75Se]Selenite moves into red cells *in vitro* by an oxygen-dependent, active transport mechanism[438]. The magnitude of ^{75}Se uptake is inversely proportional to the prior intake of dietary selenium and an assay based on this observation has been used to assess the dietary status of lambs[441].

After absorption, selenium is transported in the plasma and is soon widely distributed in all tissues. Greatest concentration occurs in liver and kidney[436]. 1 h after administration of [75Se]selenite to mice, 30% of the ^{75}Se was present in liver. Organic selenium, as it occurs in seleniferous cereal grains, is retained more efficiently than inorganic forms but prolonged intake of selenium does not result in appreciably greater accumulation[436]. When physiological levels of selenite are administered to rats fed a deficient Torula yeast diet, retention is nearly complete, only 5% of a 0.02 μg dose appearing in the urine[442] in 24 h. Rats fed a selenium-supplemented diet retain less than those fed the basal diet. The retention and excretion of selenium has also been studied in chickens, dogs, pigs, and sheep. In general the concentration in kidney is influenced more by a single dose of selenium than that in other organs. Selenium levels in liver remain somewhat more stable, but there is not uniform distribution within the cells. In liver homogenates prepared 6 h after selenite injection, 56% of the total is in the soluble fraction, 25% in mitochondria, 11% in microsomes and 2% in nuclei[436].

In addition to the usual pathways, selenium may be excreted *via* the lungs as a volatile compound, dimethyl selenide. Under physiological conditions this constitutes a small or negligible fraction of that excreted, less than 2%, but the proportion increases with dose. Mice injected with single doses of selenite, 0.1–0.7 mg of Se per kg, excreted 63% in the urine, 8% in the feces and 6% in the expired air[436] within 24 h. Biliary excretion is usually a minor pathway of excretion so that, for the most part, fecal selenium represents the unabsorbed portion of that consumed. Fecal selenium is greater in ruminants and this has been attributed to reduction of dietary selenium to non-absorbable forms by rumen microorganisms. In the rabbit, a large proportion of urinary selenium is present in a form that behaves like the ethereal or neutral sulfur fraction. Recent evidence shows that in the rat trimethylselenide, present as the basic selenonium ion, accounts for up to 60% of the urinary selenium after a subtoxic dose of selenite[443] or when the animals consume seleniferous wheat[444].

As might be predicted from their chemical similarity, sulfate and selenate constitute a pair of competitive antagonists. This has been observed in microorganisms[445] as well as in the rat[436]. Sulfate increases urinary excretion of selenate in the rat but has little or no effect on selenite. Endogenous sulfate produced from ingested protein probably contributes to the beneficial effect of high protein in alleviating selenium toxicity. By the same token sulfate may aggravate selenium deficiency when the element is limiting.

Arsenic alleviates chronic selenium toxicity but does not decrease selenium retention by tissues. Arsenite inhibits selenium volatilization and increases fecal excretion of injected selenite. Cadmium is as effective as arsenic for the inhibition of volatile selenium production, but it accentuates selenium retention[436]. A physiological antagonism between selenium and cadmium is dramatically demonstrated by another animal model. Mason[446] has reviewed the literature and reported data that extend the original observation of Parizek that cadmium produces a highly selective testicular injury. One-half equimolar selenium dioxide injected at the same time as cadmium chloride affords protection. Clearly there is a cadmium–selenium interaction but the biochemical mechanism is obscure. Among other metal ions that interact with selenium, silver is notable because its toxicity signs are highly similar to those of selenium and vitamin E deficiency. Rats develop liver necrosis and chicks develop exudative diathesis and muscular dystrophy[447]. Minute quantities of selenium afford protection as does vitamin E. An attractive hypothesis is that heavy metals complex an essential selenium compound and thus produce a selenium deficiency at that site.

(i) Physiological function

Selenium and vitamin E are closely related physiologically. Factor 3, the selenium-containing compound in biological materials, is one of three substances that prevent liver necrosis in rats fed diets based on Torula yeast[433]. The other two factors are α-tocopherol and the sulfur amino acids. Other deficiency diseases, such as exudative diathesis and muscular dystrophy in avian species are similarly related. There are some vitamin E deficiency disorders which do not respond to selenite. These include resorption sterility in rats, increased erythrocyte hemolysis, encephalomalacia in chicks, dental depigmentation in rats and increased rate of kidney autolysis[448]. Cystine and methionine protect against muscular degeneration and delay the onset of liver necrosis.

Although not well characterized, there are selenium responsive diseases that do not respond to vitamin E. A muscle degeneration in lambs and calves, known as white muscle disease, responds to selenium but not completely to vitamin E[449]. Until recently there was scant evidence for selenium-responsive disease in small animals that received the normal allowance of vitamin E. Scott and coworkers[450] have shown that chicks fed diets based on crystalline amino acids and containing less than 0.005 p.p.m. of selenium, exhibit high mortality and a poor growth rate even when the diet contained 220 p.p.m. of α-tocopheryl acetate[450]. When the diet contained 100 p.p.m. of vitamin E, the selenium requirement was less than 0.01 p.p.m. and with no vitamin E it was approximately 0.05 p.p.m. Rats fed a low selenium diet with adequate vitamin E grow and reproduce normally, but their offspring are almost hairless, grow slowly and fail to reproduce when fed the same diet[451]. Selenium supplements restore the haircoat, growth rate and reproductive capacity. The deficient animals are also afflicted with an eye injury which can be reversed by selenium supplementation. These observations strongly support the view that selenium has specific and distinct biochemical functions aside from sparing the vitamin E requirement.

(b) Chemistry

Selenium belongs to the Group VI elements and is thus closely related to sulfur, tellurium and polonium. It has a $4s^2$ and $4p^4$ electronic configuration and commonly exists in the Se^{6+}, Se^{4+} or Se^{2-} oxidation states. The higher oxidation states usually exist as oxy acids or their salts, while Se^{2-} tends to form compounds that contain two electron pair bonds, e.g., $(CH_3)_2Se$ and H_2Se which are volatile gases[80]. The chemical similarity of sulfur and selenium plays a major role in interrelationships of sulfo- and selenocompounds in biological systems.

(c) Biochemical function

There are strongly divergent points of view as to the mechanism of action[437] of selenium and vitamin E. This lack of agreement emphasizes the fact that little is actually known about the biochemical role of these catalysts. The tocopherols and certain seleno-organic compounds possess the common property of being good antioxidants but their physiological activities can not be directly related to their antioxidant potencies. This may relate in part to differences in membrane transport and in part to the lack of knowledge as to the active forms and their sites of action. Any consideration of biochemical function must encompass the sulfur amino acids which are also antioxidants and exert beneficial effects in the same deficiency diseases as selenium and vitamin E. Sulfhydryl compounds possess antioxidant activity and the similarity of selenium and sulfur allows selenium analogs of biochemically important sulfur compounds to be incorporated into organisms[434].

The toxicity of selenium has commonly been attributed to its interference with sulfur metabolism and function. This is a reasonable but not a proven hypothesis. Seleno-amino acids occur in plant and animal tissue as well as in bacteria but in many cases proof of their identity has been based on chromatographic techniques that lack rigor. According to Shrift[445] there is little doubt that selenomethionine and Se-adenosylselenomethionine are present in bacteria but the presence of selenocysteine is less well authenticated. If seleno-amino acids arise in animal tissues from selenite there must be enzyme systems, separate from those involved in sulfur metabolism, that reduce selenium and incorporate it into organic molecules. In support of this possibility, Ganther[452] has shown that mammalian tissues convert selenite into dimethyl selenide by a process requiring glutathione.

Particular caution must be exercised in concluding that selenite is incorporated into amino acids in proteins because at low pH selenite complexes strongly with sulfur compounds. By dialysis at high pH it is possible to effectively remove selenite from protein. This procedure has led to the conclusion that, at least, the rabbit cannot convert selenite to selenocystine and selenomethionine[453]. Selenite reacts with sulfhydryl compounds, such as cysteine, glutathione and coenzyme A, to form selenotrisulfides that are moderately stable except at high pH[454].

$$4\,RSH + H_2SeO_3 \rightarrow RSSeSR + RSSR + 3\,H_2O$$

Sulfhydryl groups in protein react in the same manner and change the conformation of the protein, probably forming selenotrisulfide cross-links[455]. Reduced ribonuclease, which contains 8 thiol groups, incorporates

2 g.atoms of selenium per mole when treated with selenious acid at pH 2. Under these conditions enzymatic activity is largely lost. There is lack of substantial evidence that selenite selenium is converted to seleno-amino acids by animals, but clearly animals can incorporate seleno-methionine into protein[456].

There is a large body of evidence to support the concept that the primary function of vitamin E is to inhibit lipid peroxidation. Whether or not this is its sole function, α-tocopherol is an excellent antioxidant, a property shared by both sulfo- and seleno-amino acids. Tappel and coworkers[457] are proponents of the view that selenium, as well as vitamin E, exerts its biochemical function as an antioxidant. As antioxidants, selenium compounds are 2–3 times as effective as the corresponding sulfur compounds, but the ratio of organic sulfur to α-tocopherol to selenium in the average tissue[457] is approximately 2000:25:1. Thus, if selenium's role is to prevent peroxidation of unsaturated lipids in membranes, it must function catalytically and not as a general scavenger of free radicals. One possible mechanism by which selenium compounds might function at catalytic levels is in protection of sulfhydryl-dependent enzymes which are themselves catalysts. By preventing free-radical damage or repairing such damage to critical enzymes, selenium could amplify its biological effect. Dickson and Tappel[457] observed that selenocystine increases the rate of activation of two sulfhydryl enzymes, papain (EC 3.4.4.10) and glyceraldehyde-3-phosphate dehydrogenase (EC 1.2.1.9, 12), by small sulfhydryl compounds. Papain activated in the presence of selenomethionine was also partially protected against oxidative inactivation. These seleno-amino acids bind to the sulfhydryl group of papain to form a substrate displaceable complex which serves to protect against oxidation. It was postulated that selenocystine catalyzes sulfhydryl–disulfide exchange reactions such as indicated below for papain.

$$CySH + PSSCy \rightleftharpoons PSH + CySSCy$$

The mechanism visualized follows:

(a) $$CySe^- + H^+ + PSSCy \rightleftharpoons PSH + CySeSCy$$

(b) $$CySeSCy + CySH \rightleftharpoons CySe^- + H^+ + CySSCy$$

Since the catalytic effect of selenocystine occurred at pH 7 and at physiological concentrations, this mechanism offers a plausible explanation for the physiological effect of this trace element.

References p. 254

10. Iodine

So far as presently known, the biochemical significance of iodine relates entirely to its presence in the thyroid hormones. Inasmuch as these hormones are concerned with a variety of vital metabolic functions, they clearly establish iodine as an essential trace element. Land plants and animals contain, on the average, less than 0.5 p.p.m. of iodine[7]. While soil generally contains about 5 p.p.m. there are many regions on the earth's surface that are so deficient as to cause goiter in mammals[6]. The metabolism and function of iodine have been thoroughly reviewed by several authors[6,458-460].

(a) Metabolism

Iodine occurs in foods primarily as the iodide ion, a form which is readily absorbed throughout the length of the small intestine except for a small portion near the midpoint. This portion and parts of the gastric mucosa actually secrete iodide into the lumen of the respective organs[458]. In spite of the tendency of the stomach and salivary glands to concentrate iodide in their secretions as much as 40-fold over the level found in plasma, orally administered iodide, or iodate after reduction to iodide, is nearly completely absorbed from the gastrointestinal tract. Membrane transport of iodide by everted gut sacs is reduced by cyanide and perchlorate as well as by anaerobic conditions[459].

Absorbed iodide is transported in the plasma as the free ion and quickly becomes part of the total iodide pool. The pool size is much larger than suggested by plasma levels because of the tendency of several extrathyroidal tissues, notably salivary, gastric, and mammary tissues, to concentrate iodide[459]. When the iodide concentration is high, many membranes do not distinguish between chloride and iodide but at physiological levels there is preferential concentration of iodide in the parotids, stomach, mammary gland, placenta and ovaries as well as in the thyroid gland.

The kidney is virtually the only excretory pathway of iodide besides the exocrine glands and the latter excrete only a minute proportion of the total. Organically bound iodine, primarily thyroxine, is secreted into the intestine and undergoes enterohepatic recycling. Certain natural products, such as hemoglobin, water-insoluble liver residue, soybean meal and cottonseed meal, decrease absorption and recycling of thyroxine and thus increase fecal excretion[461]. Such products have been described as antithyrotoxic because of their protective action against toxic level of thyroactive compounds. The protective component is probably not the same in all products inasmuch as hemin, the active component of hemoglobin, is not common to all active products, but all must function by impairing the absorption of a thyroactive compound[462].

Iodide is rapidly cleared from plasma by the thyroid gland and the kidneys. Of the total iodine in the adult human, 70–80% is present in the thyroid which, on a dry basis, contains 0.2–0.5% or 8–10 mg of iodine. Iodide exists in other tissues, such as muscle and blood at extremely low levels, 0.01–0.02 p.p.m. Organic iodine levels in extrathyroidal tissues are also low, approximately 0.05 p.p.m. in muscle and serum. Thyroxine (T_4) constitutes 90% of the plasma iodine, the remainder being mostly 3,5,3'-triiodothyronine (T_3) and 3,3'-diiodothyronine. These compounds are bound to plasma proteins and are precipitated with trichloroacetic acid giving rise to what is commonly termed protein-bound iodine, PBI. Thyroxine is bound more firmly than T_3 and this relationship is believed to explain the fact that T_3 has greater physiological activity[6] than T_4.

The metabolic fate of iodine is determined largely by the thyroid gland and the metabolism of the thyroid hormones. Although the mechanism by which the thyroid concentrates iodine is not clearly established, it has been referred to as the iodide trap or the iodide pump. This pump is stimulated by the thyrotropic hormone of the pituitary and is inhibited by excess iodide and organic iodine as well as by such inorganic ions as thiocyanate, perchlorate and nitrate[458].

Iodide trapped by the thyroid is rapidly incorporated into thyroglobulin, a protein found in the "colloid" within the follicles. Thyroglobulin has a molecular weight of 660 000 and an iodine content of 0.5–1%, is composed of 4 subunits, and upon hydrolysis yields, besides the usual amino acids, various iodinated derivatives of tyrosine and thyronine. Only the 3,5,3'-tri-(T_3) and the 3,5,3',5'-tetraiodothyronines (T_4) are of physiological importance, but 3,5-diiodotyrosine accounts for 25–40% of the total iodine in the protein. Formation and metabolism of the thyroid hormones is depicted by the simplified scheme shown in Fig. 6. Iodide trapped by the thyroid must be oxidized to an active form which iodinates a tyrosine residue in thyroglobulin to form mono- or more commonly di-iodotyrosine. Possibly by a free-radical mechanism, two of these residues suitably located in the molecule react non-enzymatically to form an iodinated thyronine and a three-carbon unit both of which remain in peptide linkage[463].

Iodinated thyronine, mostly thyroxine, is released from thyroglobulin by a proteolytic enzyme. It then moves into the plasma where it is bound to a specific glycoprotein. Deiodination may occur in the thyroid gland, catalyzed by a dehalogenase, or in extrathyroidal tissues, catalyzed by a deiodinase of different specificity. The oxidative step for activation of

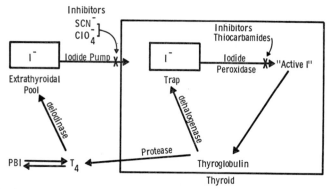

Fig. 6. Metabolic pathways of iodine. The iodide pump is inhibited by thiocyanate and iodide peroxidase by thiocarbamides.

iodine is inhibited by thiocarbamides such as thiourea, thiouracil and related compounds. Goitrogenic compounds of this type also occur naturally in various plants of the Brassica family. The best characterized is L-5-vinyl-2-thioxazolidone, a goitrin isolated from rutabagas, cabbage and rape-seed[460,463].

Deficiency of iodine results in goiter and eventually a decrease in the concentration of circulating thyroxine. Lack of thyroxine results in failure of growth and differentiation as shown by failure of metamorphosis in tadpoles and the occurrence of cretinism in children reared in iodine-deficient areas. In domestic animals iodine deficiency results in reproductive failure and the birth of hairless, non-viable offspring. Among survivors, in cases of less severe deficiency in all species, the skin and its appendages are adversely affected. There is some evidence that the central nervous system is permanently affected in children so as to reduce intelligence[6].

(b) Chemistry

Iodine belongs to the Group VII halogen family, members of which are only one electron short of the inert gas configuration. It is found in nature primarily as iodide and iodate salts. In hexane solution iodine is violet in color but in ether it is brown. This change in absorption maximum is attributed to the formation of a charge-transfer complex. Aromatic hydrocarbons can be iodinated directly in the presence of a catalyst and the reaction is reversible[80].

(c) Biochemical function

Discussion of the biochemical function of iodine entails essentially a consideration of thyroid hormone function. Since a detailed treatment of hormone action is outside the scope of this chapter, the reader is referred to other reviews[460,464,465]. In general thyroxine accelerates all cellular reactions and this is manifested, and mostly commonly measured, by increased oxygen consumption under standard conditions, the so-called basal metabolic rate.

The primary action of thyroid hormones is unknown despite the considerable research activity expended in that direction. Much of the effort has focused on the electron-transport system and oxidative phosphorylation since these processes are closely related to oxygen consumption and metabolic control. Thyroxine uncouples or at least "loosens" the couple of oxidation and phosphorylation. Although the evidence for this conclusion is not beyond question, the hypothesis affords the best model for unifying all observations made to date. The activities of over 100 enzymes are affected by administration of thyroid hormones, but it is doubtful that any reflects a primary action. Induction of mitochondrial α-glycerophosphate dehydrogenase activity is increased up to 20-fold, more than any other enzyme studied. Its activity has served as assay for thyroid hormone and antithyrotoxic activity[462]. This inductive effect may reflect the ability of thyroxine, both *in vivo* and *in vitro*, to stimulate protein synthesis by liver-cell particulates[460,463].

11. Fluorine

Fluorine does not meet all of the criteria usually required to establish dietary essentiality because it has not been demonstrated to be vital for life. Nevertheless, trace quantities of fluoride are of such benefit to human health that the element is classified here as essential. Even if it were not so categorized, fluoride deserves inclusion because of its dramatic physiological effects and the voluminous literature which refers to its toxicity when consumed in excess. Although there is no clear evidence that fluoride is essential for survival, it is also true that no one has produced animals without fluoride in their bones. There is no doubt that an optimal intake of fluoride increases the resistance of human teeth to dental caries. Several reviews relating to fluoride toxicity have appeared, but relatively few are concerned with its metabolism[6,466].

(a) Metabolism

Fluoride enters the animal body primarily by absorption from the gastro-intestinal tract. As regards biological availability, it makes little difference whether fluoride is present in food or water but its chemical form affects absorption. Essentially all of the fluoride in soluble compounds, such as sodium fluoride and sodium silicofluoride, is physiologically available; whereas only one-third to two-thirds of the fluoride in calcium fluoride and sodium aluminum fluoride is available[466]. In general, the fluoride in domestic water makes the greatest contribution to the total fluoride absorbed by man.

After absorption, fluoride quickly equilibrates with the extracellular fluids, a behavior similar to that of chloride ion. There is no evidence that fluoride ion is bound to protein in the blood or elsewhere and consequently fluoride is rapidly cleared from the blood. In sheep, 41% of a parenteral fluoride dose disappears per minute, and in cows 32% per minute[467]. Concentrations of fluoride in soft tissue generally parallel, but remain lower than those of blood. The concentration in kidney tends to be higher than that in other organs during the first few hours after fluoride dosing and before excretion in the urine. Bones and teeth accumulate fluoride selectively so that the concentration in the bones of rats given radiofluorides doubles during the period of 10–90 min after administration. Only traces of fluoride appear in milk.

Storage of fluoride in soft tissue is so low that analysis is difficult and few reliable data on tissue concentrations are available. Bone fluoride data are numerous but it is difficult if not impossible to define "normal" values since the concentration is directly related to fluoride intake. A common human diet supplies 0.5–1.5 mg of fluoride per day and the long bones of such individuals may contain from 100 to 200 p.p.m. Soft tissues[466] usually contain less than 1 p.p.m. Fluoride moves freely across the placenta so that fetal blood contains essentially the same concentration as maternal blood. Once it is absorbed into the blood, fluoride is rapidly deposited into the skeleton. During tooth formation fluoride is incorporated into both enamel and dentine. The bone mineral surface that contacts extracellular fluid constitutes the site of deposition and the latter involves the exchange of fluoride for hydroxyl ion in the bone salt. In adult animals only 1–2% of the skeleton is available for rapid exchange and the rate and quantity of uptake of a given dose depends upon previous exposure to fluoride. When the fluoride intake is constant, a steady state is established, but equilibrium is largely dependent upon the quantity of fluoride consumed daily since there seems to be a simple competition between fluoride and hydroxyl ions for a site in the apatite crystal lattice. The limit of fluoride storage as fluoroapatite has not been established experimentally but theoretically it could approach 35 000 p.p.m. fluoride[466]. Since fluoride in bone is in equilibrium with that in extracellular fluids, the ion is readily mobilized from bone when the intake is decreased. The biological half-life of fluoride in the adult human skeleton has been estimated to be approximately 2 years[466,468].

Fluoride is rapidly excreted by way of the kidneys. Approximately one-fourth of a 1-mg oral dose given to young adult men is excreted[466] in 4 h. Both fluoride and chloride ions are filtered through the glomerulus but the reabsorption of chloride by the tubules is the more efficient. Considering that 1 mg of fluoride is mixed with a body pool of 10 000 times that quantity of chloride and that, percentagewise, approximately 8 times as much fluoride as chloride is excreted, there is clearly high selectivity for elimination of fluoride from plasma. After skeletal equilibrium is established, the concentration of fluoride in the urine is almost the same as that in the drinking water. The concentration in urine may be more or less than that in the water, depending on previous fluoride intake of the individual[466].

Although it is generally outside of the scope of this review to discuss the effect of excesses of the trace elements, fluoride toxicity deserves at least brief mention. With increasing levels of fluoride consumption, there appears increasingly severe pathology, mottled teeth, osteosclerosis and crippling fluorosis. Mottled enamel, which may arise from many causes, results from failure of the enamel-forming cells to elaborate and deposit enamel during tooth development. Osteosclerosis is characterized by hypercalcification of and does not occur until the bone fluoride reaches approximately 6000 p.p.m. Crippling fluorosis involves calcification of ligaments and new bone

References p. 254

growth from surface of long bones, formation of exostoses. The pathology of excess fluoride is no more onerous than other trace-element toxicities, but it has received more public attention because of water fluoridation[466].

An optimal intake of fluoride is highly beneficial for dental health because teeth low in fluoride are highly susceptible to caries[466]. Furthermore, in an epidemiological study[469], it was observed that older women who had consumed low fluoride water (0.15–0.30 p.p.m.) most of their lives had a higher incidence of osteoporosis and collapsed vertebrae than those that consumed a higher fluoride water (4–6 p.p.m.). Of equal, if not greater, interest is the fact that men in the low fluoride area showed a significantly higher incidence of calcified aortas in the abdominal region.

(b) Chemistry

Fluorine is widespread in nature, occurring as insoluble fluorides such as CaF_2, Na_3AlF_6 and fluorapatite, $3Ca_3(PO_4)_2 \cdot CaF_2$. Compounds of fluorine are of two types, ionic and covalent, in all of which fluorine has a complete electronic octet. The ionic fluorides are of greatest biological interest because they give rise to the fluoride ion. This ion has a relatively small radius, 1.36 Å, which is almost the same as that of the hydroxide and oxide (1.40 Å) ions. Fluorine is the most electronegative of all the elements, but its electron affinity, 83.5 kcal/g.atom is intermediate between those of chlorine and bromine[80].

(c) Biochemical function

The well-established beneficial functions of fluoride relate to the stability of bones and particularly of teeth. The mechanism by which fluoride reduces dental caries is obscure but various hypotheses have been offered. These include: (1) A tooth more perfect in form and structure results when fluoride is present during development. (2) The solubility of the tooth minerals is reduced and hence the solution rate of the enamel. (3) The chemical composition makes for greater resistance to abrasion. (4) Bacterial production of organic acid is diminished[466]. All of these factors may play a role in strengthening the defenses of the teeth and they all depend on the incorporation of fluoride into the apatite crystal lattice, replacing the hydroxyl ion. Clearly, "fluorapatite" crystals are more perfect than those without fluoride and the mineral is less soluble[470]. Decreased solubility may explain the lowered caries susceptibility but it does not adequately explain the greater stability of bone as indicated by the role of fluoride in the prevention of osteoporosis.

If fluoride, at levels found in domestic water supplies, prevents soft-tissue calcification as indicated by an epidemiological study[469], this function would be as significant a contribution to human health as the prevention of dental caries. By what mechanism might fluoride prevent metastatic calcification? Soft-tissue calcification is a cardinal sign of magnesium deficiency in most animals[471] and dietary supplements of fluoride reduce or prevent calcinosis in dogs[472] and guinea pigs[473,474] fed magnesium-deficient diets. In the guinea pig, fluoride ameliorates the growth depression, hypomagnesemia and loss of bone magnesium[474], but in the dog there is no effect on growth rate or on serum and bone magnesium[472]. Conceivably there are cases of borderline magnesium deficiency in the human population that may benefit from traces of fluoride in the food or water. While fluoride supplementation of magnesium-deficient diets has proven beneficial, high levels of both fluoride (0.08%) and magnesium (0.45%) in chick diets results in reduced growth rate and bone mineralization[475]. The activity of pyrophosphatase in bone is increased by high dietary levels of both magnesium and fluoride but not by either element alone. Addition of magnesium to control bone homogenates stimulates pyrophosphatase activity in the absence of fluoride but inhibits its activity in the presence of added fluoride[476]. These results clearly show an interaction between magnesium and fluoride at the physiological and biochemical levels.

ACKNOWLEDGEMENTS

The authors are grateful to Dr. Bert Vallee for the opportunity to read a manuscript which will be part of a book, *The Role of Zinc in Medicine and Biology*, to be published by Academic Press and Dr. Howard Ganther for the opportunity to read a manuscript, *Selenium*, which is to be published as a monograph by Reinhold.

REFERENCES

1   B. L. VALLEE AND J. E. COLEMAN, in M. FLORKIN AND E. H. STOTZ (Eds.), *Comprehensive Biochemistry*, Vol. 12, Elsevier, Amsterdam, 1964, p. 165.
2   B. L. VALLEE AND R. J. P. WILLIAMS, *Proc. Natl. Acad. Sci. (U.S.)*, 59 (1968) 498.
3   R. J. P. WILLIAMS, *Endeavour*, 26 (1967) 96.
4   F. R. N. GURD AND P. E. WILCOX, *Advan. Protein Chem.*, 11 (1956) 311.
5   H. R. MAHLER, in C. L. COMAR AND F. BRONNER (Eds.), *Mineral Metabolism*, Vol. 1 B, Academic Press, New York, 1961.
6   E. J. UNDERWOOD, *Trace Elements*, 2nd ed., Academic Press, New York, 1962.
7   J. J. M. BOWEN, *Trace Elements in Biochemistry*, Academic Press, New York, 1966.
8   F. GROSS (Ed.), *Iron Metabolism: An International Symposium*, Springer, Berlin, 1964.
9   S. GRANICK, in C. A. LAMB, O. G. BENTLEY AND J. M. BEATTIE (Eds.), *Trace Elements*, Academic Press, New York, 1958.
10 C. V. MOORE AND R. DUBACK, in C. L. COMAR AND F. BRONNER (Eds.), *Mineral Metabolism, An Advanced Treatise*, Vol. 2, Academic Press, New York, 1962, p. 287.
11 C. V. MOORE, *Harvey Lectures*, 55 (1961) 67.
12 T. H. BOTHWELL AND C. A. FINCH, *Iron Metabolism*, Little and Brown Co., Boston, 1962.
13 E. B. BROWN, *Am. J. Clin. Nutr.*, 12 (1963) 205.
14 C. M. COONS, *Ann. Rev. Biochem.*, 33 (1964) 459.
15 M. ROCHE (Moderator), *Iron Metabolism and Anemia*, Proceedings of a Symposium of PAHO Advisory Committee, Scientific Publication No. 184, Washington, D.C., 1969.
16 H. W. JOSEPHS, *Blood*, 13 (1968) 1.
17 E. B. BROWN AND B. W. JUSTUS, *Am. J. Physiol.*, 194 (1958) 319.
18 S. GRANICK, *J. Biol. Chem.*, 164 (1946) 737.
19 C. V. MOORE, R. DUBACH, V. MINNICH AND H. K. ROBERTS, *J. Clin. Invest.*, 23 (1944) 755.
20 H. BRISE AND L. HALLBERG, *Acta Med. Scand.*, Suppl. 376 (1962) 1.
21 R. SANFORD, *Nature*, 185 (1960) 533.
22 L. M. SHARPE, W. C. PEACOCK, R. COOKE AND R. S. HARRIS, *J. Nutr.*, 41 (1950) 433.
23 W. M. BALFOUR, P. F. HAHN, W. F. BALE, W. J. POMMERENKE AND G. H. WHIPPLE, *J. Exptl. Med.*, 76 (1942) 15.
24 T. H. BOTHWELL, G. PIRZIO-BIROLE AND C. A. FINCH, *J. Lab. Clin. Med.*, 51 (1958) 24.
25 S. KRANTZ, E. GOLDWASSER AND L. O. JACOBSON, *Blood*, 14 (1959) 654.
26 G. PIRZIO-BIROLI AND C. A. FINCH, *J. Lab. Clin. Med.*, 55 (1960) 216.
27 C. J. GUBLER, G. E. CARTWRIGHT AND M. M. WINTROBE, *J. Biol. Chem.*, 184 (1950) 563.
28 R. DUBACH, C. V. MOORE AND S. T. CALLENDER, *J. Lab. Clin. Med.*, 45 (1955) 599.
29 C. V. MOORE AND R. DUBACH, *J. Am. Med. Ass.*, 162 (1956) 197.
30 R. A. MCCANCE AND E. M. WIDDOWSON, *Lancet*, 233 (1937) 680.
31 E. B. DOWDLE, D. SCHACHTER AND H. SCHENKER, *Am. J. Physiol.*, 198 (1960) 609.
32 J. M. JOB AND D. A. HOWELL, *J. Clin. Invest.*, 41 (1962) 1368.
33 P. F. HAHN, W. F. BALE, J. F. ROSS, W. M. BALFOUR AND G. H. WHIPPLE, *J. Exptl. Med.*, 78 (1943) 169.
34 M. AWAI AND E. B. BROWN, *J. Lab. Clin. Med.*, 61 (1963) 363.
35 M. E. CONRAD, L. R. WEINTRAUB AND W. H. CROSBY, *J. Clin. Invest.*, 43 (1964) 963.
36 J. FLETCHER AND E. R. HUEHNS, *Nature*, 218 (1968) 1211.

37 S. CROMWELL, in H. PEETERS (Ed.), *Protides of the Biological Fluids*, Vol. 11, Elsevier, Amsterdam, 1964, p. 484.
38 R. S. LANE, *Brit. J. Haematol.*, 15 (1968) 355.
39 L. D. HAMILTON, C. J. GUBLER, G. E. CARTWRIGHT AND M. M. WINTROBE, *Proc. Soc. Exptl. Biol. Med.*, 75 (1950) 65.
40 G. E. HANKS, M. CASSELL, R. N. RAY AND J. CHAPLIN JR., *J. Lab. Clin. Med.*, 56 (1960) 486.
41 C. G. HOLMBERG AND C. B. LAURELL, *Acta Physiol. Scand.*, 10 (1945) 307.
42 A. L. SCHADE AND L. CAROLINE, *Science*, 104 (1946) 340.
43 P. SALTMAN AND P. J. CHARLEY, in M. J. SEVEN (Ed.), *Metal-Binding in Medicine*, Lippincott, Philadelphia, 1960, p. 241.
44 C. E. RATH AND C. A. FINCH, *J. Clin. Invest.*, 28 (1949) 79.
45 A. MAZUR, S. GREEN AND A. CARLETON, *J. Biol. Chem.*, 235 (1960) 595.
46 J. H. JANDL, J. K. INMAN, R. L. SIMMONS AND D. W. ALLEN, *J. Clin. Invest.*, 38 (1959) 161.
47 J. FLETCHER AND E. R. HUEHNS, *Nature*, 215 (1967) 584.
48 J. H. JANDL AND H. J. KATZ, *J. Clin. Invest.*, 42 (1963) 314.
49 J. H. KATZ AND J. H. JANDL, in F. GROSS (Ed.), *Iron Metabolism, An International Symposium*, Springer, Berlin, 1964.
50 A. SCHODEN, B. W. GAKRIO AND C. A. FINCH, *J. Biol. Chem.*, 204 (1953) 823.
51 W. MEIER-RUGE, G. BENEKE AND G. AHLERT, *Nature*, 199 (1963) 175.
52 D. N. S. KERR AND A. R. A. MUIR, *J. Ultrastruct. Res.*, 3 (1960) 313.
53 P. M. HARRISON, *J. Mol. Biol.*, 6 (1963) 404.
54 A. ROTHEN, *J. Biol. Chem.*, 152 (1944) 678.
55 A. SHODEN AND P. STURGEON, *Nature*, 189 (1961) 846.
56 R. H. MCKAY AND R. A. FINEBERG, *Arch. Biochem. Biophys.*, 104 (1964) 496.
57 G. W. RIDITER, *J. Exptl. Med.*, 112 (1960) 551.
58 R. A. FINEBERG AND D. M. GREENBERG, *J. Biol. Chem.*, 214 (1955) 97.
59 R. A. FINEBERG AND D. M. GREENBERG, *J. Biol. Chem.*, 214 (1955) 107.
60 R. SADDI AND A. VON DER DECKEN, *Biochim. Biophys. Acta*, 90 (1964) 196.
61 G. W. RICHTER, *Lab. Invest.*, 12 (1963) 1026.
62 P. STURGEON AND A. SHODEN, in F. GROSS (Ed.), *Iron Metabolism: An International Symposium*, Springer, Berlin, 1964.
63 G. T. MATIOLI AND R. F. BAKER, *J. Ultrastruct. Res.*, 8 (1963) 477.
64 R. A. MCCANCE AND E. M. WIDDOWSON, *J. Physiol. (London)*, 94 (1938) 138.
65 R. A. MCCANCE AND E. M. WIDDOWSON, *Nature*, 152 (1943) 326.
66 C. P. LEBLOND AND B. E. WALKER, *Physiol. Rev.*, 36 (1956) 255.
67 R. GREEN, R. CHARLTON, H. SEFTEL, T. BOTHWELL, F. MAYET, B. ADAMS, C. FINCH AND M. LAYRISSE, *Am. J. Med.*, 45 (1968) 336.
68 H. FOY AND A. KONDI, *J. Trop. Med. Hyg.*, 60 (1957) 105.
69 C. A. FINCH, Iron Loses, in M. ROCHE (Moderator), *Iron Metabolism and Anemia*, Proceedings of a Symposium PAHO Advisory Committee, Scientific Publication No. 184, Washington, D.C., 1969.
70 T. H. BOTHWELL, B. C. ELLIS, H. W. VAN DOORN-WITTKAMPF AND O. L. ABRAHAMS, *J. Lab. Clin. Med.*, 45 (1955) 167.
71 E. J. FREIREICH, J. F. ROSS, T. B. BAYLES, C. P. EMERSON AND S. C. FINCH, *J. Clin. Invest.*, 36 (1957) 1043.
72 J. DAVIES, D. B. BROWN JR., D. STEWART, C. W. TERRY AND J. SISSON, *Am. J. Physiol.*, 197 (1959) 87.
73 H. THEORELL, M. BEZNAK, R. BONNICHSEN, K. G. PAUL AND V. AKESON, *Acta Chem. Scand.*, 5 (1951) 445.

74 L. PAPE, J. S. MULTANI, C. STITT AND P. SALTMAN, *Biochemistry*, 7 (1969) 606.
75 A. MAZUR, S. BAERZ AND E. SHORR, *J. Biol. Chem.*, 213 (1955) 147.
76 S. GREEN AND A. MAZUR, *J. Biol. Chem.*, 277 (1957) 653.
77 A. MAZUR, S. GREEN, A. SAHA AND A. CARLETON, *J. Clin. Invest.*, 37 (1958) 1809.
78 W. J. DARBY, *Handbook of Nutrition*, 2nd ed., McGraw-Hill (Blakiston), New York, 1951.
79 J. SEAMER, *Vet. Rev. Annotations*, 2 (1956) 79.
80 F. A. COTTON AND G. WILKINSON, *Advanced Inorganic Chemistry*, Interscience, New York, 1962.
81 W. S. CAUGHEY, *Ann. Rev. Biochem.*, 36 (1967) 611.
82 B. CHANCE AND R. W. ESTABROOK, *Hemes and Hemoproteins*, Academic Press, New York, 1966.
83 H. R. ITANO, *Advan. Protein Chem.*, 12 (1957) 215.
84 A. ROSSI FANELLI, E. ANTONINI AND A. CAPUTO, *Advan. Protein Chem.*, 19 (1964) 73.
85 J. WYMAN JR., *Advan. Protein Chem.*, 19 (1964) 223.
86 T. E. KING, H. S. MASON AND M. MORRISON (Eds.), *Oxidases and Related Redox Systems*, Vol. 2, Wiley, New York, 1965.
87 E. MARGOLIASH AND A. SCHEJTER, *Advan. Protein Chem.*, 21 (1966) 113.
88 R. E. DICKERSON, M. L. KOPKA, J. WEINZIERL, J. VARNUM, D. EISENBERG AND E. MARGOLIASH, *J. Biol. Chem.*, 242 (1967) 3015.
89 R. H. RINGLER, S. MINAKAMI AND T. P. SINGER, *J. Biol. Chem.*, 238 (1963) 801.
90 T. E. KING AND R. L. HOWARD, *Biochim. Biophys. Acta*, 37 (1960) 557.
91 A. SAN PIETRO (Ed.), *Non-Heme Iron Proteins; Role in Energy Conversion*, Antioch Press, Yellow Springs, Ohio, 1965.
92 B. B. BUCHANAN, *Struct. Bonding (Berlin)*, 1 (1966) 109.
93 P. NICHOLLS AND G. R. SCHONBAUM, Catalases, in P. D. BOYER, H. LARDY AND K. MYRBACK (Eds.), *The Enzymes*, Vol. 8, Academic Press 1963, New York, p. 147.
94 K. G. PAUL, Peroxidases, in P. D. BOYER, H. LARDY AND K. MYRBACK (Eds.), *The Enzymes*, Vol. 8, Academic Press, New York, 1963, p. 227.
95 E. BOERI, in M. FLORKIN AND E. H. STOTZ (Eds.), *Comprehensive Biochemistry*, Vol. 8, Elsevier, Amsterdam, 1963, p. 38.
95a S. KERESZTES-NAGY AND I. M. KLOTZ, *Biochemistry*, 4 (1965) 919.
96 K. OKUNUKI, in M. FLORKIN AND E. H. STOTZ (Eds.), *Comprehensive Biochemistry*, Vol. 14, Elsevier, Amsterdam, 1966, p. 232.
97 R. C. BRAY, in P. D. BOYER, H. LARDY AND K. MYRBACK (Eds.), *The Enzymes*, Vol. 7, Academic Press, New York, 1963, p. 533.
98 C. A. ELVEHJEM, *Physiol. Rev.*, 15 (1935) 471.
99 S. J. ADELSTEIN AND B. L. VALLEE, *New Engl. J. Med.*, 265 (1961) 892.
100 G. E. CARTWRIGHT AND M. M. WINTROBE, *Am. J. Clin. Nutr.*, 14 (1964) 224.
101 I. H. SCHEINBERG AND I. STERNLIEB, *Pharmacol. Rev.*, 12 (1960) 355.
102 A. SASS-KORTSAK, *Advan. Clin. Chem.*, 8 (1965) 1.
103 B. C. STARCHER, *J. Nutr.*, 97 (1969) 321.
104 P. FARRER AND S. P. MISTILIS, *Nature*, 213 (1967) 291.
105 D. R. VAN CAMPEN AND P. U. SCAIFE, *J. Nutr.*, 91 (1967) 473.
106 M. KIRCHGESSNER, U. WESER AND H. L. MUELLER, *Futtermittelk.*, 22 (1967) 76.
107 C. A. OWEN JR., *Am. J. Physiol.*, 209 (1965) 900.
108 B. SARKAR AND T. P. A. KRUCK, in J. PEISACH, P. AISEN AND W. E. BLUMBERG (Eds.), *The Biochemistry of Copper*, Academic Press, New York, 1966, p. 183.
109 G. W. EVANS AND R. E. WIEDERANDERS, *Am. J. Physiol.*, 213 (1967) 1183.
110 B. STARCHER AND C. H. HILL, *Comp. Biochem. Physiol.*, 15 (1965) 429.
111 C. H. HILL, G. MATRONE, W. L. PAYNE AND C. W. BARBER, *J. Nutr.*, 80 (1963) 227.

112 C. H. HILL, B. STARCHER AND G. MATRONE, *J. Nutr.*, 83 (1964) 107.

113 L. F. GRAY AND L. J. DANIEL, *J. Nutr.*, 84 (1964) 31.

114 R. P. DOWDY AND G. MATRONE, *J. Nutr.*, 95 (1968) 191.

115 G. R. LEE, S. NACHT, J. N. LUKENS AND G. E. CARTWRIGHT, *J. Clin. Invest.*, 47 (1968) 2058.

116 G. J. EVERSON, R. E. SHRADER AND T. WANG, *J. Nutr.*, 96 (1968) 115.

117 C. H. GALLAGHER, J. D. JUDAH AND K. R. REES, *Proc. Roy. Soc. Med.*, 145 (1956) 134.

118 G. S. SHIELDS, W. F. COULSON, D. A. KIMBALL, W. H. CARNES, G. E. CARTWRIGHT AND M. M. WINTROBE, *Am. J. Pathol.*, 41 (1962) 603.

119 B. L. O'DELL, B. C. HARDWICK, G. REYNOLDS AND J. E. SAVAGE, *Proc. Soc. Exptl. Biol. Med.*, 108 (1961) 402.

120 C. H. HILL, B. STARCHER AND C. KIM, *Federation Proc.*, 26 (1967) 129.

121 B. L. O'DELL, D. W. BIRD, D. L. RUGGLES AND J. E. SAVAGE, *J. Nutr.*, 88 (1966) 9.

122 S. M. PARTRIDGE, D. F. ELSDEN AND J. THOMAS, *Biochem. J.*, 93 (1964) 30 C.

123 E. J. MILLER, G. R. MARTIN, C. E. MECCA AND K. A. PIEZ, *J. Biol. Chem.*, 240 (1965) 3623.

124 W. S. CHOU, J. E. SAVAGE AND B. L. O'DELL, *Proc. Soc. Exptl. Biol. Med.*, 128 (1968) 948.

125 W. S. CHOU, J. E. SAVAGE AND B. L. O'DELL, *J. Biol. Chem.*, 244 (1969) 5785.

126 R. B. RUCKER, H. E. PARKER AND J. C. ROGLER, *J. Nutr.*, 98 (1969) 57.

127 K. A. PIEZ, *Ann. Rev. Biochem.*, 37 (1968) 547.

128 D. W. BIRD, J. E. SAVAGE AND B. L. O'DELL, *Proc. Soc. Exptl. Biol. Med.*, 123 (1966) 250.

129 G. E. CARTWRIGHT AND M. M. WINTROBE, *Am. J. Clin. Nutr.*, 15 (1964) 94.

130 J. M. WALSCHE, in J. PEISACH, P. AISEN AND W. E. BLUMBERG (Eds.), *The Biochemistry of Copper*, Academic Press, New York, 1966, p. 475.

131 H. BEINERT AND G. PALMER, *J. Biol. Chem.*, 239 (1964) 1221.

132 I. M. KLOTZ, G. H. CZERLINSKI AND H. A. FRESS, *J. Am. Chem. Soc.*, 80 (1958) 2920.

133 I. M. KLOTZ AND B. J. CAMPBELL, *Arch. Biochem. Biophys.*, 96 (1962) 92.

134 C. J. GUBLER, G. E. CARTWRIGHT AND M. M. WINTROBE, *J. Biol. Chem.*, 244 (1957) 658.

135 D. KEILIN AND E. F. HARTREE, *Proc. Roy. Soc. (London) Ser. B*, 127 (1939) 167.

136 B. EICHEL, W. W. WAINIO, P. PERSON AND S. J. COOPERSTEIN, *J. Biol. Chem.*, 183 (1950) 89.

137 E. C. SLATER, B. F. VAN GELDER AND K. MINNAERT, in T. E. KING, H. S. MASON AND M. MORRISON (Eds.), *Oxidases and Related Redox Systems*, Wiley, New York, 1965, p. 667.

138 H. BEINERT, in J. PEISACH, P. AISEN AND W. E. BLUMBERG (Eds.), *Biochemistry of Copper*, Academic Press, New York, 1966, p. 213.

139 K. OKUNUKI, in O. HAYAISHI (Ed.), *Oxygenases*, Academic Press, New York, 1962, p. 409.

140 K. OKUNUKI, R. SEKUZU, Y. ORII, T. TSUDZUKI AND Y. MATSUMURA, in K. OKUNUKI, M. D. KAMEN AND I. SELKUZU (Eds.), *Structure and Function of Cytochromes*, Univ. of Tokyo Press, Tokyo, 1968, p. 35.

141 M. MORRISON, in K. OKUNUKI, M. D. KAMEN AND I. SEKUZU (Eds.), *Structure and Function of Cytochromes*, Univ. Tokyo Press, Tokyo, 1968, p. 89.

142 H. BEINERT AND G. PALMER, in T. E. KING, H. S. MASON AND M. MORRISON (Eds.), *Oxidases and Related Redox Systems*, Wiley, New York, 1965, p. 567.

143 P. M. NAIR AND H. S. MASON, *J. Biol. Chem.*, 242 (1967) 1406.

144 I. H. SCHEINBERG, in J. PEISACH, P. AISEN AND W. E. BLUMBERG (Eds.), *Biochemistry of Copper*, Academic Press, New York, 1966, p. 513.

145 E. B. KASPER AND H. F. DEUTSCH, *J. Biol. Chem.*, 238 (1963) 2325.

146 B. STARCHER AND C. H. HILL, *Biochim. Biophys. Acta*, 127 (1966) 400.

147 S. OSAKI, D. A. JOHNSON AND E. FRIEDEN, *J. Biol. Chem.*, 241 (1966) 2746.

148 D. W. BROOKS AND C. R. DAWSON, in J. PEISACH, P. AISEN AND W. E. BLUMBERG (Eds.), *Biochemistry of Copper*, Academic Press, New York, 1966, p. 343.

149 K. KERTSZ, in J. PEISACH, P. AISEN AND W. E. BLUMBERG (Eds.), *Biochemistry of Copper*, Academic Press, New York, 1966, p. 359.

150 H. R. MAHLER, H. M. BAUM AND G. HUEBSCHER, *Science*, 124 (1956) 705.

151 S. FRIEDMAN AND S. KAUFMAN, *J. Biol. Chem.*, 240 (1965) 4763.

152 E. A. ZELLER, in P. D. BOYER, H. LARDY AND K. MYRBACK (Eds.), *The Enzymes*, Vol. 8, 2nd ed., Academic Press, New York, 1963, p. 313.

153 H. BLASCHKO, in P. D. BOYER, H. LARDY AND K. MYRBACK (Eds.), *The Enzymes*, Vol. 8, 2nd ed., Academic Press, New York, 1963, p. 337.

154 J. LEVY, *Therapie*, 23 (1968) 1.

155 B. MONDOVI, G. ROTILIO, M. T. COSTA, A. FINAZZI-AGRO, E. CHIANCONE, R. E. HANSEN AND H. BEINERT, *J. Biol. Chem.*, 242 (1967) 1160.

156 H. YAMADA, H. KUMAGAI, H. KAWASAKI, H. MATSUI AND K. OGATA, *Biochem. Biophys. Res. Commun.*, 29 (1967) 723.

157 K. F. TIPTON, *Biochim. Biophys. Acta*, 159 (1968) 451.

158 V. G. ERWIN AND L. HELLERMAN, *J. Biol. Chem.*, 242 (1967) 4230.

159 B. GOMES, I. IGAUE, H. G. KLOEPFER AND K. T. YASUNOBU, *Arch. Biochem. Biophys.*, 132 (1969) 16.

160 S. NARA, B. GOMES AND K. T. YASUNOBU, *J. Biol. Chem.*, 241 (1966) 2774.

161 H. YAMADA AND K. T. YASUNOBU, *J. Biol. Chem.*, 237 (1962) 1511.

162 F. BUFFONI AND H. BLASCHKO, *Proc. Roy. Soc. (London)*, *Ser. B.*, 161 (1964) 153.

163 H. BLASCHKO, F. BUFFONI, N. WEISSMAN, W. H. CARNES AND W. F. COULSON, *Biochem. J.*, 96 (1965) 4 C.

164 C. F. MILLS, A. C. DALGARNO AND R. B. WILLIAMS, *Biochem. Biophys. Res. Commun.*, 24 (1966) 537.

165 C. H. HILL AND C. S. KIM, *Biochem. Biophys. Res. Commun.*, 27 (1968) 94.

166 H. KOMAI AND J. B. NEILANDS, *Biochim. Biophys. Acta*, 171 (1969) 311.

167 D. W. FANSHIER AND E. KUN, *Biochim. Biophys. Acta*, 58 (1962) 266.

168 S. KATOH, *Nature*, 186 (1960) 533.

169 H. PORTER, in J. PEISACH, P. AISEN AND W. E. BLUMBERG (Eds.), *Biochemistry oɟ Copper*, Academic Press, New York, 1966, p. 159.

170 F. HAUROWITZ AND R. L. HARDIN, in H. NEURATH AND K. BAILEY (Eds.), *The Proteins*, Vol. II, Part A, Academic Press, New York, 1954, p. 336.

171 A. TZAGOLOFF, C. P. YANG, D. C. WHARTON AND J. S. RIESKE, *Biochim. Biophys. Acta*, 96 (1965) 1.

172 C. G. HOLMBERG AND C. B. LAURELL, *Acta Chem. Scand.*, 2 (1948) 550.

173 C. R. DAWSON, in J. PEISACH, P. AISEN AND W. E. BLUMBERG (Eds.), *Biochemistry of Copper*, Academic Press, New York, 1966, p. 305.

174 W. G. LEVINE, in J. PEISACH, P. AISEN AND W. R. BLUMBERG (Eds.), *Biochemistry of Copper*, Academic Press, New York, 1966, p. 371.

175 D. AMARAL, L. BERNSTEIN, D. MORSE AND B. L. HORECKER, *J. Biol. Chem.*, 238 (1963) 2281.

176 J. M. HILL AND P. J. G. MANN, *Biochem. J.*, 91 (1964) 171.

177 T. MANN AND D. KEILIN, *Proc. Roy. Soc. (London)*, *Ser. B*, 126 (1938) 303.

178 H. MARKOWITZ, G. E. CARTWRIGHT AND M. M. WINTROBE, *J. Biol. Chem.*, 234 (1959) 40.
179 J. R. KIMMEL, H. MARKOWITZ AND D. M. BROWN, *J. Biol. Chem.*, 234 (1959) 46.
180 M. S. MOHAMED AND P. M. GREENBERG, *J. Gen. Physiol.*, 37 (1954) 433.
181 H. PORTER, M. SWEENEY AND E. PORTER, *Arch. Biochem. Biophys.*, 105 (1964) 319.
182 H. PORTER AND J. FOLCH, *J. Neurochem.*, 1 (1957) 260.
183 R. LOUTRE AND R. WITTERS, in J. PEISACH, P. AISEN AND W. E. BLUMBERG (Eds.), *Biochemistry of Copper*, Academic Press, New York, 1966, p. 455.
184 G. BERTRAND AND R. C. BHATTACHERJEE, *Compt. Rend.*, 198 (1934) 1823.
185 W. R. TODD, C. A. ELVEHJEM AND E. G. HART, *Am. J. Physiol.*, 107 (1934) 146.
186 B. L. O'DELL AND J. E. SAVAGE, *Poultry Sci.*, 36 (1957) 489.
187 A. S. PRASAD, J. A. HALSTEAD AND M. NADIMI, *Am. J. Med.*, 31 (1961) 532.
188 D. KEILIN AND T. MANN, *Biochem. J.*, 34 (1940) 1163.
189 B. L. VALLEE, in C. L. COMAR AND F. BRONNER (Eds.), *Mineral Metabolism, An Advanced Treatise*, Vol. 2, Academic Press, New York, 1962, p. 443.
190 T. K. LI, in A. S. PRASAD (Ed.), *Zinc Metabolism*, Thomas, Springfield, Ill., 1966, p. 48.
191 A. S. PRASAD, in A. S. PRASAD (Ed.), *Zinc Metabolism*, Thomas, Springfield, Ill., 1966, p. 250.
192 R. M. FORBES, in A. A. ALBANESE (Ed.), *New Methods of Nutritional Biochemistry with Applications and Interpretations*, Academic Press, New York, 1967, p. 339.
193 R. A. McCANCE AND E. M. WIDDOWSON, *Biochem. J.*, 36 (1942) 692.
194 W. N. PEARSON, T. SCHWINK AND M. REICH, in A. S. PRASAD (Ed.), *Zinc Metabolism*, Thomas, Springfield, Ill., 1966, p. 239.
195 R. M. FORBES, *J. Nutr.*, 83 (1964) 225.
196 F. A. SUSO AND H. M. EDWARDS, *Poultry Sci.*, 47 (1968) 991.
197 W. J. MILLER, *Am. J. Clin. Nutr.*, 22 (1969) 1323.
198 H. M. TRIBBLE AND F. I. SCOULAR, *J. Nutr.*, 52 (1954) 209.
199 H. SPENCER, B. ROSOFF, I. LEWIN AND J. SAMACHSON, in A. S. PRASAD (Ed.), *Zinc Metabolism*, Thomas, Springfield, Ill., 1966, p. 339.
200 D. OBERLEAS, M. E. MUHRER AND B. L. O'DELL, *J. Nutr.*, 90 (1966) 56.
201 D. R. VAN CAMPEN AND E. A. MITCHELL, *J. Nutr.*, 86 (1965) 120.
202 B. L. O'DELL, *Am. J. Clin. Nutr.*, 22 (1969) 1315.
203 B. L. O'DELL AND J. E. SAVAGE, *Proc. Soc. Exptl. Biol. Med.*, 103 (1960) 304.
204 H. J. A. LIKUSKI AND R. M. FORBES, *J. Nutr.*, 85 (1965) 230.
205 M. P. PLUMLEE, D. R. WHITAKER, W. H. SMITH, J. H. CONRAD, H. E. PARKER AND W. M. BEESON, *J. Anim. Sci.*, 19 (1960) 1285.
206 W. G. HOEKSTRA, *Am. J. Clin. Nutr.*, 22 (1969) 1268.
207 I. VIKBLADH, *Scand. J. Clin. Lab. Invest.*, 2 (1950) 143.
208 H. WOLFF, *Klin. Wochschr.*, 34 (1956) 409.
209 E. DENNES, R. TUPPER AND A. WORMALL, *Biochem. J.*, 82 (1962) 466.
210 W. BURGIE AND K. SCHMID, *J. Biol. Chem.*, 236 (1961) 1066.
211 D. M. SURGENOR, B. A. KOECHLIN AND L. E. STRONG, *J. Clin. Invest.*, 28 (1949) 73.
212 B. L. VALLEE, F. L. HOCH, S. J. ADELSTEIN AND W. E. C. WACKER, *J. Am. Chem. Soc.*, 78 (1956) 5879.
213 B. L. VALLEE, F. L. HOCH AND W. L. HUGES JR., *Arch. Biochem. Biophys.*, 48 (1954) 347.
214 S. E. HERRECK, J. R. OKUNEWICK AND T. G. HENNESSY, *Radiation Res.*, 22 (1964) 196.
215 E. M. WIDDOWSON, R. A. McCANCE AND C. M. SPRAY, *Clin. Sci.*, 10 (1951) 113.
216 B. L. VALLEE AND M. D. ALTSCHULE, *Physiol. Rev.*, 29 (1949) 370.

217 G. Weitzel, E. Buddecke, A. M. Fretzdorff, E. J. Strecker and U. Roester, *Z. Physiol. Chem.*, 299 (1955) 193.
218 C. A. Mawson and M. I. Fischer, *Can. J. Med. Sci.*, 20 (1952) 336.
219 C. A. Mawson and M. I. Fischer, *Biochem. J.*, 55 (1953) 696.
220 S. A. Gunn, T. C. Gould, S. S. Ginori and J. C. Morse, *Proc. Soc. Exptl. Biol. Med.*, 88 (1955) 556.
221 E. P. Reaven and A. J. Cox, *J. Histochem. Cytochem.*, 11 (1963) 782.
222 W. H. Strain and W. J. Pories, in A. S. Prasad (Ed.), *Zinc Metabolism*, Thomas, Springfield, Ill., 1966, p. 363.
223 M. J. Millar, N. R. Vincent and C. A. Mawson, *J. Histochem. Cytochem.*, 9 (1961) 111.
224 M. E. Rubini, G. Montaho, C. P. Lockhart and C. R. Johnson, *Am. J. Physiol.*, 200 (1961) 1345.
225 R. E. Thiers and B. L. Vallee, *J. Biol. Chem.*, 266 (1957) 911.
226 C. Edwards, K. B. Olson, G. Heggen and J. Glenn, *Proc. Soc. Exptl. Biol. Med.*, 107 (1961) 94.
227 G. C. Cotzias and P. S. Papavasiliow, *Am. J. Physiol.*, 206 (1964) 787.
228 G. E. Sheline, I. L. Chaikoff, H. B. Jones and M. L. Montgomery, *J. Biol. Chem.*, 147 (1943) 409.
229 I. G. F. Gilbert and D. M. Taylor, *Biochim. Biophys. Acta*, 21 (1956) 545.
230 G. C. Cotzias, D. C. Borg and B. Selleck, *Am. J. Physiol.*, 202 (1962) 359.
231 J. E. Savage, J. M. Yohe, E. E. Pickett and B. L. O'Dell, *Poultry Sci.*, 43 (1964) 420.
232 B. L. Vallee, R. G. Fluharty and J. G. Givson, *Acta Unio Intern. Contra Cancrum*, 6 (1949) 869.
233 J. P. Okunewick, B. Pond and T. G. Hennessy, *Am. J. Physiol.*, 202 (1962) 926.
234 J. P. Feaster, S. L. Hansard, J. T. McCall, F. H. Skipper and G. K. Davis, *J. Anim. Sci.*, 13 (1954) 781.
235 J. C. Heath and J. Liqiuer-Milward, *Biochim. Biophys. Acta*, 5 (1950) 404.
236 J. C. Wakeley, B. Moffat, A. Crook and J. R. Mallard, *Intern. J. Appl. Radiat. Isotop.*, 7 (1960) 225.
237 R. M. Forbes, *Federation Proc.*, 19 (1960) 643.
238 C. A. Byrd and G. Matrone, *Proc. Soc. Exptl. Biol. Med.*, 119 (1965) 347.
239 D. R. van Campen, *J. Nutr.*, 97 (1969) 104.
240 H. D. Ritchie, R. W. Luecke, B. V. Baltzer, E. R. Miller, D. E. Ullrey and J. A. Hoefer, *J. Nutr.*, 79 (1963) 117.
241 W. G. Hoekstra, *Federation Proc.*, 23 (1964) 1068.
242 P. J. O'Hara, H. P. Newman and R. Jackson, *Australian Vet. J.*, 36 (1960) 225.
243 J. Parizek and Z. Zahor, *Nature*, 177 (1956) 1036.
244 S. A. Gunn, T. C. Gould and W. A. D. Anderson, *Arch. Pathol.*, 71 (1961) 274.
245 W. C. Supplee, *Science*, 139 (1963) 119.
246 S. A. Gunn, T. C. Gould and W. A. D. Anderson, *Proc. Soc. Exptl. Biol. Med.*, 111 (1962) 559.
247 S. A. Gunn, T. C. Gould and W. A. D. Anderson, *Acta Endocrinol.*, 37 (1961) 24.
248 R. H. Follis, H. G. Day and E. V. McCollum, *J. Nutr.*, 22 (1941) 223.
249 H. G. Day and B. E. Skidmore, *J. Nutr.*, 33 (1947) 27.
250 H. F. Tucker and W. D. Salmon, *Proc. Soc. Exptl. Biol. Med.*, 88 (1955) 613.
251 J. K. Miller and W. J. Miller, *J. Nutr.*, 76 (1962) 467.
252 C. F. Mills, A. C. Dalgarno, R. B. Williams and J. Quarterman, *Brit. J. Nutr.*, 21 (1967) 751.

253 E. A. OTT, W. H. SMITH, M. STAB, H. E. PARKER, R. B. HARRINGTON AND W. M. BEESON, *J. Nutr.*, 87 (1965) 495.

254 W. J. MILLER, D. M. BLACKWON, R. P. GENTRY, W. H. PITTS AND G. W. POWELL, *J. Nutr.*, 92 (1967) 71.

255 B. T. ROBERTSON AND M. J. BURNS, *Am. J. Vet. Res.*, 24 (1963) 997.

256 M. P. MACAPINLAC, G. H. BARNEY, W. N. PEARSON AND W. J. DARBY, *J. Nutr.*, 93 (1967) 499.

257 B. L. O'DELL, P. M. NEWBERNE AND J. E. SAVAGE, *J. Nutr.*, 65 (1958) 503.

258 F. H. KRATZER, P. VOHRA, J. B. ALLRED AND P. N. DAVIS, *Proc. Soc. Exptl. Biol. Med.*, 98 (1958) 205.

259 M. R. S. FOX AND B. N. HARRISON, *Proc. Soc. Exptl. Biol. Med.*, 116 (1964) 256.

260 E. W. KIENHOLZ, D. E. TURK, M. L. SUNDE AND W. G. HOEKSTRA, *J. Nutr.*, 75 (1961) 211.

261 A. S. PRASAD, A. MIALE, Z. FARID, H. H. SANDSTEAD AND A. SCHULERT, *J. Lab. Clin. Med.*, 61 (1963) 537.

262 L. S. HURLEY, *Am. J. Clin. Nutr.*, 22 (1969) 1332.

263 J. APGAR, *Am. J. Physiol.*, 215 (1968) 160.

264 B. L. VALLEE, *The Role of Zinc in Medicine and Biology*, Academic Press, New York, in the press.

265 K. S. V. SAMPATH-KUMAR, K. A. WALSH, J. P. BARGETZI AND H. NEURATH, *Biochemistry*, 2 (1963) 1475.

266 B. L. VALLEE, J. A. RUPLEY, T. L. COOMBS AND H. NEURATH, *J. Biol. Chem.*, 233 (1960) 64.

267 J. E. COLEMAN AND G. L. VALLEE, *J. Biol. Chem.*, 235 (1960) 390.

268 J. E. COLEMAN AND B. L. VALLEE, *J. Biol. Chem.*, 236 (1966) 2244.

269 B. L. VALLEE, J. F. RIORDAN AND J. E. COLEMAN, *Proc. Natl. Acad. Sci. (U.S.)*, 49 (1963) 109.

270 J. E. COLEMAN AND B. L. VALLEE, *J. Biol. Chem.*, 237 (1962) 3430.

271 J. E. COLEMAN AND B. L. VALLEE, *Biochemistry*, 1 (1962) 1083.

272 N. NEURATH, R. A. BRADSHAW, L. H. ERICSSON, D. R. BABBIN, P. H. PETRA AND K. A. WALSH, *Brookhaven Symp. Biol.*, 21 (1968).

273 H. NEURATH, R. A. BRADSHAW, P. H. PETRA AND K. A. WALSH, *Proc. Roy. Soc. (London) Ser. A*, in the press.

274 W. N. LIPSCOMB, J. C. COPPOLA, J. A. HARTSUCK, M. L. LUDWIG, H. MUIRHEAD, J. SEARL AND T. A. STEITZ, *J. Mol. Biol.*, 19 (1966) 423.

275 M. L. LUDWIG, J. A. HARTSUCK, T. A. STEITZ, M. MUIRHEAD, J. C. COPPOLA, G. N. REEKE AND W. N. LIPSCOMB, *Proc. Natl. Acad. Sci. (U.S.)*, 57 (1967) 511.

276 T. A. STEITZ, M. L. LUDWIG, F. A. QUIOCHO AND W. N. LIPSCOMB, *J. Biol. Chem.*, 242 (1967) 4662.

277 G. N. REEKE, J. A. HARTSUCK, M. L. LUDWIG, F. A. QUIOCHO, T. A. STEITZ AND W. N. LIPSCOMB, *Proc. Natl. Acad. Sci. (U.S.)*, 58 (1967) 2220.

278 W. N. LIPSCOMB, J. A. HARTSUCK, G. N. REEKE, F. A. QUIOCHO, P. H. BETHGE, M. L. LUDWIG, T. A. STEITZ, H. MUIRHEAD AND J. C. COPPOLA, *Brookhaven Symp. Biol.*, 21 (1968) 24.

279 W. N. LIPSCOMB, G. N. REEKE, J. A. HARTSUCK, F. A. QUIOCHO AND P. H. BETHAGE, *Proc. Roy. Soc. (London) Ser. B*, in the press.

280 R. E. DICKERSON AND I. GEIS, *The Structure and Action of Proteins*, Harper and Row, New York, 1969.

280a W. N. LIPSCOMB, *Accounts Chem. Res.*, 3 (1970) 81.

281 J. E. FOLK, K. A. PIEZ, W. R. CARROLL AND J. A. GLADNER, *J. Biol. Chem.*, 235 (1960) 2272.

282 E. WINTERSBERGER, D. J. COX AND H. NEURATH, *Biochemistry*, 1 (1962) 1069.
283 J. E. FOLK AND J. A. GLADNER, *J. Biol. Chem.*, 235 (1960) 60.
284 E. WINTERSBERGER, H. NEURATH, T. L. COOMBS AND B. L. VALLEE, *Biochemistry*, 4 (1965) 1526.
285 J. M. ARMSTRONG, D. V. MEYERS, J. A. VERPOORTE AND J. T. EDSALL, *J. Biol. Chem.*, 241 (1966) 5137.
286 S. LINDSKOG, *J. Biol. Chem.*, 238 (1963) 945.
287 Y. POCKER AND J. T. STONE, *J. Am. Chem. Soc.*, 87 (1965) 5479.
288 S. LINDSKOG, *Biochemistry*, 5 (1966) 2641.
289 J. C. KERNOHAN, *Biochim. Biophys. Acta*, 81 (1964) 346.
290 J. C. KERNOHAN, *Biochim. Biophys. Acta*, 96 (1965) 304.
291 S. LINDSKOG, *Biochim. Biophys. Acta*, 122 (1966) 534.
292 J. E. COLEMAN, *J. Biol. Chem.*, 242 (1967) 5212.
293 M. E. RIEPE AND J. H. WANG, *J. Am. Chem. Soc.*, 89 (1967) 4229.
294 P. L. WHITNEY, P. O. NYMAN AND B. G. MALMSTROM, *J. Biol. Chem.*, 242 (1967) 4212.
295 S. L. BRADBURY AND J. T. EDSALL, *Federation Proc.*, 27 (1968) 292.
296 S. LINDSKOG AND B. G. MALMSTROM, *J. Biol. Chem.*, 237 (1962) 1129.
297 S. LINDSKOG AND P. O. NYMAN, *Biochim. Biophys. Acta*, 85 (1964) 462.
298 L. M. SMILLIE, A. FURKHA, N. NAGABLUSKAN, K. J. STEVENSON AND C. O. PARKES, *Nature*, 218 (1968) 343.
299 F. ROTHMAN AND R. BYRNE, *J. Mol. Biol.*, 6 (1963) 330.
300 C. LEVINTHAL, E. R. SIGNER AND K. FETHEROLF, *Proc. Natl. Acad. Sci. (U.S.)*, 48 (1962) 1230.
301 M. J. SCHLESINGER AND K. BARRETT, *J. Biol. Chem.*, 204 (1965) 4284.
302 R. T. SIMPSON AND B. L. VALLEE, *Federation Proc.*, 27 (1968) 291.
303 J. H. SCHWARTZ AND F. LIPMANN, *Proc. Natl. Acad. Sci. (U.S.)*, 47 (1961) 1996.
304 C. LAZDUNSKI AND M. LAZDUNSKI, *Europ. J. Biochem.*, 7 (1969) 294.
305 G. H. TAIT AND B. L. VALLEE, *Proc. Natl. Acad. Sci. (U.S.)*, 56 (1966) 1247.
306 B. J. CAMPBELL, Y. C. LIN, R. V. DAVIS AND E. BALLEW, *Biochim. Biophys. Acta*, 118 (1966) 371.
307 A. M. RENE AND B. J. CAMPBELL, *J. Biol. Chem.*, 244 (1969) 1445.
308 D. E. DRUM, T. K. LI AND B. L. VALLEE, *Biochemistry*, 8 (1969) 3783.
309 J. H. R. KAGI AND B. L. VALLEE, *J. Biol. Chem.*, 235 (1960) 3188.
310 R. DRUYAN AND B. L. VALLEE, *Federation Proc.*, 21 (1962) 247.
311 D. E. DRUM, J. H. HARRISON, T. K. LI, J. L. BETHUNE AND B. L. VALLEE, *Proc. Natl. Acad. Sci. (U.S.)*, 57 (1967) 1434.
312 H. THEORELL AND J. S. MCKINLEY-MCKEE, *Acta Chem. Scand.*, 16 (1961) 1811.
313 A. D. WINES AND H. THEORELL, *Acta Chem. Scand.*, 14 (1960) 1729.
314 T. K. LI AND B. L. VALLEE, *Biochemistry*, 4 (1965) 1195.
315 N. EVANS AND B. R. RABIN, *Europ. J. Biochem.*, 4 (1968) 548.
316 T. K. LI AND B. L. VALLEE, *Biochemistry*, 3 (1964) 896.
317 J. I. HARRIS, *Nature*, 203 (1964) 30.
318 C. H. REYNOLDS AND J. S. MCKINLEY-MCKEE, *Europ. J. Biochem.*, 10 (1969) 474.
319 H. SUND AND W. BURCHARD, *Europ. J. Biochem.*, 6 (1968) 202.
320 S. J. ADELSTEIN AND B. L. VALLEE, *J. Biol. Chem.*, 233 (1958) 589.
321 L. E. ORGEL, in K. BLOCH AND O. HAYAISHI (Eds.), *Biological and Chemical Aspects of Oxygenases*, Maruzen, Tokyo, 1966.
322 S. J. ADELSTEIN AND B. L. VALLEE, *J. Biol. Chem.*, 234 (1959) 824.
323 B. M. ANDERSON, M. L. REYNOLDS AND C. D. ANDERSON, *Biochim. Biophys. Acta*, 113 (1966) 235.

324   J. H. HARRISON, *Federation Proc.*, 22 (1963) 493.
325   R. L. WARD AND J. A. HAPPE, *Biochem. Biophys. Res. Commun.*, 28 (1967) 785.
326   W. E. C. WACKER AND B. L. VALLEE, *J. Biol. Chem.*, 234 (1959) 3257.
327   M. FUJIOKA AND I. LIEBERMAN, *J. Biol. Chem.*, 239 (1964) 1164.
328   H. SANDSTEAD AND R. A. RINALDI, *J. Cell. Physiol.*, 73 (1969) 81.
329   M. P. MACAPINLAC, W. N. PEARSON, G. H. BARNEY AND W. J. DARBY, *J. Nutr.*, 95 (1968) 569.
330   D. J. PLOCKE AND B. L. VALLEE, *Biochemistry*, 1 (1962) 1039.
331   D. J. PLOCKE, C. LEVINTHAL AND B. L. VALLEE, *Biochemistry*, 1 (1962) 373.
332   J. D. MCCONN, D. TSURU AND K. T. YASUNOBU, *J. Biol. Chem.*, 239 (1964) 3706.
333   D. TSURU, J. D. MCCONN AND K. T. YASUNOBU, *J. Biol. Chem.*, 240 (1965) **2415.**
334   B. J. CAMPBELL, Y. C. LIN AND E. BALLEW, *J. Biol. Chem.*, 242 (1967) 930.
335   B. L. VALLEE AND F. L. HOCH, *Proc. Natl. Acad. Sci. (U.S.)*, 41 (1955) 327.
336   B. L. VALLEE, in P. D. BOYER, H. LARDY AND K. MYRBACK (Eds.), *The Enzymes*, Vol. 3, Academic Press, New York, 1960, p. 241.
337   H. KUBO, T. YAMANO, M. DWATSUBO, H. WATERI, T. SOYAMA, J. SHIRAISHI, S. SAWADA, N. KAWASHIMA, S. MITANI AND K. ITO, *Bull. Soc. Chim. Biol.*, 40 (1958) 431.
338   A. CURDEL, *Biochem. Biophys. Res. Commun.*, 22 (1966) 357.
339   T. CREMONA AND T. P. SINGER, *J. Biol. Chem.*, 239 (1964) 1466.
340   A. F. WAGNER AND K. FOLKERS, in M. FLORKIN AND E. H. STOTZ (Eds.), *Comprehensive Biochemistry*, Vol. 11, Elsevier, Amsterdam, 1963, p. 103.
341   E. LESTER SMITH, in C. L. COMAR AND F. BRONNER (Eds.), *Mineral Metabolism*, Vol. II B, Academic Press, New York, 1961, p. 349.
342   F. A. SUSO AND H. M. EDWARDS, *Poultry Sci.*, 47 (1968) 1417.
343   E. LESTER SMITH, *Vitamin B_{12}*, 3rd ed., Wiley, New York, 1965, p. 127.
344   A. CARLIER, C. BOULANGER, R. HAVEX AND G. BISERTE, *Compt. Rend.*, 264 (1967) 2240.
345   R. F. SCHILLING, *Vitamins Hormones*, 26 (1968) 547.
346   R. M. HEYSSEL, R. C. BOZIAN, W. J. DARBY AND M. C. BELL, *J. Clin. Nutr.*, 18 (1966) 176.
347   P. G. REIZENSTEIN, G. EK AND C. M. E. MATHEWS, *Phys. Med. Biol.*, 11 (1966) 295.
348   W. H. FLEMING AND E. R. KING, *Gastroenterology*, 42 (1962) 164.
349   C. C. BOOTH AND D. L. MOLLIN, *Lancet*, (1959) 118.
350   O. N. MILLER, J. L. RANEY, H. J. HANSEN AND F. J. TRONCALE, *Arch. Biochem. Biophys.*, 100 (1963) 223.
351   H. L. ROSENTHAL, *J. Nutr.*, 78 (1962) 348.
352   H. A. BARKER, *Biochem. J.*, 105 (1967) 1.
353   B. L. O'DELL, *Federation Proc.*, 27 (1968) 199.
354   G. N. SCHRAUZER, *Ann. N. Y. Acad. Sci.*, 158 (1969) 526.
355   H. WEISSBACH, A. PETERKOFSKY AND H. A. BARKER, in M. FLORKIN AND E. H. STOTZ (Eds.), *Comprehensive Biochemistry*, Vol. 16, Elsevier, Amsterdam, 1965, p. 189.
356   H. P. C. HOGENKAMP, *Federation Proc.*, 25 (1966) 1623.
357   H. A. O. HILL, J. M. PRATT AND R. J. P. WILLIAMS, *Chem. Britain*, 5 (1969) 156.
358   H. WEISSBACH AND H. DICKERMAN, *Physiol. Rev.*, 45 (1965) 80.
359   R. T. TAYLOR AND H. WEISSBACH, *Arch. Biochem. Biophys.*, 129 (1969) 745.
360   R. L. BLAKLEY, *Federation Proc.*, 25 (1966) 1633.
361   R. H. ABELES AND W. S. BECK, *J. Biol. Chem.*, 242 (1967) 3589.
362   M. AKHTAR, *Comp. Biochem. Physiol.*, 38 (1969) 1.

363 G. C. Cotzias, in C. L. Comar and F. Bronner (Eds.), *Mineral Metabolism*, Vol. II B, Academic Press, New York, 1961.
364 S. J. Bertinchamps, S. T. Miller and G. C. Cotzias, *Am. J. Physiol.*, 211 (1966) 217.
365 P. S. Papavasiliou, S. T. Miller and G. C. Cotzias, *Am. J. Physiol.*, 211 (1966) 211.
366 H. S. Wilgus and A. R. Patton, *J. Nutr.*, 18 (1939) 35.
367 A. A. Britton and G. C. Cotzias, *Am. J. Physiol.*, 211 (1966) 203.
368 R. M. Leach, *Federation Proc.*, 26 (1967) 118.
369 H. C. C. Tsai and G. J. Everson, *J. Nutr.*, 9 (1967) 447.
370 L. S. Hurley, *Federation Proc.*, 27 (1968) 193.
371 R. E. Shrader and G. J. Everson, *J. Nutr.*, 91 (1967) 453.
372 L. Erway, L. S. Hurley and A. Fraser, *Science*, 152 (1966) 1766.
373 M. Cohn, *Biochemistry*, 2 (1963) 623.
374 M. C. Scrutton, M. F. Utter and A. S. Mildvan, *J. Biol. Chem.*, 241 (1966) 3480.
375 A. S. Mildvan, M. C. Scrutton and M. F. Utter, *J. Biol. Chem.*, 241 (1966) 3488.
376 A. S. Mildvan and M. C. Scrutton, *Biochemistry*, 6 (1967) 2978.
377 R. S. Miller, A. S. Mildvan, H. C. Chang, R. L. Easterday, H. Maruyama and M. D. Lane, *J. Biol. Chem.*, 243 (1968) 6030.
378 H. Hirsch-Kolb and D. M. Greenberg, *J. Biol. Chem.*, 243 (1968) 6123.
379 R. C. Valentine, B. M. Shapiro and E. R. Stadtman, *Biochemistry*, 7 (1968) 2143.
380 M. D. Denton and A. Ginsberg, *Biochemistry*, 8 (1969) 1714.
381 E. M. Martin and J. A. Sonnabend, *J. Virol.*, 1 (1967) 97.
382 U. Maitra, Y. Nakata and J. Hurwitz, *J. Biol. Chem.*, 242 (1967) 4908.
383 R. M. Leach, A. M. Muenster and E. M. Wien, *Arch. Biochem. Biophys.*, 133 (1969) 22.
384 K. Schwarz and W. Mertz, *Arch. Biochem. Biophys.*, 85 (1959) 292.
385 H. A. Schroder, J. J. Balassa and I. H. Tipton, *J. Chronic Dis.*, 15 (1962) 941.
386 W. Mertz, *Physiol. Rev.*, 49 (1969) 163.
387 L. L. Hopkins and K. Schwarz, *Biochim. Biophys. Acta*, 90 (1964) 484.
388 P. Aisen, R. Aasa and A. G. Redfield, *J. Biol. Chem.*, 244 (1969) 4628.
389 S. J. Gray and K. Sterling, *J. Clin. Invest.*, 29 (1950) 1604.
390 L. L. Hopkins, *Am. J. Physiol.*, 209 (1965) 731.
391 W. Mertz, E. E. Roginski, F. J. Feldman and D. E. Thurman, *J. Nutr.*, 99 (1969) 363.
392 I. W. F. Davidson and W. L. Blackwell, *Proc. Soc. Exptl. Biol. Med.*, 127 (1968) 66.
393 W. Mertz and E. E. Roginski, *J. Nutr.*, 97 (1969) 531.
394 E. E. Roginski and W. Mertz, *J. Nutr.*, 93 (1967) 249.
395 E. E. Roginski and W. Mertz, *J. Nutr.*, 97 (1969) 525.
396 H. A. Schroeder, *J. Nutr.*, 88 (1966) 439.
397 T. G. Farkas and S. L. Roberson, *Exptl. Eye Res.*, 4 (1965) 124.
398 L. H. Stickland, *Biochem. J.*, 44 (1949) 190.
399 W. S. Ferguson, A. H. Lewis and S. J. Watson, *Nature*, 141 (1938) 553.
400 D. A. Richert and W. W. Westerfeld, *J. Biol. Chem.*, 203 (1953) 915.
401 E. C. DeRenzo, E. Kaleita, P. Heytler, J. J. Oleson, B. L. Hutchings and J. H. Williams, *Arch. Biochem. Biophys.*, 45 (1953) 247.
402 J. T. Spence, *Z. Naturwiss. Med. Grundlagenforsch.*, 2 (1965) 267.
403 E. C. DeRenzo, in C. L. Comar and F. Bronner (Eds.), *Mineral Metabolism, An Advanced Treatise*, Vol. 2, Academic Press, New York, 1962, p. 483.
404 A. T. Dick, *Soil Sci.*, 81 (1956) 229.

405 A. NASON, in C. A. LAMB, O. G. BENTLEY AND J. M. BEATTIE (Eds.), *Trace Elements*, Academic Press, New York, 1958.
406 L. T. FAIRHALL, R. C. DUNN, N. E. SHARPLESS AND E. A. PRITCHARD, *U.S. Public Health Serv., Public Health Bull.*, 293 (1945) 1.
407 I. J. CUNNINGHAM, in W. D. MCELROY AND B. GLASS (Eds.), *Symposium on Copper Metabolism*, Johns Hopkins, Baltimore, Md., 1950.
408 A. T. DICK, in W. D. MCELROY AND B. GLASS (Eds.), *Inorganic Nitrogen Metabolism*, Johns Hopkins, Baltimore, Md., 1956.
409 J. F. SCAIFE, *New Zealand J. Sci. Technol., A*, 38 (1956) 285.
410 I. H. TIPTON, in M. J. SEVEN (Ed.), *Metal Binding in Medicine*, Lippincott, Philadelphia, 1960, p. 27.
411 R. E. DAVIS, B. L. REID, A. A. KURNICK AND J. R. COUCH, *J. Nutr.*, 70 (1960) 193.
412 C. L. COMAR, in W. D. MCELROY AND B. GLASS (Eds.), *Symposium on Copper Metabolism*, Johns Hopkins, Baltimore, Md., 1950.
413 B. ROSOFF AND H. SPENCER, *Nature*, 202 (1964) 410.
414 J. B. NEILANDS, R. M. STRONG AND C. A. ELVEHJEM, *J. Biol. Chem.*, 172 (1948) 431.
415 C. L. COMAR, K. SINGER AND G. K. DAVIS, *J. Biol. Chem.*, 180 (1949) 913.
416 E. S. HIGGINS, D. A. RICHERT AND W. W. WESTERFELD, *J. Nutr.*, 59 (1956) 539.
417 R. M. LEACH AND L. C. NORRIS, *Poultry Sci.*, 36 (1957) 1136.
418 S. S. GABALLAH, L. G. ABOOD, G. T. CALEEL AND A. KAPSALIS, *Proc. Soc. Exptl. Biol. Med.*, 120 (1965) 733.
419 S. S. GABALLAH, A. KAPSALIS AND D. STURDIVANT, *J. Am. Osteopath. Ass.*, 64 (1966) 945.
420 N. A. MARCILESE, C. B. AMMERMAN, R. M. VALSECCHI, B. G. DUNAVANT AND G. K. DAVIS, *J. Nutr.*, 99 (1969) 177.
421 R. F. MILLER AND R. W. ENGEL, *Federation Proc.*, 19 (1960) 666.
422 W. C. ELLIS, W. H. PFANDER, M. E. MUHRER AND E. E. PICKETT, *J. Anim. Sci.*, 17 (1958) 180.
423 G. PALMER, R. C. BRAY AND H. BEINERT, *J. Biol. Chem.*, 239 (1964) 2657.
424 R. C. BRAY, G. PALMER AND H. BEINERT, *J. Biol. Chem.*, 239 (1964) 2667.
425 K. V. RAJAGOPALAN AND P. HANDLER, *J. Biol. Chem.*, 242 (1967) 4097.
426 S. T. SMITH, K. V. RAJAGOPALAN AND P. HANDLER, *J. Biol. Chem.*, 242 (1967) 4108.
427 M. P. COUGHLAN, D. V. RAJAGOPALAN AND P. HANDLER, *J. Biol. Chem.*, 244 (1969) 2658.
428 P. HANDLER AND K. V. RAJAGOPALAN, *J. Biol. Chem.*, 239 (1964) 2027.
429 J. T. SPENCE AND H. CHANG, *Inorg. Chem.*, 2 (1963) 319.
430 R. H. GARRETT AND A. NASON, *J. Biol. Chem.*, 244 (1969) 2870.
431 D. J. D. NICHOLAS AND C. G. WALKER, *Biochim. Biophys. Acta*, 49 (1961) 350.
432 D. J. D. NICHOLAS, P. W. WILSON, W. HEINEN AND G. PALMER, *Nature*, 196 (1962) 433.
433 K. SCHWARZ AND C. M. FOLTZ, *J. Am. Chem. Soc.*, 79 (1957) 3293.
434 I. ROSENFELD AND O. A. BEATH, *Selenium*, Academic Press, New York, 1964.
435 O. H. MUTH, J. E. OLDFIELD AND P. H. WESWIG (Eds.), *Selenium in Biomedicine*, Avi Publishing Co., Westport, Conn., 1967.
436 H. E. GANTHER, *World Rev. Nutr. Diet.*, 5 (1965) 338.
437 R. E. OLSON, *Federation Proc.*, 24 (1965) 55.
438 P. L. WRIGHT, in O. H. MUTH, J. E. OLDFIELD AND P. H. WESWIG (Eds.), *Selenium in Biomedicine*, Avi Publishing Co., Westport, Conn., 1967.
439 C. F. EHLIG, D. E. HOGUE, W. H. ALLAWAY AND D. J. HAMM, *J. Nutr.*, 92 (1967) 121.
440 K. P. MCCONNELL AND G. J. CHO, in O. H. MUTH, J. E. OLDFIELD AND P. H. WESWIG (Eds.), *Selenium in Biomedicine*, Avi Publishing Co., Westport, Conn., 1967.

441 P. L. LOPEZ, R. L. PRESTON AND W. H. PFANDER, *J. Nutr.*, 94 (1968) 219.
442 L. L. HOPKINS, A. L. POPE AND C. A. BAUMANN, *J. Nutr.*, 88 (1966) 61.
443 J. L. BYARD, *Arch. Biochem. Biophys.*, 130 (1969) 556.
444 I. S. PALMER, D. D. FISHER, A. W. HALVERSON AND O. E. OLSON, *Biochim. Biophys. Acta*, 177 (1969) 336.
445 A. SHRIFT, in O. H. MUTH, J. E. OLDFIELD AND P. H. WESWIG (Eds.), *Selenium in Biomedicine*, Avi Publishing Co., Westport, Conn., 1967.
446 K. E. MASON AND J. O. YOUNG, in O. H. MUTH, J. E. OLDFIELD AND P. H. WESWIG (Eds.), *Selenium in Biomedicine*, Avi Publishing Co., Westport, Conn., 1967.
447 J. GREEN AND J. BUNYAN, *Nutr. Abstr. Rev.*, 39 (1969) 321.
448 E. SONDEGAARD, in O. H. MUTH, J. E. OLDFIELD AND P. H. WESWIG (Eds.), *Selenium in Biomedicine*, Avi Publishing Co., Westport Conn., 1967.
449 J. R. SCHUBERT, O. H. MUTH, J. E. OLDFIELD AND L. F. REMMERT, *Federation Proc.*, 20 (1961) 689.
450 J. N. THOMPSON AND M. L. SCOTT, *J. Nutr.*, 97 (1969) 335.
451 K. E. M. McCOY AND P. H. WESWIG, *J. Nutr.*, 98 (1969) 383.
452 H. E. GANTHER, *Biochemistry*, 5 (1966) 1089.
453 L. M. CUMMINS AND J. L. MARTIN, *Biochemistry*, 6 (1967) 3162.
454 H. E. GANTHER, *Biochemistry*, 7 (1968) 2898.
455 H. E. GANTHER AND C. CORCORAN, *Biochemistry*, 8 (1969) 2557.
456 A. OCHOA-SOLANO AND C. GITLER, *J. Nutr.*, 94 (1968) 243.
457 R. C. DICKSON AND A. L. TAPPEL, *Arch. Biochem. Biophys.*, 131 (1969) 100.
458 J. GROSS, in C. L. COMAR AND F. BRONNER (Eds.), *Mineral Metabolism*, Vol. II B, Academic Press, New York, 1962, p. 221.
459 K. BROWN-GRANT, *Physiol. Rev.*, 41 (1961) 189.
460 R. PITT-RIVERS AND W. R. TROTTER, *The Thyroid Gland*, Vol. I, Butterworth, London, 1964.
461 W. R. RUEGAMER, B. J. WAGNER, M. BARSTOW AND E. E. KERAN, *Endocrinology*, 81 (1967) 49.
462 B. E. BAKER, D. A. RICHERT, D. MIRISOLOFF AND W. W. WESTERFELD, *J. Nutr.*, 87 (1965) 93.
463 A. WHITE, P. HANDLER AND E. L. SMITH, *Principles of Biochemistry*, 4th ed., McGraw-Hill, New York, 1968.
464 S. B. BARKER, *Federation Proc.*, 21 (1962) 635.
465 F. L. HOCH, *Physiol. Revs.*, 42 (1962) 605.
466 H. C. HODGE, in C. L. COMAR AND F. BRONNER (Eds.), *Mineral Metabolism*, Vol. II A, Academic Press, New York, 1964, p. 573.
467 J. D. PERKINSON, I. B. WHITNEY, R. A. MONROE, W. E. LOTZ AND C. L. COMAR, *Am. J. Physiol.*, 182 (1955) 383.
468 E. J. LARGENT, in J. H. SHAW (Ed.), *Fluoridation as a Public Health Measure*, Am. Ass. Advancement Science, Washington, D.C., 1954, p. 49.
469 D. S. BERNSTEIN, N. SADOWSKY, D. M. HEGSTED, C. D. GURI AND F. J. STARE, *J. Am. Med. Ass.*, 198 (1966) 499.
470 I. ZIPKIN, A. S. POSNER AND E. D. EANES, *Biochim. Biophys. Acta*, 59 (1962) 255.
471 B. L. O'DELL, *Federation Proc.*, 19 (1960) 648.
472 Y. CHIMCHAISRI AND P. H. PHILLIPS, *J. Nutr.*, 86 (1965) 23.
473 D. J. THOMPSON, J. F. HEINTZ AND P. H. PHILLIPS, *J. Nutr.*, 84 (1964) 27.
474 R. E. PYKE, W. G. HOEKSTRA AND P. H. PHILLIPS, *J. Nutr.*, 92 (1967) 311.
475 E. E. GARDINER, J. C. ROGLER AND H. E. PARKER, *J. Nutr.*, 75 (1961) 270.
476 F. SPIERTO, J. C. ROGLER AND H. E. PARKER, *Proc. Soc. Exptl. Biol. Med.*, 132 (1969) 568.

SUBJECT INDEX

[267]

Cerebrocuprein, mol. wt., Cu content, Cu
storage, 202, 203
Ceruloplasmin, absence in turkey and
peacock blood, 193
—, apo-, formation and reconstitution,
199
—, and biliary excretion of Cu, 236
— copper, in plasma, 192
—, and Cu balance and transport, 198
—, and Cu utilization by tissues, 236
—, levels in blood of mammals, 193
—, mol. wt., Cu content, 197, 198
—, oxidase activity, inhibition by
transferrin and apotransferrin, 199
—, and promotion of Fe utilization by
stimulation of Fe oxidation, 199
—, sialic acid containing carbohydrate
component, 199
—, stimulation by Fe^{2+}, 199
—, synthesis, interference by Mo and
sulphate, 236
Chicken, biotin catabolism, 101
—, Co metabolism, 217, 219, 220
—, Cu metabolism, 191, 193, 194, 199
— embryos, control of vitamin B_6
metabolism, 58
—, Mn metabolism, 227, 228
—, Mo metabolism, 236, 238
— pancreas, conjugase, 127, 128
—, pantothenate and pantothenylcysteine
as nutritional factor, 77
—, Se metabolism, 243
—, vitamin B_6 required in diet, 48
—, vitamin B_{12} concns. in kidney and
liver, 167, 168
—, Zn metabolism, 204, 207, 208
Chimpansee, vitamin B_{12} binding proteins
from gastric mucosa, 163
Chondrodystrophy, chick embryos, in Mn
deficiency, 225
Chromate, as erythrocyte label in
haematology, 230
Chromatium strain D, photoreduction of
dihydrobiopterin by chromatophore
preparation, 145
Chromic oxide, as non-digestible marker
for absorption studies, 230
Chromium, absorption from intestine, of
Cr^{2+} and Cr^{3+}, 230
—, and activation of phosphoglucomutase,
233

Chromium, (continued)
—, biological function, analogy with
fat-soluble vitamins, 230
—, concn. in lung and brain, 230
—, — in soil, land plants, animals and
tissues, 230
—, and control mechanisms in
multicellular organisms, 230
—, corneal vascularization and opacity in
rats on Cr-deficient diets, 232
—, Cr^{3+} transport in blood by transferrin,
231
—, Cr^{6+} transport in blood, absorption by
erythrocytes, 231
—, and effect of injected insulin, 232
—, electronic structure, oxidation states,
complexes, olation, 232
—, essentiality for higher animals, 230
—, excretion in urine and faeces, 231
—, and glucose tolerance factor, 230–232
—, metabolism and metabolic function,
230–233
—, and nucleic acids, interaction, 233
—, requirement of a specific complex at
the functional level, 231
—, transport across the placenta, 231
—, uptake by mature testes after
intravenous injection, 231
Ciliapterin, origin of side-chain, 131–134
—, structure, 133
Cinnamic acid hydroxylation,
tetrahydropteridine-dependent, 144
Clostridia, anaerobic, absence of
pyridoxine-P oxidase, 55
—, —, pyridoxamine-P:α-ketoglutarate
aminotransferase, 56
Clostridium acetobutylicum, riboflavin
synthesis, commercial production, 11
— cylindrosporum, pteroylglutamates
containing additional amino acids, 125
— pasteurianum, ferredoxin, mol. wt., Fe
content, 190
— sticklandii, serine hydroxymethylase,
requirement for
tetrahydropteroyltriglutamate, 126, 127
CO_2 fixation, role of Mn in enzymes
concerned, 227, 228
Cobalamase, lack of, as primary cause of
pernicious anaemia, 163
Cobalamin(s), see also Cobalt,
Coenzyme B_{12} and Vitamin B_{12}

DNA, (*continued*)
—, synthesis in regenerating rat liver, Zn requirement, 216
Dog, cobalamin metabolism, 158, 159, 169, 171
—, Co metabolism, 218
—, Cu metabolism, 193
—, F metabolism, 253
—, Se metabolism, 241
—, vitamin B_6 absorption, 50
—, — requirement in diet, 48
—, Zn metabolism, 205, 206
Dopamine-β-hydroxylase, Cu^{2+} content, 200
—, hydroxylation of 3,4-dihydroxyphenylethylamine to norepinephrine, role of ascorbate, 200
Drosophila melanogaster, folate compounds, biosynthesis, 112, 120
—, unconjugated pteridines, biosynthesis, 130, 135, 137, 139, 140
Drosopterins, formation from biopterin-sepiapterin pool, 139
—, occurrence, structure, 130
Duck, vitamin B_6 requirement in diet, 48

Ehrlich ascites tumour cells, concentration of pyridoxal and of pyridoxal-P from media, 49
Endomycopsis, bisnorbiotin and bisnorbiotin d-sulphoxide in culture filtrates, 102
Entasis, inactive enzyme–substrate complex formation, role of metals, 181
Eremothecium ashbyii, riboflavin synthesis, 11–13, 39, 41–43, 112
Erythrocuprein, mol. wt., Cu content, Cu storage, 202, 203
Erythrocytes, concentration of vitamins B_6 and vitamins B_6 phosphates from media, 49
—, vitamin B_{12} absorption, 166
Erythrogenesis, effect of Cu, 194
Escherichia coli, alkaline phosphatase, Zn content and function, 213
—, biotin (vitamers) metabolism, 85, 86, 88–90, 96–98
—, folate compounds, biosynthesis, 116–118, 121–124
—, glutamine synthetase, Mn^{2+} requirement for activity, 229

Escherichia coli, (*continued*)
—, neopterin and neopterin phosphates in, 132
—, pantoate synthesis from α-ketoisovaleric acid, 73, 74
—, pteroylpolyglutamates, biosynthesis, 126
—, riboflavin synthesis, 12, 16, 22
—, thiamine biosynthesis, 5, 7, 8
—, vitamin B_6 metabolism, 48, 49, 52, 53, 56–58, 60
Ethanolamine deaminase, coenzyme B_{12} dependent, 223
Ethylenediamine tetraacetate, effect on Zn absorption from phytate-containing diets, 204
Euglena, pteridine content, increase in light, 144

FAD, effect on riboflavin synthetase, 23
Fatty acids, desaturation, unconjugated pteridines in, 144
Ferredoxin, *Clostridium pasteurianum*, mol. wt., Fe content, 190
—, as electron acceptor for xanthine oxidase, 238
Ferritin stores, formation, composition, mol. wt., 186, 190
Fish, vitamin B_{12} concns. in liver and kidney, 167, 168
Flavobacterium sp., conjugases, 128
Flour beetle, pimelic acid as vitamin for larvae, 83
Fluorescyanine, *see* Ichtyopterin
Fluoride, absorption from gastrointestinal tract, 250
—, beneficial effects, 250, 252
—, biological availability, 250
—, — half-life in adult human skeleton, 251
—, clearance from the blood, concn. in soft tissues, 250
—, concn. in long bones, 250
—, content in common human diet, 250
—, in domestic water, contribution to total F absorbed by man, 250
—, exchange for hydroxyl ion in bone salt, 251
—, excretion by the kidneys, selectivity for elimination of fluoride from the plasma, 251

Phosphatase, alkaline, (*continued*)
—, —, Zn atoms for activity and for
maintenance of quaternary structure,
213
—, non-specific hydrolysis of vitamin B$_6$
phosphates, 56
Phosphates, in diet, effect on intestinal iron
absorption, 183
—, effect on Zn absorption from intestine,
204
Phosphodoxin, stimulation of
photosynthetic phosphorylation, 145
Phosphoenolpyruvate carboxykinase,
pig-liver mitochondria, role of Mn^{2+},
228
— carboxylase, peanut cotyledons, role of
Mn^{2+}, 228
Phosphoglucomutase, activation by Cr,
233
Phosphopantetheine, in CoA synthesis, 76
Phosphopantothenic acid, in CoA
synthesis, 77
Phosphopantothenylcysteine pathway, in
CoA synthesis from pantothenic acid,
76, 77
—, role of ATP and CTP in
phosphopantothenylcysteine formation
from phosphopantothenic acid, 77
Photosynthetic bacteria, neopterin, 131
— —, role of pteridines in photosynthesis,
144, 145
— phosphorylation, role of pteridines, 145
Phycomyces blakesleeanus, biotin
metabolism, 84, 97
Physalia physalis, conjugase, 128
Phytate(s), in diet, effect on intestinal Fe
absorption, 183
—, effect on Zn absorption from intestine,
204
Pigs, Cu metabolism, 194, 201
—, Fe metabolism, 188
—, intrinsic factor, 158
—, Mn metabolism, 225, 228
—, Se metabolism, 241
—, vitamin B$_6$ metabolism, 48, 56, 57
—, Zn metabolism, 208, 211, 212, 214,
228
Pigeon liver, kinase for phosphorylation
of pantothenylcysteine, 79
Pimelic acid, growth-stimulating effect of
— *per se*, 83

Pimelic acid, (*continued*)
—, incorporation as a unit in biosynthesis
of biotin, evidence from isotope
experiments, 83, 84
—, odd-numbered and even-numbered
higher homologues, in biotin
biosynthesis stimulation, 82
—, organisms in which biotin requirement
is not satisfied by —, 83
—, role in biotin biosynthesis, 82–84
—, urinary secretion in herbivores, 83
Plants, higher, origin of vitamin B$_{12}$ in —,
154
—, tissues, Cr metabolism, 230
—, —, Fe metabolism, 189
—, —, Se metabolism, 240
Plasma, (blood), and cobalamin
metabolism, 157, 163–165, 169, 170
Plastocyanin, 203
Plaut *et al.*, mechanism of enzymatic
riboflavin synthesis from
6,7-diMe-8-ribityllumazine, 35–38
Pneumococci, folate compound
biosynthesis, 122
—, pteroylpolyglutamates, 126
Polycythaemia vera, vitamin B$_{12}$ binding
protein, 164, 165
γ-Polyglutamic acid peptidases, in
Flavobacterium from soil, 128
γ-Polyglutamylglutamic acid peptidases,
from animals, use in assay for total
folates, limitations, 128
Polyphenoloxidase, synonyms, 199
Propionate metabolism, in ruminants and
Co deficiency, 221
— —, and vitamin B$_{12}$ deficiency, 221
Pseudomonads, biotin and biotin in
sulphoxide degradation, 101
—, catabolism of vitamins B$_6$ to
open-chain compounds, inducible
pathways, (Fig.), 61, 62
—, conjugases, 128
—, metabolism of *d*-biotinol, 104
—, neopterin, 131
—, (soil), biotin metabolism, 83, 85,
99–101, 104
Pseudomonas aeruginosa, respiratory
nitrate reductase, cofactors, 239
— AM-1, serine hydroxymethylase,
requirement for
tetrahydropteroylglutamate, 126, 127

Riboflavin synthetase, (*continued*)
— —, 7-(2-OH-2-Me-3-oxobutyl)-6-Me-8-
(1′-D-ribityl)lumazine as intermediate,
38
— —, pH effect on activity at the
acceptor site, 35–37
— —, — on substrate binding to donor
site, 35–37
— —, positioning of ribityl groups on
synthetase and positioning of
heterocyclic ring portions of lumazine,
38
— — protein, binding of lumazine
derivatives, (table), 25
— — —, catalytic activity, 24
— — —, complexes with lumazines, 29–32
— — —, lumazine binding with different
affinities to donor and acceptor sites,
28, 29
— —, purification, 22
— —, pyrazine ring opening, in formed
dimeric lumazine intermediates, 37
— —, rate-limiting effect of the presence
of deuterium in the 6-Me group of
substrate on riboflavin formation, and
velocity of H–T exchange at the 7-Me
group, 37
— —, rates of enzymatic conversion of
non-deuterated and of 6- and
7-Me-deuterated 6,7-diMe-8-
ribityllumazine, 34
— —, reaction intermediates, 31–38
— —, — and Rowan and Wood reaction,
32, 33, 38
— —, — stoichiometry, 17, 18
— —, requirement for the configuration
of riboflavin side-chain at position-8,
18
— —, riboflavin as competitive inhibitor,
23
— —, role of SH groups in catalysis, 23
— —, stabilization by reducing
compounds, 23
— —, substrate specificity, 18–22
— —, ternary complex formation in
reaction, 31
— —, tritium–hydrogen exchange
reaction, of lumazines, relationship
with substrate specificity, 35
— —, — —, and riboflavin synthesis, by
the same enzyme protein, 35

Riboflavin synthetase, (*continued*)
— —, yeast, binding of 4-carbon donor
and acceptor, 28–31
— —, —, complex with riboflavin, removal
of riboflavin with charcoal, 24
— —, —, —, spectral properties, 24
— —, —, mechanism, 24–38
— —, —, reaction kinetics, 27, 28
— —, —, riboflavin-free, *see* Riboflavin
synthetase protein
— —, —, role of bound riboflavin in
synthetase reaction, 24
—, *o*-xylene portion, origin, 12, 17, 37, 41,
42
Riboflavinimine, effect on riboflavin
synthetase, 23
Ribonucleic acid, polymerase systems,
Mn^{2+} requirement, Mg^{2+} substitution,
229
—, Zn in, and configuration of RNA
molecule, 216
Ribonucleoproteins, Cr content, 233
Ribonucleotide reductase, coenzyme B_{12}
requiring, in bacteria and animals, 222,
223
Rowan and Wood mechanism, chemical
conversion of 6,7-dimethyl-8-
substituted lumazine to flavin, 31–33
Royal jelly, biopterin, 130
Rumen microorganisms, Se, dietary
reduction to non-absorbable forms, 252
Ruminants, Co deficiency, 219, 220
—, Co metabolism, 217, 219, 221
—, Cu metabolism, 201
—, Mo, metabolic functions, interrelation
with Cu and sulphate, 236
—, Mo metabolism, 236
—, propionate metabolism, and Co
deficiency, 221
—, Se metabolism, 240, 242
—, vitamin B_{12} production by rumen
flora, 217
—, — synthesis by microorganisms in
fore-stomach, 154
—, Zn metabolism, 204, 207

Saccharomyces cerevisiae, riboflavin
synthesis, 16, 17, 39
Salmonella typhimurium, biosynthesis of
pyrimidine moiety of thiamine, 5–7
—, folate compounds, biosynthesis, 114, 122